Communications in Computer and Information Science 629

Commenced Publication in 2007
Founding and Former Series Editors:
Alfredo Cuzzocrea, Dominik Ślęzak, and Xiaokang Yang

More information about this series at http://www.springer.com/series/7899

Chrisina Jayne · Lazaros Iliadis (Eds.)

Engineering Applications of Neural Networks

17th International Conference, EANN 2016
Aberdeen, UK, September 2–5, 2016
Proceedings

Editors
Chrisina Jayne
Robert Gordon University
Aberdeen
UK

Lazaros Iliadis
Lab of Forest Informatics (FiLAB)
Democritus University of Thrace
Orestiada
Greece

ISSN 1865-0929 ISSN 1865-0937 (electronic)
Communications in Computer and Information Science
ISBN 978-3-319-44187-0 ISBN 978-3-319-44188-7 (eBook)
DOI 10.1007/978-3-319-44188-7

Library of Congress Control Number: 2016947184

Printed on acid-free paper

This Springer imprint is published by Springer Nature
The registered company is Springer International Publishing AG Switzerland

Preface

The 17th International Conference on Engineering Applications of Neural Networks (EANN) was held at Robert Gordon University in Aberdeen, UK, during September, 2016. The supporters for the conference were the International Neural Network Society (INNS), The Scottish Informatics and Computer Science Alliance (SICSA), Visit Aberdeen, Visit Scotland, and Robert Gordon University in Aberdeen, UK. The 17th EANN 2016 attracted delegates from 12 countries across the world: Czech Republic, China, Chile, Colombia, Greece, Italy, Japan, Poland, Portugal, Russia, the UK, and USA.

The volume includes 22 full papers, three short papers, and two tutorial papers. All papers were subject to a rigorous peer-review process by at least two independent academic referees. EANN 2016 accepted approximately 53 % of the submitted papers as full papers. The authors of the best 10 papers were invited to submit extended contributions for inclusion in a special issue of *Neural Computing and Applications* (Springer). The papers demonstrate a variety of novel neural network and other computational intelligence approaches applied to challenging real-world problems. The papers cover topics such as: convolutional neural networks and deep learning applications, real-time systems, ensemble classification, chaotic neural networks, self-organizing maps applications, intelligent cyber physical systems, text analysis, emotion recognition, and optimization problems.

The following keynote speakers were invited and gave lectures on exciting neural network application topics:

- Professor Nikola Kasabov, Director and Founder, Knowledge Engineering and Discovery Research Institute (KEDRI), Chair of Knowledge Engineering, Auckland University of Technology, New Zealand
- Professor Marley Vellasco, Head of the Electrical Engineering Department and the Applied Computational Intelligence Laboratory (ICA) at PUC-Rio, Brazil
- Professor John MacIntyre, Dean of the Faculty of Appliced Sciences, Pro Vice Chancellor Director of Research, Innovation and Employer Engagement, University of Sunderland, UK

On behalf of the conference Organizing Committee we would like to thank all those who contributed to the organization of this year's program, and in particular the Program Committee members.

September 2016

Chrisina Jayne
Lazaros Iliadis

Organization

General Chair

Chrisina Jayne	Robert Gordon University, UK

Advisory Chair

Nikola Kasabov	Auckland University of Technology, New Zenland

Program Chairs

Chrisina Jayne	Robert Gordon University, UK
Lazaros Iliadis	Democritus University of Thrace, Greece

Local Organizing Committee Chair

Michael Heron	Robert Gordon University, UK

Program Committee

A. Canuto	Federal University of Rio Grande do Norte, Brazil
A. Petrovski	Robert Gordon University, UK
B. Beliczynski	Institute of Control and Industrial Electronics, Poland
D. Coufal	Czech Academy of Sciences
D. Pérez	University of Oviedo, Spain
E. Kyriacou	Frederick University, Cyprus
H. Leopold	Austrian Institute of Technology GmbH, Austria
I. Bukovsky	Czech Technical University in Prague, Czech Republic
J.F. De Canete Rodriguez	University of Malaga, Spain
K.L. Kermanidis	Ionian University, Greece
K. Margaritis	University of Macedonia, Greece
M. Holena	Academy of Sciences of the Czech Republic
M. Fiasche	Politecnico di Milano, Italy
M. Trovati	Derby University, UK
N. Wiratunga	Robert Gordon University, UK
P. Hajek	University of Pardubice, Czech Republic
P. Kumpulainen	Tempere University of Technology, Finland
S. Massie	Robert Gordon University, UK
V. Kurkova	Czech Academy of Sciences
Z. Ding	Hewlett Packard Enterprise, USA

A. Papaleonidas	Democritus University of Thrace, Greece
A. Kalampakas	Democritus University of Thrace, Greece
B. Ribeiro	University of Coimbra, Portugal
D. Gorse	University College London, UK
E. Elyan	Robert Gordon University, UK
F. Marcelloni	University of Pisa, Italy
I. Bougoudis	Democritus University of Thrace, UK
I. Stephanakis	Hellenic Telecommunication Organization SA, Greece
K. Demertzis	Democritus University of Thrace, Greece
K. Koutroumbas	National Observatory of Athens, Greece
M. Gaber	Robert Gordon University, UK
M. Kolehmainen	Environmental Science University of Eastern, Finland
M. Tauber	Austrian Institute of Technology GmbH, Austria
N. Nicolaou	Imperial College London, UK
P. Gastaldo	Università degli Studi di Genoa, Italy
P. Vidnerov	Czech Academy of Sciences
R. Tanscheit	PUC-Rio, Brazil
S. Sani	Robert Gordon University, UK
Y. Manolopoulos	Aristotle University of Thessaloniki, Greece

Supporting Organizations

International Neural Network Society (INNS)
The Scottish Informatics and Computer Science Alliance (SICSA)
Visit Aberdeen
Visit Scotland
Robert Gordon University, Aberdeen, UK

Contents

Active Learning and Dynamic Environments

Semi-supervised Modeling

Classification Applications

Clustering Applications

Active Learning and Dynamic Environments

Deep Active Learning for Autonomous Navigation

Ahmed Hussein(✉), Mohamed Medhat Gaber, and Eyad Elyan

School of Computing, Robert Gordon University,
Garthdee Road, Aberdeen AB10 7QB, UK
a.s.h.a.hussein@rgu.ac.uk

Abstract. Imitation learning refers to an agent's ability to mimic a desired behavior by learning from observations. A major challenge facing learning from demonstrations is to represent the demonstrations in a manner that is adequate for learning and efficient for real time decisions. Creating feature representations is especially challenging when extracted from high dimensional visual data. In this paper, we present a method for imitation learning from raw visual data. The proposed method is applied to a popular imitation learning domain that is relevant to a variety of real life applications; namely navigation. To create a training set, a teacher uses an optimal policy to perform a navigation task, and the actions taken are recorded along with visual footage from the first person perspective. Features are automatically extracted and used to learn a policy that mimics the teacher via a deep convolutional neural network. A trained agent can then predict an action to perform based on the scene it finds itself in. This method is generic, and the network is trained without knowledge of the task, targets or environment in which it is acting. Another common challenge in imitation learning is generalizing a policy over unseen situation in training data. To address this challenge, the learned policy is subsequently improved by employing active learning. While the agent is executing a task, it can query the teacher for the correct action to take in situations where it has low confidence. The active samples are added to the training set and used to update the initial policy. The proposed approach is demonstrated on 4 different tasks in a 3D simulated environment. The experiments show that an agent can effectively perform imitation learning from raw visual data for navigation tasks and that active learning can significantly improve the initial policy using a small number of samples. The simulated testbed facilitates reproduction of these results and comparison with other approaches.

1 Introduction

One of the important aspects of artificial intelligence is the ability of autonomous agents to behave effectively and realistically in a given task. There is a rising demand for applications in which agents can act and make decisions similar to human behavior in order to achieve a goal. Imitation learning is a paradigm in which an agent learns how to behave by observing demonstrations of correct

C. Jayne and L. Iliadis (Eds.): EANN 2016, CCIS 629, pp. 3–17, 2016.
DOI: 10.1007/978-3-319-44188-7_1

behavior provided by a teacher. In contrast to explicit programming, learning from demonstrations does not require knowledge of the task to be integrated in the learning process. It favors a generic learning process where the task is learned completely from observing the demonstrations. Thus, an intelligent agent can be trained to perform a new task simply by providing examples. Since an agent is able to learn complex tasks by mimicking a teacher's behavior, imitation learning is relevant to many robotic applications [2,4,6,11,13,24,29,36] and is considered an integral part in the future of intelligent robots [31].

One of the biggest challenges in imitation learning is finding adequate representations for the state of the agent in its environment. The agent should be able to extract meaningful information from sensing of its surroundings, and utilize this information to perform actions in real time. Deep learning methods have recently been applied in a wide array of applications and are especially successful in handling raw data. One of the most popular deep learning techniques is *Convolutional Neural Networks* (CNNs). CNNs are particularly popular in vision applications due to their ability to extract features from high dimensional visual data. The ability of deep networks to automatically discover patterns provides a generic alternative to engineered features which have to be designed for each specific task. For instance traditional planning approaches that use computer-vision methods of object recognition and localization need to tailor the methods for every individual target and task. CNNs achieve results competitive with the state of the art in many image classification tasks [8,17] and have been recently used to learn Atari 2600 games from raw visual input [20,21]. These and other recent attempts have shown that deep learning can be successful in teaching an agent to perform a task from visual data. However, most studies focus on 2D environments with stationary views; which does not reflect real world applications. Moreover, direct imitation is performed without considering refining the policy based on the agent's performance. To the best of our knowledge, training an agent from raw visual input using deep networks and active learning in a 3D environment has not been done.

In this paper we present a novel method that utilizes deep learning and active learning to train agents in a 3D setting. The method is demonstrated on several navigation tasks in a 3D simulated environment. Navigation is one of the most explored domains in imitation learning due to its relevance to many robotic applications, such as flying [1,23,29] and ground vehicles [7,26,27,32]. Navigation is also an essential base task in high degree of freedom robots (e.g. humanoid robots) [7,30]. We propose a generic method for learning navigation tasks from demonstrations that does not require any prior knowledge of the task's goals, environment or possible actions. A training set is gathered by having a teacher control the agent to successfully perform the task. The controlled agent's view of the 3D environment is captured along with the actions performed in each frame. A deep convolution network is used to learn visual representation from the captured video footage and learn a policy to mimic the teacher's behavior. We also employ active learning to improve the agents policy by emphasizing situations in which it is not confident. We show that active learning can significantly improve the policy with a limited number of queried instances. Once trained, the agent is able to extract features from the scene and predict actions in real time. We conduct our experiments on benchmark testbed that makes it seamless to replicate our results and compare with other approaches.

Benchmark environments are useful tools for evaluating intelligent agents. A few benchmarks are available for 2D tasks such as [3,15,25] and are being increasingly employed in the literature. 3D environments however have not been as widely explored, although they provide a closer simulation to real robotic applications. We use mash-simulator [19] as our testbed to facilitate the evaluation and comparison of learning methods. It is also convenient for extending the experiments to different navigation tasks within the same framework.

In the next section we review related work. Section 3 describes the proposed methods. Section 4 details our experiments and results. Finally we present our conclusions and discuss future steps in Sect. 5.

2 Related Work

2.1 Navigation

Navigation tasks have been of interest in AI in general and imitation learning specifically from an early stage. Sammut et al. [29] provides an early example of an aircraft learning autonomous flight from demonstrations provided via remote control. Later research tackle more elaborate navigation problems including obstacles and objects of interest. Chernova et al. [7] use Gaussian mixture models to teach a robot to navigate through a maze. The robot is fitted with an IR sensor to provide information about the proximity of obstacles. This data coupled with input from a teacher controlling the robot is used to learn a policy. The robot is then able to make a decision to execute one of 4 motion primitives(unit actions) based on its sensory readings. In [10] the robot uses a laser sensor to detect and recognize objects of interest. A policy is learned to predict subgoals associated with the detected objects rather than directly predicting the motion primitives. Such sensing methods provide an abstract view of the environment, but can't convey visual details that might be needed for intelligent agents to mimic human behavior. [22] use neural networks to learn a policy for driving a car in racing game using features extracted from the game engine (such as position of the car relative to the track). Driving is a complex task compared to other navigation problems due to the complexity of the possible actions. The outputs of the neural network in [22] are high DOF low level actions. However, the features extracted from the game engine to train the policy would be difficult to extract in the real world. Advances in computational resources have prompted the use of visual data over simpler sensory data. Visual sensors provide detailed information about the agents surrounding and are suitable to use in real world applications. In [28] a policy for a racing game is learned from visual data. Demonstrations are provided by capturing the games video stream and the controller input. The raw frames (downsampled) without extracting engineered features are used as input to train a neural network.

2.2 Deep Learning

Deep learning methods are highly effective in problems that don't have established sets of engineered features. CNNs have been used with great success to extract features from images. In recent studies [20,21] CNNs are coupled with reinforcement learning to learn several Atari games. A sequence of raw frames is used as input to the network and trial and error is used to learn a policy. Trial and error methods such as reinforcement learning have been extensively used to learn policies for intelligent agents [16]. However, providing demonstrations of correct behavior can greatly expedite the learning rate. Moreover, learning through trial and error can lead the agent to learn a way of performing the task that doesn't seem natural or intuitive to a human observer. In [12] learning from demonstrations is applied on the same Atari benchmark. A supervised network is used to train a policy using samples from a high performing but non real time agent. This approach is reported to outperform agents that learn from scratch through reinforcement learning. Other examples of using deep learning to play games include learning the game of 'GO' using supervised convolution networks [9] and a combination of supervised and reinforcement learning [33]. These examples all focus on learning 2D games that have a fixed view. However in real applications, visual sensors would capture 3D scenes, and the sensors would most likely be mounted on the agent which means it is unrealistic to have a fixed view of the entire scene at all times.

In [18] a robot is trained to perform a number of object manipulation tasks. First a trajectory is learned using reinforcement learning with the position of the objects and targets known to the robot. These trajectories then serve as demonstrations train a supervised convolutional neural network. In this case no demonstrations are needed to be provided by a teacher. However, this approach requires expert knowledge for the initial setup of the reinforcement learning phase. Compared to related work that employs deep learning to teach an intelligent agent, this is a realistic application implemented with a physical robot. However, the features are extracted from a set scene with small variations. This is different from applications where the agent moves and turns around, and with that completely altering it's view.

2.3 Active Learning

In many imitation learning applications direct imitation is not sufficient for robust behavior. One of the common challenges facing direct imitation is that the training set doesn't fully represent the desire task. The collected demonstrations only include optimal actions performed by the teacher. If the agent makes an error it arrives at a state that was not represented in its learned policy [35]. It is therefore necessary in many cases to provide further training to an agent based on its own performance of the task. One of the methods to enhance a trained agent is active learning. Active learning relies on querying a teacher for the correct decision in cases where the trained model performs poorly. The teacher's answers are used to improve the model in its weakest areas. In [7] active learning is used to teach a robot navigation tasks. The agent estimates a confidence measure for its prediction and queries a teacher for the correct action when the confidence is low. Erroneous behavior may also be identified by the teacher. In [5] the robot is allowed to perform the task while a human teacher physically adjusts its actions, which in turn provides corrected demonstrations. Some imitation learning tasks involve actions that are performed continuously over a period of time (i.e. an action is comprised of a series of motions performed in sequence). In such cases a correction can be provided by the teacher at any point in the action trajectory [14,28]. This way the agent is able to adapt to errors in the trajectory.

3 Proposed Method

In this section we detail our proposed method for learning navigation tasks from demonstrations. The source code for this work can be accessed at: https://github.com/ahmedsalaheldin/ImitationMASH.git

3.1 Collecting Demonstrations

In imitation learning it is assumed that a human teacher is following an unknown optimal policy. It is therefore possible to use an optimal policy if it exists to collect demonstrations. To collect a training set we use a deterministic automated teacher that has access to information hidden from a human or intelligent playing agent such as position of targets and obstacles in a 3D space. Each training instance consists of a raw 120×90 image of the rendered 3D scene and the action performed by the teacher. We only use the current frame (not a sequence of previous frames) in an instance because for the navigation tasks investigated here adhere to the Markov property. That is, that current state is sufficient to make a decision. And any previous actions and states need not be included in the representation of the current state. In that case training an imitation learning policy is reduced to a supervised image classification problem; where the current view of the agent is the image and the action chosen by the teacher is the label. Subsequently the trained agent will be able to predict a decision (as it would be taken by the teacher) given its current view. More formally, the agent learns a policy π from a set of demonstrations $D = (x_i, y_i)$ such that $u = \pi(x, \alpha)$. Where x_i is a 120×90 image, y is the action performed by the teacher at frame i, u is the action predicted by policy π for input x and α is the set of policy parameters that are changed through learning.

3.2 Deep Learning

To learn the policy we employ a deep convolutional neural network. The proposed network uses several convolution layers to automatically extract features from the raw visual footage. Then a fully connected layer is used to map the learned features to actions. Each convolution layer is followed by a pooling layer that down-samples the output of the convolution layer. The convolution layers take advantage of spacial connection between visual features to reduce connections in the network. The pooling layers reduce the dimensionality to further alleviate the computations needed. Our network follows the pattern in [21]. It consists of 3 convolution layers each followed by a pooling layer. The input to the first layer is a frame of 120×90 pixels. We apply a luminance map to the colored images to obtain one value for each pixel instead of 3 channels, resulting in a feature vector of size 10,800. Figure 1 shows the architecture of the network. The filter sizes for the three layers are 7×9, 5×5 and 4×5 respectively; and the number of filters are 20, 50 and 70 respectively. The pooling layers all use maxpool of shape (2,2). Following the last convolution layer is a fully connected hidden layer with rectifier activation function and fully connected output layer with three output nodes representing the 3 possible actions. Table 1 summarizes the architecture of the network.

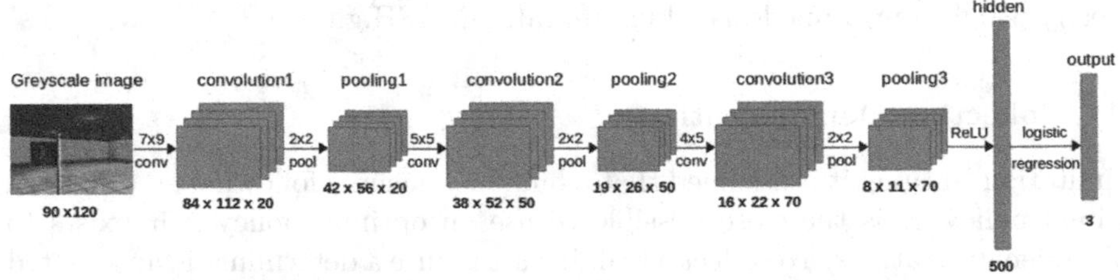

Fig. 1. Architecture of the neural network used to train the agent

Table 1. Neural network architecture

Layer	Size of activation volume
Input	120 * 90
Conv1	7 * 9 * 20
Conv2	5 * 5 * 50
Conv3	4 * 5 * 70
FC	500
Output(FC)	3

3.3 Active Learning

Active learning is employed to improve the initial policy learned from demonstrations. This is achieved by acquiring a new data set to train the agent that emphasizes the weaknesses of the initial policy. The agent is allowed to perform the task for a number of rounds. For each prediction the network's confidence is calculated, and if the confidence is low the optimal policy is queried for the correct action. The action provided by the teacher is performed by the agent and is recorded along with the frame image. The confidence is measured as the entropy of the output of the final layer in the network. The entropy $H(X)$ is calculated as:

$$H(X) = -\sum_{i} P(x_i) \log_2 P(x_i) \tag{1}$$

Where X is the prediction of the network, $P(x_i)$ is the probability distribution produced by the network for action i.

The active samples are added to the training set and used to update the initial policy. We find that updating a trained network using only the active samples results in forgetting the initial policy in favor of an inadequate one rather than complementing it. Therefore the training set is augmented with the active samples collected from the playing agent. The augmented dataset is used to update the network that was previously trained. We find that it is easier and faster for the network to converge if it is pre-trained with the initial dataset than training from scratch. Algorithm 1 shows the steps followed to perform active learning.

Low confidence predictions are mainly caused by situations that were not covered by the training data. Therefore, for active learning to be effective, it is important that it is performed in the simulation rather than on a collected dataset. Because by performing its current policy in the simulation, the agent arrives at unfamiliar situations where it is not confident in its behavior and thus utilize active learning.

Algorithm 1. Active Learning Algorithm

Given: A policy π trained on a Data set $D = (x_i, y_i)$
    Confidence threshold β
while Active_Learning **do**
    x = current_frame
    $u = \pi(x, \alpha)$
    $H(X) = -\sum_{i} P(u_i) \log_2 P(u_i)$
    if $H(X) < \beta$ **then**
        $y = Query(x)$
        perform action y
        add (x, y) to D
    else
        perform $max(u)$
Update π using D

4 Experiments

We conduct our experiments in the framework of mash-simulator [19]. Mash-simulator is a tool for benchmarking computer vision techniques for navigation tasks. The simulator includes a number of different tasks and environments.

As well as optimal policies for a number of tasks. All the navigation is viewed from the first person perspective. The player has 4 possible actions: 'Go forward', 'Turn left', 'Turn right' and 'Go back'. Although there are 4 possible actions, the action 'Go back' was never used in the demonstrations by the optimal policy. Therefore the network is only presented with 3 classes in the training set and thus has 3 output nodes.

4.1 Tasks

The experiments are conducted on the following 4 navigation tasks:

Reach the Flag. This task is set in a single rectangular room with a flag placed randomly in the room. The goal is to reach the flag. The task fails if the flag is not reached within a time limit.

Fig. 2. sample images from "Reach the flag"

Follow the Line. This task is set in a room with directed lines drawn on the floor. The lines show the direction to follow in order to reach the flag. The target is to follow the line to the flag, and the agent fails if it deviates from the line on the floor.

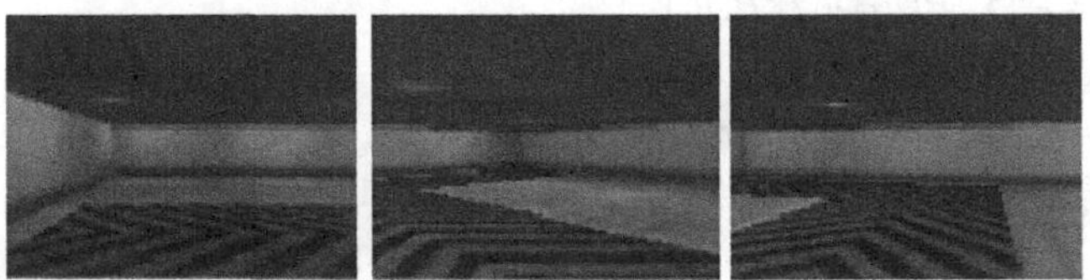

Fig. 3. sample images from "Follow the line"

Reach the Correct Object. In this task two objects are placed on pedestals in random positions in the room. The objective is to reach the pedestal with the trophy on it. The task fails if a time limit is reached or if the player reaches the wrong object. The wrong object has the same material of the trophy and can take different shapes.

Fig. 4. sample images from "Reach the correct object"

Eat All Disks. This task is set in a large room containing several black disks on the floor. The target is to keep reaching the disks. A disk is 'eaten' once the agent reaches it and dissapears. New disks appear when one is eaten. The goal of this task is to eat as many disks as possible within a time limit.

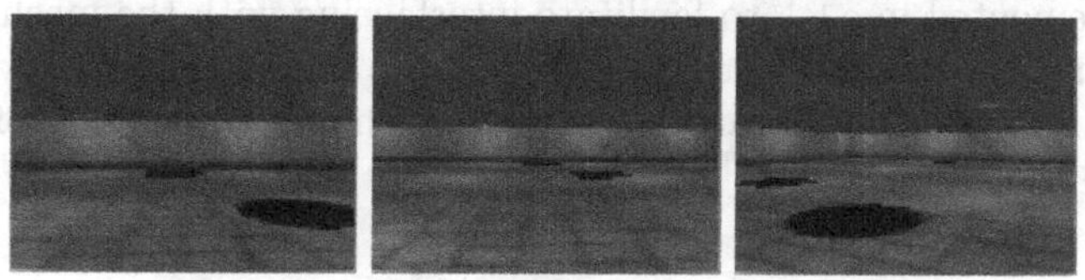

Fig. 5. sample images from "Eat all disks"

Figures 2, 3, 4 and 5 show sample images of the 4 tasks in the 120×90 size used in the experiments.

4.2 Setup

To evaluate the proposed methods, the performance of the agent is measured over 1,000 rounds. A round starts when the task is initialized and ends when the agent reaches the target or a time limit is reached. The number of frames in a round might vary depending on how fast the agent can reach the target. For all tasks, in each round the environment is randomized including room size and shape, lighting and the location of the target and the agent. A time limit is set for each round and the round fails if the limit is reached before the agent reaches the target. The time limit is measured in frames to avoid any issues with different frame rates. The time limit is set as the maximum time needed for the optimal policy to finish the task; which is 500 frames for "Reach the flag" and "Reach the correct object" and 5000 frames for "Follow the line". In "Eat all disks" the task is continuous, so a time limit was set to match the total number of frames in the other tasks.

4.3 Implementation Details

Inter-process communication is used to communicate data across the different components of the testbed. The agent acts as a client and communicates with the simulator via a TCP connection as follows: The agent requests a task from the server, the server initiates a round and sends an image to the client. The client sends an action to the server. The server calculates the simulations and responds with a new image. Figure 6 shows a flowchart of the data collection process.

The network used for prediction is also decoupled from the agent. The network acts as a predicting server where an agent sends frames that it receives from the simulator and in return receives a decision from the network. The entire process of communication with both servers occurs in real time. This implementation facilitates experimentation, as making changes to the network doesn't affect the client or the simulator server. Moreover, it is easier to extend this system to physical robots. A predicting server can be located on the robot or on another machine if the robot's computational capabilities are not sufficient. A predicting server can also serve multiple agents simultaneously. The agent client is implemented in c++ to facilitate interfaceing with the mash-simulator. The predicting server and the training process are implemented in python using the Theano deep learning library [34]. Figure 7 shows a flowchart of the agent performing a task.

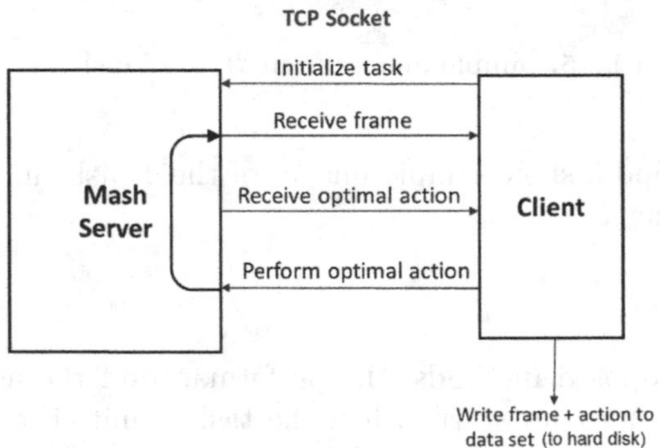

Fig. 6. Dataset collection flowchart

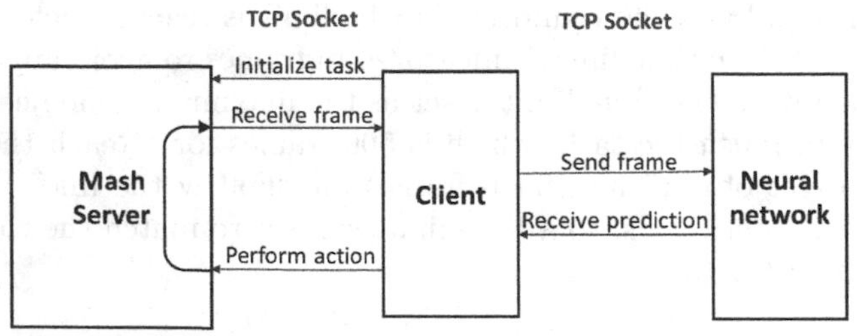

Fig. 7. Imitation agent playing flowchart

4.4 Results

In this section we present the results of the proposed method. The same network and parameters are used to learn all tasks. For each task 20,000 images are used for training. Testing is conducted by allowing an agent to attempt the tasks in the mash-simulator and recording the number of successful attempts. An agent's performance for the first 3 tasks is evaluated as the percentage of times it reaches the target in 1,000 rounds. For "Eat all disks", the performance is measured as the number of disks eaten in 1,000 rounds. We also report the classification error on an unseen test set of 20,000 images collected from the teacher's demonstrations.

Table 2 shows the results for the first 3 tasks. The success measure is the percentage of rounds (out of 1000) in which the agent reached the target. While error is the classification error on the test set collected from the teacher's demonstrations. The agent performs well on "Reach the flag" and is significantly less successful in the other two tasks. "Follow the line" is considerably less fault tolerant than "Reach the flag". As a small error can result in the agent deviating from the line and subsequently failing the round. Whereas in "Reach the flag" the agent can continue to search for the target after a wrong prediction. In "Reach the correct object" the agent is not able to effectively distinguish between the two objects. This could be attributed to insufficient visual details in the training set, as the teacher avoids the wrong object from a distance. Qualitative analysis of "Reach the flag" shows that the agent aims towards corners as they resemble the erect flag from a distance. Upon approaching the corner, as the details of the image become clearer, the agent stops recognizing it as the target and continues its search. While this did not pose a big problem in the agent's ability to execute the task it is interesting to examine the ability of CNNs to distinguish small details in such environments. It is also worth noting that the teacher's policy for "Reach correct object" does not avoid the wrong object if it is in the way of the target and achieves 80.2 % success rate.

Table 2. Direct imitation results

Task	Reach the flag	Reach object	Follow the line
success	96.20 %	53.10 %	40.70 %
error	2.48 %	4.06 %	0.86 %

Table 3 shows results for the 4th task "Eat all disks". The table shows the score of the agent compared to the score achieved using the optimal policy. The agent is shown to achieve 97.9 % of the score performed by the optimal policy.

To improve the agent's ability to adapt to wrong predictions and unseen situations, active learning is used to train the agent on "Follow the line". In the other tasks where the agent searches for the target, the optimal policy remembers the location of the target even if it goes out of view due to agent error.

Table 3. "Eat all disks" results

Task	Agent	Optimal policy
score	1051	1073
error	1.70 %	-

Therefore active learning samples include information that is not represented in the visual data available to the agent and thus degrade the performance. This can be rectified by devising a teaching policy that does not use historical information, or by incorporating past experience in the learned model.

Figure 8 shows the results of active learning on the "Follow the line" task. Active learning is demonstrated to significantly improve the performance of the agent using a relatively small number of samples. Comparing the classification error with success rate emphasizes the point that the errors come from situations that are not represented in the teacher's demonstrations.

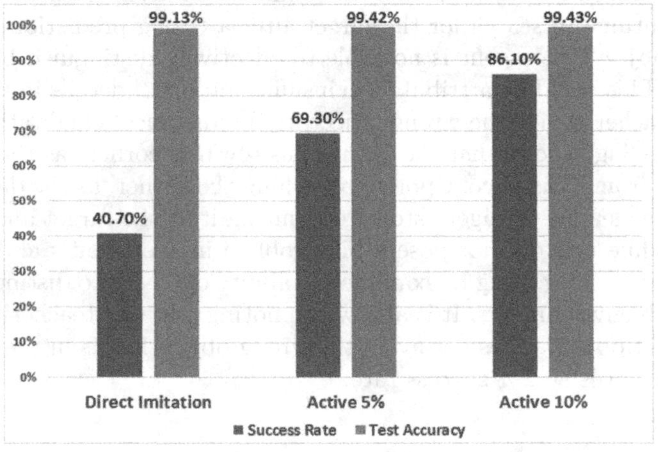

Fig. 8. Results for active learning on "follow the line" task

The task in which the time limit affected the performance was "Reach the flag". As the agent continues to follow its policy in search of the flag even after performing wrong predictions. The effect of the time limit is evaluated in Fig. 9 which presents the success rate of "reach the flag" task with different time limits. The horizontal axis represents the time limit as a percentage of the maximum time needed by the teacher. The graph shows that the longer the agent is allowed to look for the target the higher the success rate.

Overall the results show good performance on 3 out of the 4 tasks. They demonstrate the effectiveness of active learning to significantly improve a weak policy with a limited number of samples. Even without active learning the agent can learn a robust policy for simple navigation tasks.

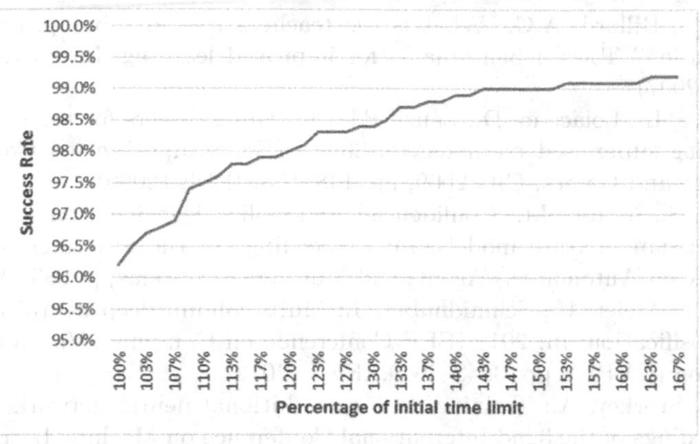

Fig. 9. Results for "reach the flag" task with increasing time limits

5 Conclusion and Future Directions

In this paper, we propose a framework for learning autonomous policies for navigation tasks from demonstrations. A generic learning process is employed to learn from raw visual data without integrating any knowledge of the task. The experiments are conducted on a testbed that facilitates reproduction, comparison and extension of this work. The results show that CNNs can learn meaningful features from raw images of 3D environments and learn a policy from demonstrations. They also show that active learning can significantly improve a learned policy with a limited number of samples.

Our next step is to conduct an investigation of the proposed approach in more visually cluttered environments to further evaluate the ability of convolution networks to create adequate representations from (relatively) low resolution 3D scenes. As well as extend active learning experiments to more tasks. We also aim to integrate reinforcement learning with learning from demonstrations to improve the learned policies through trial and error. This allows the agent to generalize its policy to unseen situations and adapt to changes in the task without requiring to query the teacher.

References

1. Abbeel, P., Coates, A., Quigley, M., Ng, A.Y.: An application of reinforcement learning to aerobatic helicopter flight. Adv. Neural Inf. Process. Syst. **19**, 1 (2007)
2. Argall, B.D., Chernova, S., Veloso, M., Browning, B.: A survey of robot learning from demonstration. Robot. Auton. Syst. **57**(5), 469–483 (2009)
3. Bellemare, M.G., Naddaf, Y., Veness, J., Bowling, M.: The arcade learning environment: an evaluation platform for general agents (2012). arXiv preprint arXiv:1207.4708
4. Bemelmans, R., Gelderblom, G.J., Jonker, P., De Witte, L.: Socially assistive robots in elderly care: a systematic review into effects and effectiveness. J. Am. Med. Direct. Assoc. **13**(2), 114–120 (2012)

5. Calinon, S., Billard, A.G.: What is the teachers role in robot programming by demonstration? Toward benchmarks for improved learning. Interact. Stud. **8**(3), 441–464 (2007)
6. Cardamone, L., Loiacono, D., Lanzi, P.L.: Learning drivers for torcs through imitation using supervised methods. In: 2009 IEEE Symposium on Computational Intelligence and Games, CIG 2009, pp. 148–155. IEEE (2009)
7. Chernova, S., Veloso, M.: Confidence-based policy learning from demonstration using Gaussian mixture models. In: Proceedings of the 6th International Joint Conference on Autonomous Agents and Multiagent Systems, p. 233. ACM (2007)
8. Ciresan, D., Meier, U., Schmidhuber, J.: Multi-column deep neural networks for image classification. In: 2012 IEEE Conference on Computer Vision and Pattern Recognition (CVPR), pp. 3642–3649. IEEE (2012)
9. Clark, C., Storkey, A.: Training deep convolutional neural networks to play go. In: Proceedings of the 32nd International Conference on Machine Learning (ICML 2015), pp. 1766–1774 (2015)
10. Dixon, K.R., Khosla, P.K.: Learning by observation with mobile robots: a computational approach. In: Proceedings 2004 IEEE International Conference on Robotics and Automation, ICRA 2004, vol. 1, pp. 102–107. IEEE (2004)
11. Gorman, B.: Imitation learning through games: theory, implementation and evaluation. Ph.D. thesis, Dublin City University (2009)
12. Guo, X., Singh, S., Lee, H., Lewis, R.L., Wang, X.: Deep learning for real-time Atari game play using offline monte-carlo tree search planning. In: Proceedings of Advances in Neural Information Processing Systems, pp. 3338–3346 (2014)
13. Ijspeert, A.J., Nakanishi, J., Schaal, S.: Learning rhythmic movements by demonstration using nonlinear oscillators. In: Proceedings of the IEEE/RSJ International Conference on Intelligent Robots and Systems (IROS 2002), pp. 958–963 (2002). No. BIOROB-CONF-2002-003
14. Judah, K., Fern, A., Dietterich, T.G.: Active imitation learning via reduction to IID active learning (2012). arXiv preprint arXiv:1210.4876
15. Karakovskiy, S., Togelius, J.: The mario AI benchmark and competitions. IEEE Trans. Comput. Intell. AI Games **4**(1), 55–67 (2012)
16. Kober, J., Bagnell, J.A., Peters, J.: Reinforcement learning in robotics: a survey. Int. J. Robot. Res. **32**, 1238 (2013). 0278364913495721
17. Krizhevsky, A., Sutskever, I., Hinton, G.E.: Imagenet classification with deep convolutional neural networks. In: Proceedings of Advances in Neural Information Processing Systems, pp. 1097–1105 (2012)
18. Levine, S., Finn, C., Darrell, T., Abbeel, P.: End-to-end training of deep visuomotor policies (2015). arXiv preprint arXiv:1504.00702
19. Mash-simulator (2014). https://github.com/idiap/mash-simulator
20. Mnih, V., Kavukcuoglu, K., Silver, D., Graves, A., Antonoglou, I., Wierstra, D., Riedmiller, M.: Playing Atari with deep reinforcement learning (2013). arXiv preprint arXiv:1312.5602
21. Mnih, V., Kavukcuoglu, K., Silver, D., Rusu, A.A., Veness, J., Bellemare, M.G., Graves, A., Riedmiller, M., Fidjeland, A.K., Ostrovski, G., et al.: Human-level control through deep reinforcement learning. Nature **518**(7540), 529–533 (2015)
22. Munoz, J., Gutierrez, G., Sanchis, A.: Controller for torcs created by imitation. In: 2009 IEEE Symposium on Computational Intelligence and Games, CIG 2009, pp. 271–278. IEEE (2009)

23. Ng, A.Y., Coates, A., Diel, M., Ganapathi, V., Schulte, J., Tse, B., Berger, E., Liang, E.: Autonomous inverted helicopter flight via reinforcement learning. In: Ang Jr., M.H., Khatib, O. (eds.) Experimental Robotics IX. Springer Tracts in Advanced Robotics, vol. 21, pp. 363–372. Springer, Heidelberg (2006)
24. Nicolescu, M.N., Mataric, M.J.: Natural methods for robot task learning: instructive demonstrations, generalization and practice. In: Proceedings of the Second International Joint Conference on Autonomous Agents and Multiagent Systems, pp. 241–248. ACM (2003)
25. Noda, I., Matsubara, H., Hiraki, K., Frank, I.: Soccer server: a tool for research on multiagent systems. Appl. Artif. Intell. **12**(2–3), 233–250 (1998)
26. Ollis, M., Huang, W.H., Happold, M.: A Bayesian approach to imitation learning for robot navigation. In: 2007 IEEE/RSJ International Conference on Intelligent Robots and Systems, IROS 2007, pp. 709–714. IEEE (2007)
27. Ratliff, N., Bradley, D., Bagnell, J.A., Chestnutt, J.: Boosting structured prediction for imitation learning. In: Proceedings of Robotics Institute, p. 54 (2007)
28. Ross, S., Bagnell, D.: Efficient reductions for imitation learning. In: International Conference on Artificial Intelligence and Statistics, pp. 661–668 (2010)
29. Sammut, C., Hurst, S., Kedzier, D., Michie, D., et al.: Learning to fly. In: Proceedings of the Ninth International Workshop on Machine Learning, pp. 385–393 (1992)
30. Saunders, J., Nehaniv, C.L., Dautenhahn, K.: Teaching robots by moulding behavior and scaffolding the environment. In: Proceedings of the 1st ACM SIGCHI/SIGART Conference on Human-robot Interaction, pp. 118–125. ACM (2006)
31. Schaal, S.: Is imitation learning the route to humanoid robots? Trends Cogn. Sci. **3**(6), 233–242 (1999)
32. Silver, D., Bagnell, J., Stentz, A.: High performance outdoor navigation from overhead data using imitation learning. In: Proceedings of Robotics: Science and Systems IV, Zurich, Switzerland (2008)
33. Silver, D., Huang, A., Maddison, C.J., Guez, A., Sifre, L., van den Driessche, G., Schrittwieser, J., Antonoglou, I., Panneershelvam, V., Lanctot, M., et al.: Mastering the game of go with deep neural networks and tree search. Nature **529**(7587), 484–489 (2016)
34. Theano Development Team: Theano: a Python framework for fast computation of mathematical expressions. arXiv e-prints abs/1605.02688, May 2016. http://arxiv.org/abs/1605.02688
35. Togelius, J., De Nardi, R., Lucas, S.M.: Towards automatic personalised content creation for racing games. In: 2007 IEEE Symposium on Computational Intelligence and Games, CIG 2007, pp. 252–259. IEEE (2007)
36. Vogt, D., Amor, H.B., Berger, E., Jung, B.: Learning two-person interaction models for responsive synthetic humanoids. J. Virtual Real. Broadcast. **11**(1) (2014)

2D Recurrent Neural Networks for Robust Visual Tracking of Non-Rigid Bodies

G.L. Masala(✉), B. Golosio, M. Tistarelli, and E. Grosso

Department of Political Science, Communication,
Engineering and Information Technologies - Computer Vision Laboratory, University of Sassari,
Sassari, Italy
{gilmasala,golosio,tista,grosso}@uniss.it

Abstract. The efficient tracking of articulated bodies over time is an essential element of pattern recognition and dynamic scenes analysis. This paper proposes a novel method for robust visual tracking, based on the combination of image-based prediction and weighted correlation. Starting from an initial guess, neural computation is applied to predict the position of the target in each video frame. Normalized cross-correlation is then applied to refine the predicted target position.

Image-based prediction relies on a novel architecture, derived from the Elman's Recurrent Neural Networks and adopting nearest neighborhood connections between the input and context layers in order to store the temporal information content of the video. The proposed architecture, named 2D Recurrent Neural Network, ensures both a limited complexity and a very fast learning stage. At the same time, it guarantees fast execution times and excellent accuracy for the considered tracking task. The effectiveness of the proposed approach is demonstrated on a very challenging set of dynamic image sequences, extracted from the final of triple jump at the London 2012 Summer Olympics. The system shows remarkable performance in all considered cases, characterized by changing background and a large variety of articulated motions.

Keywords: Recurrent neural network · Tracking · Video analysis

1 Introduction

Motion has been one of the main cues studied in Computer Vision and Pattern Recognition. As stated by David Marr [1]: "Motion pervades the visual world, a circumstance that has not failed to influence substantially the process of evolution. The study of visual motion is the study of how information about only the organization of movement in an image can be used to make inferences about the structure and the movement of the outside world".

In many cases motion enables three-dimensional perception (consider for example the counter-rotating cylinders effect described by Ullman [2]) and, in its simplest form, explains the unrivaled ability of humans to perform scene segmentation and object tracking [3].

C. Jayne and L. Iliadis (Eds.): EANN 2016, CCIS 629, pp. 18–34, 2016.
DOI: 10.1007/978-3-319-44188-7_2

Object tracking can be defined as the estimation of the trajectory of an object in the image plane as it moves around a scene. Errors in tracking are often due to abrupt changes in object motion, changes in the objects appearance, non-rigidity of the object, occlusions and non-linear camera motion [4]. The robustness of the representation of target appearance, against these and other unpredictable events, is crucial to successfully track objects over time. Interestingly, assumptions are often made to constrain the tracking problem within the context of a particular application.

Recent tracking algorithms are classified into two major categories, based on the learning strategy adopted: generative and discriminative methods. Generative methods describe the target appearance by a statistical model estimated from the previous frames. To maintain the integrity of the target appearance model, various approaches have been proposed, including sparse representation [5, 8, 9], on-line density estimation [10]. On the other hand, discriminative methods [11, 13] directly implement classifiers to discriminate the target from the surrounding background. Several learning algorithms have been adopted, including on-line boosting [13], multiple instance learning [11], structured support vector machines [12] and random forests [14, 15]. These approaches are often limited by the adoption of hand-crafted features for target representation, such as iconic templates, Haar-like features, histograms and others, which may not generalize well to handle the challenges arising in video sequences from everyday life scenes.

In this paper, an original method is proposed for robust visual tracking, based on a combination of image prediction and weighted correlation matching techniques. Image prediction is based on a novel recurrent neural network which can be easily generalized to track any visual pattern in dynamic scenes. The proposed approach is derived from both Recurrent Neural Networks (RNNs) and Convolutional Neural Networks (CNNs).

A Recurrent Neural Network (RNN) is an artificial neural network with feedback connections between nodes, with the capability to model dynamic systems [16]. Elman's neural network, also known as Simple Recurrent Network (SRN), is a partially recurrent neural network first proposed by Elman [17]. Because of the context neurons and local recurrent connections between the context layer and the hidden layer, the Elman's neural network has several dynamic advantages over a static neural network. Training and convergence of SRNs usually take a long time, which makes them useless in time critical applications [10] and/or when dealing with high resolution images. Therefore, to efficiently process high resolution images, a compromise between the representation power and the dimension of the network must be sought.

A CNN is a feed-forward artificial neural network where individual neurons are arranged to respond to overlapping regions in the visual field. A CNN consists of multiple layers of small neuron collections taking as input small overlapping areas of the image. The outputs of each layer are tiled and overlapping to better represent the original image. This feature ensures a reasonable invariance to planar translation on the image plane [18]. Due to their representation power, Convolutional Neural Networks have recently attracted a considerable attention in the Computer Vision community [7], particularly for image- and video-based recognition. However, only few attempts can be found in the literature to employ CNNs for visual tracking. One reason is that offline classifiers require a model of the objects class. On the other hand, performing online learning based on CNNs is not straightforward, due to the large network size and the lack of sufficiently large training sets. According to Hong et al. [7], the extraction of features from the deep structure may not be appropriate for visual tracking because top layers encode semantic information and may provide a relatively poor localization accuracy.

In this paper a variation of Elman's architecture, the Two-dimensional Recurrent Neural Network (2D-RNN) is proposed. This neural architecture is derived from a CNN where the input layer captures small areas of the input image. This mapping of the image pixels allows to reduce both the training time and the network dimension, yet keeping the temporal information embedded in the video and the image details unaltered.

The paper is organized as follows: in Sect. 2, the tracking problem is analytically stated, the solution based on the novel 2D-RNN architecture is described and compared with the Elman's SRN. A case study for video tracking (triple-jumping runner and related dataset) is first introduced in Sect. 3; then experimental steps and experimental protocols are defined. Section 4 is devoted to the comparison and discussion of the experimental results. Conclusions and future developments are finally discussed in Sect. 5.

2 Object Tracking in Real-Time Video

In this paragraph we first define the tracking problem for scenes including non-rigid and articulated bodies; thence the two types of neural networks used in the experimental section, the original Elman RNN and the proposed 2D-RNN are detailed.

2.1 Tracking

In a tracking scenario, an object can be defined as "anything that is of interest for further analysis" [6]. Objects can be represented by their varying shapes and appearances; the position of a single object can be traced through a single point as the centroid or by a set of points related to a small region in the image; for example primitive geometric shapes (suitable for rigid object but also used for tracking of non rigid objects), object silhouette and contour, articulated shape models or skeletal models. In the proposed approach a primitive rectangular shape (bounding box or BB) is used. The BB has a fixed dimension for all frames of the database. Note that for the purposes of this paper, the initialization of the tracking process, for example by moving objects detection or direct object recognition, it is not explicitly considered; as a consequence the object of interest must be defined at the time step 0 by manually placing a starting BB in the first frame.

Afterward, the tracking algorithm iteratively determines the object position. At each time step t, it can be assumed that the object position has been detected in the previous $t - i$ time steps, through the centroid of the bounding box. The past i images inside the BB are fed as input to a RNN, which produces as output the prediction of the image in the bounding box for the current time step. An important outcome of such a prediction is that the expected position of the bounding box for the current time step can be evaluated and refined through the correlation between the predicted sub-image (RNN next frame prediction) and the current image. At the same time also the predicted content of the BB can be evaluated (both the dynamic background and the object of interest) by considering the residual error corresponding to the maximum of the correlation.

Figure 1 depicts in detail the tracking scheme based on the RNN next frame prediction. The correlation matrix is computed by convolution in the Fourier domain; the position of the maximum of the correlation matrix corresponds to the best prediction of the BB position for the current time step. Note that, in general, the correlation matrix can have more than one local maximum, and it can happen that the target BB position is close to a local maximum that is not the absolute maximum.

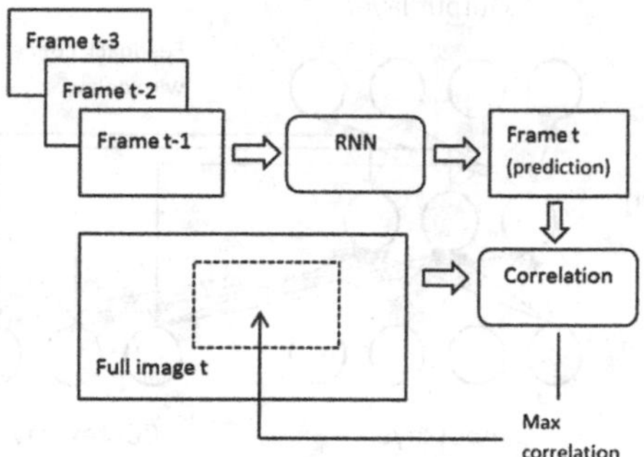

Fig. 1. The Recurrent Neural Network predicts the bounding box of the object at the next frame starting from the bounding box at the previous frame. The location refinement is performed by the correlation between the predicted bounding box and the entire image

This problem is particularly important when the moving object is subject to abrupt deformations, partial occlusions, etc. In order to deal with this problem, in our approach the correlation matrix is weighted with a Gaussian function centered on the extrapolated position of the moving object, based on the two most recent observations, i.e. the positions at time t – 1 and t – 2.

More precisely, the coordinates $X_c^{t,extr}$ and $Y_c^{t,extr}$ of the extrapolated center position are defined through:

$$X_c^{t,extr} = X_c^{t-1} + \Delta X_c^{t-1} \tag{1}$$

$$Y_c^{t,extr} = Y_c^{t-1} + \Delta Y_c^{t-1} \tag{2}$$

Where

$$\Delta X_c^{t-1} = X_c^{t-1} - X_c^{t-2} \tag{3}$$

$$\Delta Y_c^{t-1} = Y_c^{t-1} - Y_c^{t-2} \tag{4}$$

2.2 The Elman Neural Network

The Elman's Simple Recurrent Network (SRN) consists of an input layer, a hidden layer, a context layer, and an output layer. The outputs of the context neurons and the external input neurons are fed to the hidden neurons. Context neurons are known as memory units as they store the previous output of hidden neurons. At the time step t, the context layer nodes carry the output of hidden layer nodes of the time step t – 1 iteration and supply that as input during processing of the time step t data. The SRN architecture is presented in figure Fig. 2.

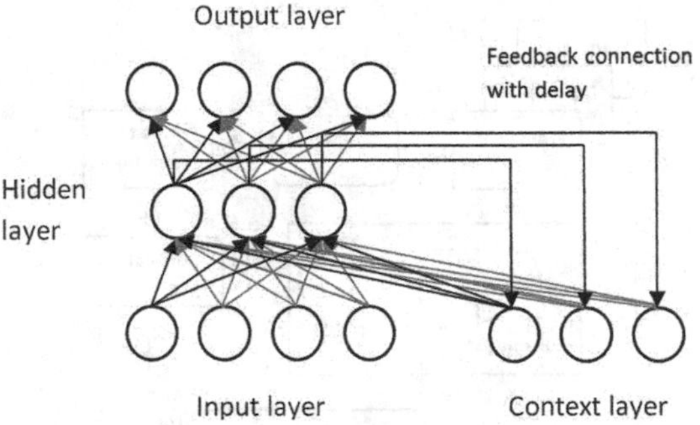

Fig. 2. Architecture of the SRN. The layers are fully connected with a feedback connection between the hidden and the context layers. The context layer provides both actual and delayed inputs to the hidden layer.

Considering I, S, C and O as input, hidden, context and output layer vectors, respectively, the vector components at the tth iteration can be written as [19]:

$$i_p^t \in I, p = 1, 2, \ldots, n \tag{5}$$

$$s_q^t \in S, q = 1, 2, \ldots, m \tag{6}$$

$$o_r^t \in O, r = 1, 2, \ldots, 1 \tag{7}$$

$$c_q^t = C \tag{8}$$

$$s_q^t = f(b_q^t) \tag{9}$$

$$c_q^t = s_q^{t-1} \tag{10}$$

In the above equations, *n*, *m* and *l* represent the numbers of nodes of input, hidden, and output layer, respectively, $f(\cdot)$ indicates the activation function of the qth hidden node at the tth iteration, while c_q^t denotes the input of the qth context layer node at the tth iteration and b_q^t is the linear output of the hidden node *q* at t^{th} iteration.

Let $W1$, $W2$ and $W3$ be the weight matrices between input and hidden layer, hidden and context layer and hidden and output layer, respectively. The output of hidden layer and output layer nodes at the tth iteration with these weight matrices can be represented by the following equations:

$$s_q^t = f\left(\sum_{p=1}^{m} w_{1qp} i_p^t + \sum_{q=1}^{m} w_{2qj} s_j^{t-1}\right) \tag{11}$$

$$o_r^t = f\left(\sum_{q=1}^{m} w_{3rq} s_q^t\right) \tag{12}$$

where the w_{1qp}, w_{2qj}, w_{3rq} are the elements of the weight matrices $W1$, $W2$, and $W3$, respectively.

The training of the network can be accomplished by exploiting the error back propagation algorithm [20]. In this algorithm, the error is minimized to converge to the target value by updating the link weights at each iteration using Eq. (13).

$$W^{new} = W^{old} + \alpha \Delta W \tag{13}$$

where α is the learning rate.

The error E expresses the difference between the set target at the output nodes and the actual output obtained as expressed in Eq. (14):

$$E(W) = \frac{1}{2} \sum_{k=1}^{e} \sum_{r=1}^{l} (O_r^t - o_r^t)^2 \tag{14}$$

where O_r^t and o_r^t represents the set target and the actual output from the network at the tth iteration, respectively, and e is the number of epochs.

2.3 Two-Dimensional Recursive Neural Networks

In the proposed 2D-RNN, hidden, context and output layers are organized in two-dimensional arrays all having the same dimensions as the input image. Unlike the Elman's network, the layers of the proposed network are not fully connected to each other. In particular, denoting by (x,y) the index of row and column of the matrix of the hidden layer, respectively, 2D-RNN uses for each element (x',y') of the input matrix also its nearest elements in the connection with the correspondent element of the hidden layer (x,y). Such type of association is replicated in the connection of the context layer with the hidden layer and in the connection between the hidden layer and the output layer, as shown in Fig. 3.

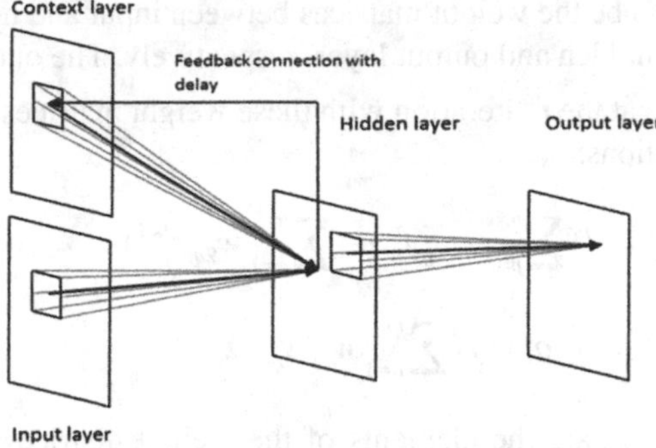

Fig. 3. Architecture of the 2D-RNN. Mapping of the image pixels from the input and context layers. Each node in the hidden layer receives input from both the actual and delayed image. Spatial information is preserved through the layers

Note that neuron (*x,y*) of the hidden layer is connected to all neurons (*x',y'*) of the input layer and to all neurons (*x'',y''*) of the context layer with:

$$x - k \leq x' \leq x + k, y - k \leq y' \leq y + k \tag{15}$$

$$x - k \leq x'' \leq x + k, y - k \leq y'' \leq y + k \tag{16}$$

In other words, the neuron at position (x,y) of the hidden layer is connected to the corresponding neuron of the input layer and to its nearest neighbors, and to the corresponding neuron of the context layer and to its nearest neighbors. Analogously, each neuron of the output layer is connected to the corresponding neuron of the hidden layer and to its nearest neighbors.

The training of the network can be accomplished again by the standard error back propagation algorithm. Note that the k parameter (dimension of the neighborhood) is also optimized and the Eqs. (11) and (12) are modified as follows:

$$s^t_{xy} = f(\sum\nolimits_{u=x-k}^{x+k} \sum\nolimits_{w=y-k}^{y+k} w_{1xyuw} i^t_{uw}) + \sum\nolimits_{u=x-k}^{x+k} \sum\nolimits_{w=y-k}^{y+k} w_{2xyuw} S^{t-1}_{uw}) \tag{17}$$

$$o^t_{xy} = f(\sum\nolimits_{u=x-k}^{x+k} \sum\nolimits_{w=y-k}^{y+k} w_{3xyuw} S^t_{uw}) \tag{18}$$

3 Experimental Results

3.1 Basic Assumptions

The proposed tracking algorithm has been validated on a limited but challenging set of sequences, extracted from the final of triple jump at the London 2012 Summer Olympics. In this case study, the computation of the runner's trajectory is subject to several critical issues such as moving background, noise, articulated motion, scene illumination changes and dynamic background, as illustrated in Fig. 4.

Fig. 4. Frames extracted from the triple jump sequence. Several visual artifacts can be noticed, such as moving background, change in the object (the runner) shape, changes in lighting and occlusions.

Several object-tracking methods impose constraints on the motion and/or the object's appearance of objects. Most tracking algorithms assume that the object motion is smooth and without abrupt changes. Some approaches constrain the object motion to be of constant velocity or constant acceleration based on a priori information. As stated in the previous section, prior knowledge about the number and the size of objects, or the object appearance and shape, have also been used to simplify the problem. The proposed method does not make assumptions. Furthermore, it does not use any pre-processing of the image to remove external objects (i.e. TV-written), it does not apply any pre-processing such as band-pass filtering or segmentation. The developed object tracker shows a bounding box that contains the athlete in all different frames of a video, as in Fig. 5. The gold standard for each frame is provided through manual labeling of the region of interest and more specifically by defining the position of the pelvic bones of the athlete.

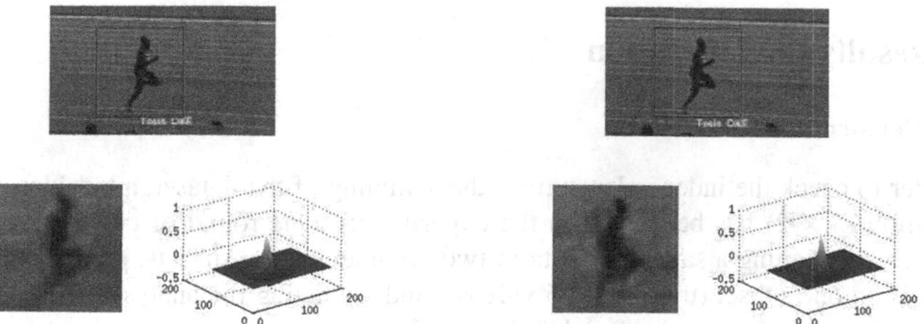

Fig. 5. Actual frame, predicted next frame and the correlation diagram computed by the SRN on the left and by the 2D-RNN on the right. The blue bounding box represents the computed gold standard while the green box represents the position computed by the RNNs.

The main processing steps for the experimental phase are the following: extraction of the single JPG frames for each sequence of the MP4 video; resizing of all frames from 1280 × 720 pixels to 128 × 72 pixels; conversion of the frames to gray levels; RNNs training and testing by applying a BB of 50 × 50 pixels to the resized images.

3.2 Dataset

The experimental dataset is composed of 10 sequences, downloaded from the YouTube platform. Each sequence relates to a different athlete in the final of the triple jump at the Olympics London 2012.

The sequences are characterized by a frame rate of 29 images/s; the dimension of each original frame is 1280 × 720 pixels. Each sequence has a duration of about 45 s; only one frame every ten is considered for further processing, therefore for each sequence the number of frames processed varies between 97 and 127.

3.3 Configuration

A comparison between the original SRN with respect to the novel 2D-RNN is performed. Input data are the same for both networks.

The SRN can identify the single-order dynamic system using fixed coefficients in the context neurons, using weight = 1 in the delay connections with the context layer; SRN best architecture needs 2500 input, 250 hidden, 250 context and 2500 output neurons. Note that the input and output layers are related to the frame input matrix (the 50 × 50 pixels bounding box) while the number of neurons of the context and hidden layers have been optimized trying several configurations.

2D-RNN is not fully connected as the SRN; it requires 2500 input, hidden, context and output neurons (the numbers of neurons for all layers is fixed with respect to the frame input matrix). Best results are obtained for a number of nearest neighbors k = 3, using weight = 1 in the delay connections with the context layer. For both RNNs and for all neurons a logistic standard transfer functions has been adopted.

4 Results and Discussion

4.1 Performances of RNNs

In order to check the independence from the sampling of the dataset, a k-folder cross validation (5 × 2) has been used in the experiments. One round of cross-validation involves partitioning a sample of data in two complementary subsets, performing the analysis on one subset (train set of 5 videos), and validating the analysis on the other subset (test set of 5 videos); after that the simulation is repeated exchanging train and test sets. To reduce variability, 5 rounds of cross-validation are performed using random different partitions, and the validation results are averaged over the 10 (5 × 2) rounds.

In Table 1 the comparison of the best configuration for both RNNs on the same random train test and blind test set is presented; learning times refer to a simple desktop architecture based on a Intel CoreTM 2 DUO CPU E 8400 @3.00 GHz and 4 GB RAM.

Table 1. Performance comparison of the of SRN and the 2D-RNN for the same blind test set

Parameters	SRN	2D-RNN
Input	2500	2500
Output	2500	2500
Hidden	250	2500
Context	250	2500
Learning rate	0.005	0.05
Epocs	280	130
Connections	1312500	367500
Learning time (s)	9230	1092
Best rmse	0.114	0.104

Table 1 clearly shows that the learning phase of 2D-RNN is faster than SRN, and 2D-RNN produces the best results. Obviously the best learning rate for both RNNs are reported,in particular, the root-mean-square deviation (rmse) is repeatedly computed on the test set after a random selection of the training set followed by the learning phase. The results for the 2D-RNN, in 5×2 cross validation, is a mean rmse $= 0.105 \pm 0.003$. In summary, Table 1 shows that 2D-RNN, compared to SRN on the same dataset, provides a better rmse; the results are stable for the 5x2 cross-validation and 2D-RNN is faster than SRN in terms of learning time and epochs. The complexity of the 2D-RNN is minor than SRN in terms of connections.

4.2 Results for Tracking

Visual tracking results can be described through the distances between the center of manual annotation (the pelvic bones) of the athlete and the center of bounding box in the 2D-RNN next frame prediction, illustrated in previous Fig. 5. Using only one frame every ten and starting from the original frame rate information, the RNN previsions correspond to one image every 0.344 s. In Fig. 5 corresponding samples for the SRN (left) with actual frames and next frame prediction are shown, together with the correlation diagram. The same results are shown for the 2D-RNN in Fig. 5 (right). In the surface plot, the peak of the cross-correlation matrix occurs where the sub images are best correlated.

It should be noted that all the next frames prediction in all figures are blurred because the RNN produces a distribution of positions related to the probability density. This distribution reflects the variability of the images used to train the RNN. Naturally, the athletes move their limbs during the run in different ways. The fact that the body image is blurred is the consequence of the inability to produce an accurate prediction. On the other hand, images with clear prediction of the part of the body with respect to blurred images, would give rise to lower correlations on the average, and consequently more average errors due to the variability and not exact predictability of the next image. This is a compromise tolerable because in the tracking problem it is only necessary to have an accurate prediction of the center of the BB to follow the object of interest.

Furthermore Fig. 6 show a qualitative comparison between the real next frame and calculated next frame prevision of the SRN and 2D-RNN.

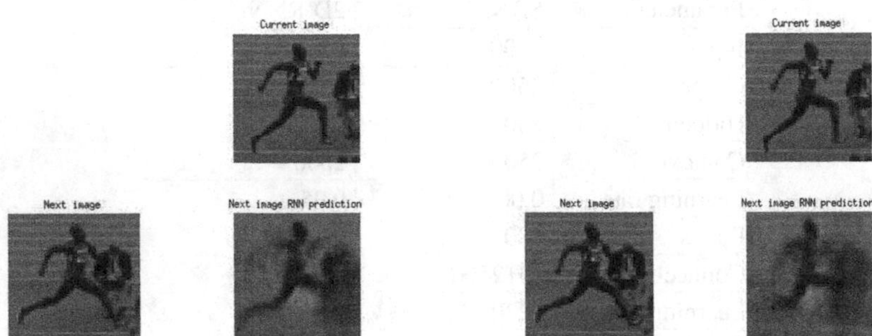

Fig. 6. Actual frame (top), next frame and predicted next frame (bottom) from the SRN on the left and 2D-RNN on the right.

In Fig. 7 the diagrams of the Euclidean distances between the gold standard and the center of the RNNs prevision are shown, normalized with respect to the dimension of the BB.

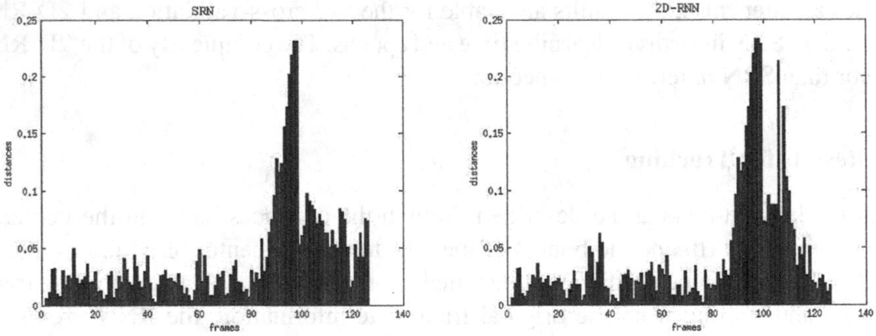

Fig. 7. Euclidean distances between the center of the computed bounding box and the gold standard for each frame, normalized with respect to the side of bounding box. The large deviation shown at about 85 frames is due to the runner landing on the sand.

The literature is divided in two types of measures for precision and recall: one is based on the localization of objects as a whole such as the F-score or other index [36] and one based on the position at pixel level. In the proposed approach object tracking is performed at pixel level. There are no lost frames in the proposed approach, therefore evaluation metrics based on accuracy is not used; however, with the aim to show the performances on the correct location of the BB, a position based measure (PBM) can be used. PBM is defined as [21]:

$$BPM = \frac{1}{N_f} \sum\nolimits_i [1 - \frac{D(i)}{T_h}] \tag{19}$$

Where

$$T_h = \frac{1}{2}[(BB_w) + (BB_h)] \tag{20}$$

depends on the dimensions (width and height) of the bounding box.

In Eq. 19, N_f is the total number of frames considered whilst $D(i)$ is the L1-norm distance between the gold standard and the BB predicted by RNN. Using such index in our dataset the resulting mean of BMP (proposed system, first 85 frames) is expressed in the Table 2. In particular it is possible to note a better performance of the 2D-RNN with respect to the SNR.

Table 2. Comparison of the two RNNs tracking system in terms of BPM

RNN	BMP (first 85 frames)
SNR	0.95 ± 0.040
2D-RNN	0.97 ± 0.002

Considering the entire test set, the distribution of the BB position errors between the predicted coordinates and the gold standard, in pixels, with respect to the original image, is shown in Fig. 8.

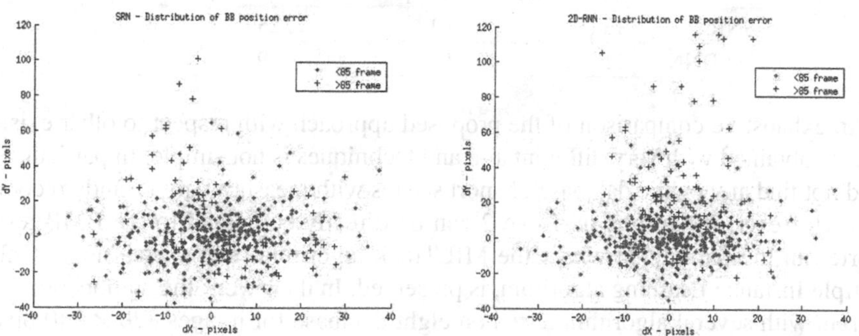

Fig. 8. Scatter plots representing the distribution of the bounding box errors, between the computed box coordinates and the gold standard for each sequential frame. The red dots represent the target positions computed before the runner landing on the sand, while the blue dots represent the target positions computed after the runner landing on the sand.

Note that the interesting part of the sequences is composed by the first 85 frames; in fact, after frame 85 we typically register a degradation of the image due to the sand effect after landing. In Fig. 9 a comparative scatter plot represents the distribution of the BB position error on the original images, for both RNNs considering only the first 85 frames.

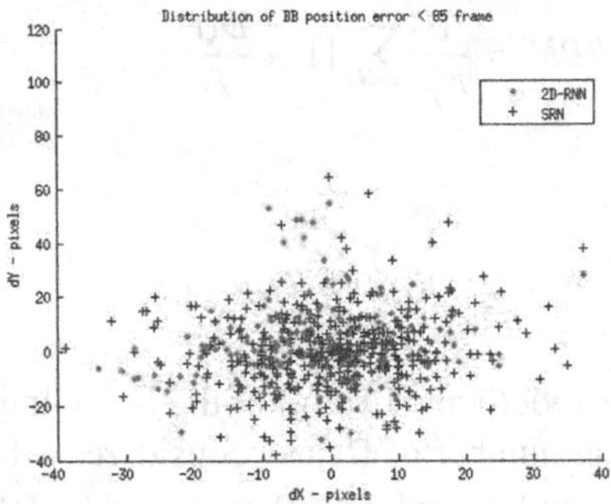

Fig. 9. Scatter plot representing the distribution of the bounding box errors, between the coordinates of the computed box and the gold standard for the first 85 frames. The colors represent the method applied to compute the target position

In particular from Fig. 9 it turns out that the most part of errors for both RNNs are within ± 30 pixels with respect to the original image of 1280X720 pixels, where the dimension of the BB is 500 × 500 pixels; in Table 3 are represented the position errors in pixels for both RNNs and the details of the coordinates of the position mean square error (Position MSE).

Table 3. Comparison of the two RNNs position error in pixels for the first 85 frames.

RNN	Position error dX (pixels)	Position error dY (pixels)	Position MSE (pixels)
SNR	13	19	23
2D-RNN	11	14	19

An exhaustive comparison of the proposed approach with respect to other existing datasets obtained with very different aim and techniques is not simple. In particular we could not find any public database of sport scenes with measured gold standard coordinates. However, the results of Table 2 can be directly compared to the BMP results reported in the paper [11] where the MILTrack algorithm, that uses a novel Online Multiple Instance Learning algorithm, is presented. In their work the authors provide a diagram with several algorithm tested on eight database for images 320 × 240 pixels. Normalizing the results to the scale of the adopted BB, it is possible to conclude that our algorithm, without lost frames, obtains similar performances of the best proposed MILTrack algorithm.

An alternative measure quite convenient for comparison is deviation. Deviation represents the capability of a tracker to determine the correct position of the target and measures the accuracy of tracking [12]. In particular, by using Deviation as the error of the center location expressed in pixels as a tracking accuracy measure:

$$Deviation = 1 - \frac{\sum_{i \in Ms} d(T^i, GT^i)}{|Ms|} \tag{21}$$

where $d(T^i, GT^i)$ is the normalized distance between the centroids of bounding box (BB) and the gold standard and Ms denotes the set of frames in a video where the tracked BB matches with the gold standard BB.

In the proposed approach, again normalizing with respect to the side of the BB and using the first 85 frames, for 10 sequences, a Deviation equal to about 0.98 for both RNNs is obtained. Taking into account all frames of the 10 sequences in the dataset the Deviation value slightly decreases to about 0.96 for both RNNs.

This result can be compared with the values reported in [21] and related to the articles [11, 12], [22–37], where the target is considered tracked correctly each time the overlap between the current forecast and the real position of the object area overlap for more than 50 %. As shown in Table 4, the proposed approach achieves the same or even better accuracy than the algorithms at the state of the art.

Table 4. Comparison of different approaches for target tracking applied to the jumping sequence.

RNN	Deviation
Elman's neural network (SRN)	0.96
2D Recurrent Neural Network (2D-RNN)	0.96
Normalized Cross-Correlation (NCC) [22]	0.95
Lucas-Kanade Tracker (KLT) [23]	0.95
Kalman Appearance Tracker (KAT) [24]	0.95
Fragments-based Robust Tracking (FRT) [25]	0.94
Mean Shift Tracking (MST) [26]	0.93
Locally Orderless Tracking (LOT) [27]	0.94
Incremental Visual Tracking (IVT) [28]	0.95
Tracking on the Affine Group (TAG) [29]	0.95
Tracking by Sampling Trackers (TST) [30]	0.94
Tracking by Monte Carlo sampling (TMC) [31]	0.96
Adaptive Coupled-layer Tracking (ACT) [32]	0.94
L1-minimization Tracker (L1T) [33]	0.95
L1 Tracker with Occlusion detection (L1O) [33]	0.95
Foreground-Background Tracker (FBT) [34]	0.95
Hough-Based Tracking (HBT) [35]	0.93
Super Pixel tracking (SPT) [36]	0.93
Multiple Instance learning Tracking (MIT) [11]	0.94
Tracking, Learning and Detection (TLD) [37]	0.93
STRuck: Structured output tracking with kernels (STR) [12]	0.94

5 Conclusion

A novel tracking algorithm has been presented, where two complementary RNN topologies are used without any pre-processing of the images. The temporal memory of the recursive neural networks is used to keep the correlation among processed pixels and to perform the next frame prediction at the temporal distances of ten frames, with respect to the frame of interest.

The novel RNN algorithm proposed performs well for generic, iconic based, image tracking. This is mainly due to the two dimensional approach where for each pixel of the input image also the information of its k nearest pixels are considered. Such kind of connection of the layers (input-hidden and hidden-output) is preferred with respect to the full connection, with great advantages in terms of rmse, learning times and BMP of the tracking.

A qualitative comparison with different approaches on different datasets is also performed, obtaining good results on measures such as deviation, that reveals an excellent performance compared to the literature.

The extension of this approach will be applied in the future to large benchmark datasets with different types of object of interest, and replacing the manual selection of the BB in the first frame with an automatic procedure designed to recognize objects belonging to predefined classes.

The results are originally measured on a triple jump dataset and could be very helpful for analysis of athlete errors in the jump in computer aided coaching or for TV highlight. However the novel method doesn't require any information related to the object of interest in the scene and it is therefore suitable for a large set of applications from sport activities to video-surveillance.

References

1. Marr, D.: Vision: A Computational Approach. Freeman & Co., San Francisco (1982)
2. Ullman, S.: The interpretation of structure from motion. Proc. Roy. Soc. Lond. B: Biol. Sci. **203**(1153), 405–426 (1979)
3. Gibson, J.J.: The Ecological Approach to Visual Perception, Classic edn. Psychology Press, New York (2014)
4. Denman, H., Rea, N., Kokaram, A.: Content-based analysis for video from snooker broadcasts. Comput. Vis. Image Underst. **92**(2), 176–195 (2003)
5. Kokaram, A., Pitie, F., Dahyot, R., Rea, N., Yeterian, S.: Content controlled image representation for sports streaming. In: Proceedings of Content-Based Multimedia Indexing (CBMI05)
6. Yilmaz, A., Javed, O., Shah, M.: Object tracking: a survey. ACM Comput. Surv. (CSUR) **38**(4), 13 (2006)
7. Hong, S., You, T., Kwak, S., Han, B.: Online tracking by learning discriminative saliency map with convolutional neural network, arXiv preprint arXiv:1502.06796
8. Bao, C., Wu, Y., Ling, H., Ji, H.: Real time robust L1 tracker using accelerated proximal gradient approach. In: 2012 IEEE Conference on Computer Vision and Pattern Recognition (CVPR), pp. 1830–1837. IEEE (2012)
9. Jia, X., Lu, H., Yang, M.-H.: Visual tracking via adaptive structural local sparse appearance model. In: 2012 IEEE Conference on Computer Vision and Pattern Recognition (CVPR), pp. 1822–1829. IEEE (2012)
10. Mei, X., Ling, H.: Robust visual tracking using L1 minimization. In: 2009 IEEE 12th International Conference on Computer Vision, pp. 1436–1443. IEEE (2009)
11. Babenko, B., Yang, M.-H., Belongie, S.: Robust object tracking with online multiple instance learning. IEEE Trans. Pattern Anal. Mach. Intell. **33**(8), 1619–1632 (2011)

12. Hare, S., Saffari, A., Torr, P.H.: Struck: structured output tracking with kernels. In: 2011 IEEE International Conference on Computer Vision (ICCV), pp. 263–270. IEEE (2011)
13. Grabner, H., Grabner, M., Bischof, H.: Real-time tracking via on-line boosting. In: BMVC, vol.1, p. 6 (2006)
14. Gall, J., Yao, A., Razavi, N., Van Gool, L., Lempitsky, V.: Hough forests for object detection, tracking, and action recognition. IEEE Trans. Pattern Anal. Mach. Intell. **33**(11), 2188–2202 (2011)
15. Schulter, S., Leistner, C., Roth, P.M., Bischof, H., Van Gool, L.J.: On-line hough forests. In: BMVC 2011, pp. 1–11 (2011)
16. Wang, X., Ma, L., Wang, B., Wang, T.: A hybrid optimization-based recurrent neural network for real-time data prediction. Neurocomputing **120**, 547–559 (2013)
17. Elman, J.L.: Finding structure in time. Cogn. Sci. **14**(2), 179–211 (1990)
18. Korekado, K., Morie, T., Nomura, O., Ando, H., Nakano, T., Matsugu, M., Iwata, A.: A convolutional neural network VLSI for image recognition using merged/mixed analog-digital architecture. In: Palade, V., Howlett, R.J., Jain, L. (eds.) KES 2003. LNCS, vol. 2774, pp. 169–176. Springer, Heidelberg (2003)
19. Şeker, S., Ayaz, E., Türkcan, E.: Elman's recurrent neural network applications to condition monitoring in nuclear power plant and rotating machinery. Eng. Appl. Artif. Intell. **16**(7), 647–656 (2003)
20. Haykin, S.: Neural networks: a comprehensive foundation, 2nd edn. Prentice Hall PTR, Upper Saddle River (1998)
21. Smeulders, A.W., Chu, D.M., Cucchiara, R., Calderara, S., Dehghan, A., Shah, M.: Visual tracking: an experimental survey. IEEE Trans. Pattern Anal. Mach. Intell. **36**(7), 1442–1468 (2014)
22. Briechle, K., Hanebeck, U.D.: Template matching using fast normalized cross correlation. In: Aerospace/Defense Sensing, Simulation, and Controls, International Society for Optics and Photonics, pp. 95–102 (2001)
23. Baker, S., Matthews, I.: Lucas-kanade 20 years on: a unifying framework. Int. J. Comput. Vis. **56**(3), 221–255 (2004)
24. Nguyen, H.T., Smeulders, A.W.: Fast occluded object tracking by a robust appearance filter. IEEE Trans. Pattern Anal. Mach. Intell. **26**(8), 1099–1104 (2004)
25. Adam, A., Rivlin, E., Shimshoni, I.: Robust fragments-based tracking using the integral histogram. In: 2006 IEEE Computer Society Conference on Computer Vision and Pattern Recognition, vol. 1, pp. 798–805. IEEE (2006)
26. Comaniciu, D., Ramesh, V., Meer, P.: Real-time tracking of non-rigid objects using mean shift. In: Proceedings of 2000 IEEE Conference on Computer Vision and Pattern Recognition, vol. 2, pp. 142–149. IEEE (2000)
27. Oron, S., Bar-Hillel, A., Levi, D., Avidan, S.: Locally orderless tracking. In: 2012 IEEE Conference on Computer Vision and Pattern Recognition (CVPR), pp. 1940–1947. IEEE (2012)
28. Ross, D.A., Lim, J., Lin, R.-S., Yang, M.-H.: Incremental learning for robust visual tracking. Int. J. Comput. Vision **77**(1–3), 125–141 (2008)
29. Kwon, J., Lee, K.M., Park, F.C.: Visual tracking via geometric particle filtering on the affine group with optimal importance functions. In: 2009 IEEE Conference on Computer Vision and Pattern Recognition CVPR 2009, pp. 991–998. IEEE (2009)
30. Kwon, J., Lee, K.M.: Tracking by sampling trackers. In: 2011 IEEE International Conference on Computer Vision (ICCV), pp. 1195–1202. IEEE (2011)

31. Kwon, J., Lee, K.M.: Tracking of a non-rigid object via patch-based dynamic appearance modeling and adaptive basin hopping monte carlo sampling. In: IEEE Conference on Computer Vision and Pattern Recognition, 2009 CVPR 2009, pp. 1208–1215. IEEE (2009)
32. Čehovin, L., Kristan, M., Leonardis, A.: An adaptive coupled-layer visual model for robust visual tracking. In: 2011 IEEE International Conference on Computer Vision (ICCV), pp. 1363–1370. IEEE (2011)
33. Mei, X., Ling, H., Wu, Y., Blasch, E., Bai, L.: Minimum error bounded efficient L1 tracker with occlusion detection. In: Proceedings of IEEE CVPR, Providence, RI, USA (2011)
34. Nguyen, H.T., Smeulders, A.W.: Robust tracking using foreground-background texture discrimination. Int. J. Comput. Vision **69**(3), 277–293 (2006)
35. Godec, M., Roth, P.M., Bischof, H.: Hough-based tracking of non-rigid objects. Comput. Vis. Image Underst. **117**(10), 1245–1256 (2013)
36. Yang, F., Lu, H., Yang, M.-H.: Robust superpixel tracking. IEEE Trans. Image Process. **23**(4), 1639–1651 (2014)
37. Kalal, Z., Matas, J., Mikolajczyk, K.: PN learning: Bootstrapping binary classifiers by structural constraints. In: 2010 IEEE Conference on Computer Vision and Pattern Recognition (CVPR), pp. 49–56. IEEE (2010)

Choice of Best Samples for Building Ensembles in Dynamic Environments

Joana Costa[1,2], Catarina Silva[1,2(✉)], Mário Antunes[1,3], and Bernardete Ribeiro[2]

[1] School of Technology and Management, Polytechnic Institute of Leiria, Leiria, Portugal
{joana.costa,catarina,mario.antunes}@ipleiria.pt

[2] Department of Informatics Engineering, Center for Informatics and Systems of the University of Coimbra (CISUC), Coimbra, Portugal
{joanamc,catarina,bribeiro}@dei.uc.pt

[3] Center for Research in Advanced Computing Systems (CRACS), INESC-TEC, University of Porto, Porto, Portugal
mantunes@dcc.fc.up.pt

Abstract. Machine learning approaches often focus on optimizing the algorithm rather than assuring that the source data is as rich as possible. However, when it is possible to enhance the input examples to construct models, one should consider it thoroughly. In this work, we propose a technique to define the best set of training examples using dynamic ensembles in text classification scenarios. In dynamic environments, where new data is constantly appearing, old data is usually disregarded, but sometimes some of those disregarded examples may carry substantial information. We propose a method that determines the most relevant examples by analysing their behaviour when defining separating planes or thresholds between classes. Those examples, deemed better than others, are kept for a longer time-window than the rest. Results on a *Twitter* scenario show that keeping those examples enhances the final classification performance.

Keywords: Dynamic environments · Ensembles · Drift · Text classification · Social networks

1 Introduction

Information spread in online scenarios has created a new information sharing paradigm. Nowadays, it is possible to create and disseminate information in numerous formats by publishing it world-wide, making the Web responsible for a deluge of data. Albeit we can undoubtedly benefit from all these data, one major drawback of such overflow is the inability to easily perceive important, significant and accurate information. This challenge arises not only because the amount of data is overwhelming to process, but also because time plays an important role by fast out-dating information.

C. Jayne and L. Iliadis (Eds.): EANN 2016, CCIS 629, pp. 35–47, 2016.
DOI: 10.1007/978-3-319-44188-7_3

In dynamic environments drift occurs and deployed models performance is reduced when changes occur between the distribution that generated the data used to define the model and the current drifted scenario. To handle such challenges, some form of model ageing must be put in place.

In [1–3] the proposed approaches try to detect that a drift occurred and react accordingly. This reaction is usually non-trivial, since the new samples can carry more relevant and new information that should probably contribute more to the final model [4]. In fact, in dynamic environments, such as in social networks we will be using as case study, effective learning requires a learning algorithm with the ability to detect context changes without being explicitly informed about them, quickly recovering from the context change and adjusting its hypothesis to the new context [1,5].

A lot of effort has been place in adapting to such news' samples. However, previous samples can also play an important role, and previously experienced situations should also be considered when old contexts and corresponding concepts reappear [6]. In this work we propose a framework for choosing the best samples for building classifiers in dynamic environments. Using a sliding time-window approach, models are retrained with updated real-time samples. Furthermore, we analyze previous samples' behaviour, e.g. when defining separating planes, to detect stronger examples and keep them in the pool. Hence, these chosen samples are kept for a longer time-window than the rest. Tests carried out on a *Twitter* stream case study support the hypothesis that keeping such chosen examples enhances the final classification performance.

The rest of the paper is organized as follows. Section 2 introduces background concepts and state of the art on dynamic environments and social networks. Section 3 presents the proposed framework for choosing the best samples for building classifiers in dynamic environments. In Sect. 4 the experimental setup is set, including the case study dataset and the performance metrics, followed by the experimental results and analysis in Sect. 5. Finally, Sect. 6 concludes the paper with conclusions and future work.

2 Background

2.1 Social Networks

Social networks have settled definitely in the daily routine of Internet users. They have also gained increasing importance and are being widely studied in many fields of research over the last years, such as computer, social, political, business and economical sciences.

There exists a wide set of distinct social networks for different purposes and scope, being *Twitter* and *Facebook* two of the most popular ones. In a purely research perspective, these social networks make accessible through their own public API a deluge of relevant data related with users' daily status, news and events. Data is produced in a non-deterministic way, turning social networks in a dynamic environment in which we may apply learning and detection strategies and algorithms.

The focus of our work is *Twitter* social media platform (www.twitter.com), more precisely on applying learning and classification strategies to cope with different types of variations of context (*drift*) through time [1,7].

2.2 Dynamic Environments

Social networks can be seen as a dynamic (also entitled as non-stationary) environment, in which information is produced by users in a timely order. Time plays a crucial role in *Twitter* information processing, as past events can give important insights to understand how previously seen information is relevant to improve learning and classification of future unseen and related events.

In that sense, learning strategies would be able to learn in dynamic environments and apply innovative strategies to dead with a *"recent memory"* of past events, in order to better identify future and unseen ones.

There can be several approaches to tackle dynamic environments [4]: instance selection, instance weighting and ensemble learning. A review of concept drift applied to intrusion detection is presented in [8].

Learn++.NSE and Learn++.CDS [3,9] are algorithms to deal with drift, namely with imbalanced datasets. An ensemble technique, DWM-WIN, was proposed in [2], to overcome the known limits of dynamic weight majority [10], namely not taking into consideration the timestamp of classifiers or the previous performances.

In this section we presented some examples of the importance of tackling drift in dynamic scenarios like social networks, and particularly in *Twitter*. Multiple applications like spam email filtering, intrusion detection, recommendation systems, event detection, or improve search capabilities are just pointed examples [1].

3 Proposed Approach

This section describes the proposed approach to define the best set of training examples using dynamic ensembles in text classification scenarios. We will firstly present our *Twitter* classification problem and then proceed with formalizing the proposed models.

3.1 Case Study: Twitter Stream

Twitter stream constitutes a paradigmatic example of a text-based scenario where drift phenomena occur commonly. *Twitter* is a micro-blogging service where users post text-based messages up to 140 characters, also known as *tweets*. It is also considered one of the most relevant social networks, along with *Facebook*, as millions of users are connected to each other by a following mechanism that allows them to read each others posts.

Twitter is also responsible for the popularization of the *'hashtag'* concept. An *hashtag* is a single word started by the symbol "#" that is used to classify the message content and to improve search capabilities. Besides improving search capabilities, *hashtags* have been identified as having multiple and relevant potentialities, like promoting the phenomenon described in [11] as *micro-meme*, i.e. an idea, behavior, or style that spreads from person to person within a culture [12]. By tagging a message with a trending topic *hashtag*, a user expands the audience of the message, compelling more users to express themselves about the subject [13].

Considering the importance of the *hashtag* in *Twitter*, it is relevant to study the possibility of evaluating message contents in order to predict its *hashtag*. If we can classify a message based on a set of *hashtags*, we are able to suggest an *hashtag* for a given *tweet*, bringing a wider audience into discussion [14], spreading an idea [15], get affiliated with a community [16], or bringing together other Internet resources [17].

This case study aims to classify *Twitter* messages. A *Twitter* classification problem can be described as a multi-class problem that can be cast as a time series of *tweets*. It consists of a continuous sequence of instances, in this case, *Twitter* messages, represented as $\mathcal{X} = \{x_1, \ldots, x_t\}$, where x_1 is the first occurring instance and x_t the latest. Each instance occurs at a time, not necessarily in equally spaced time intervals, and is characterized by a set of features, usually words, $\mathcal{W} = \{w_1, w_2, \ldots, w_{|\mathcal{W}|}\}$. Consequently, instance x_i is denoted as the feature vector $\{w_{i1}, w_{i2}, \ldots, w_{i/W/}\}$.

When x_i is a labelled instance it is represented as the pair (x_i, y_i), being $y_i \in \mathcal{Y} = \{y_1, y_2, \ldots, y_{|\mathcal{Y}|}\}$ the class label for instance x_i.

We have used a classification strategy previously introduced in [7], where the *Twitter* message *hashtag* is used to label the content of the message, which means that y_i represents the *hashtag* that labels the *Twitter* message x_i.

Notwithstanding it is a multi-class problem in its essence, it can be decomposed in multiple binary tasks in a one-against-all binary classification strategy. In this case, a classifier h^t is composed by $|Y|$ binary classifiers.

3.2 Learning Models

We are focusing on dynamic ensembles in text classification scenarios, where the ensemble must adapt to deal with changes usually dependent on hidden contexts. One of the major challenges is the amount of data, specially when dealing with streams. It is sometimes unfeasable to store all the previously seen data, but it may carry substantial information for future use.

In [18] we have studied the impact of longstanding examples in future classification time-windows. The rationale of the presented idea was to store previously seen examples for a period of time regardless the effect they might have as a solo example. The most relevant action was to keep examples for a period of time instead of discarding them for future use. We were also not dealing with ensembles but single classifiers. Differently from that approach, we are now proposing to choose examples based on the effect they might have individually.

Our baseline model, created for comparison purposes, proposes to store all the information gathered by storing models and combining them as an ensemble. For each collection of documents $\mathcal{T}$, that contain both positive and negative examples and occur in a time-window t, a classifier $\mathcal{C}^t$ is trained and stored. When a new collection of documents in the subsequent time-window occurs, all the previously trained classifiers are loaded, and the system will classify the newly seen examples. The prediction function of the ensemble, composed by the set of classifiers already created, is a combined function of the outputs of all the considered classifiers. A majority voting strategy where each model participates equally is then put forward. The documents of the previously seen time-windows are not stored in this approach even though the possible learning information is stored along in the classifier trained immediately after it.

We then propose an ensemble learning model, namely reinforced model. The main difference is that we define a collection of documents $\mathcal{R}$ that contain all the classification errors that occur in the time-windows prior to t. The classification errors are considered based on the ensemble classification and not in each model classification output. For each time-window t, a classifier $\mathcal{C}^t$ is trained with the collection of documents $\mathcal{T}$ plus the collection of documents $\mathcal{R}$ and stored. When a new collection of documents in the subsequent time-window occur, all the previously trained classifiers are loaded, and will classify the newly seen examples participating equally to the final decision of the ensemble.

The collection of documents $\mathcal{R}$ might retain the misclassified examples indefinitely or be pruned in a time-based approach. If pruned, the lifetime of an example in $\mathcal{R}$ is dependent of a pre-defined time-window size g, which means that, in time-window t, $\mathcal{R}$ contains the misclassified examples that occur in all time-windows that satisfy the condition t-$g > 0$.

Figure 1 depicts the proposed models. It is important to understand that we represented time in two different directions because we are working with a time series and the last seen scenario is the input for the new one. The major

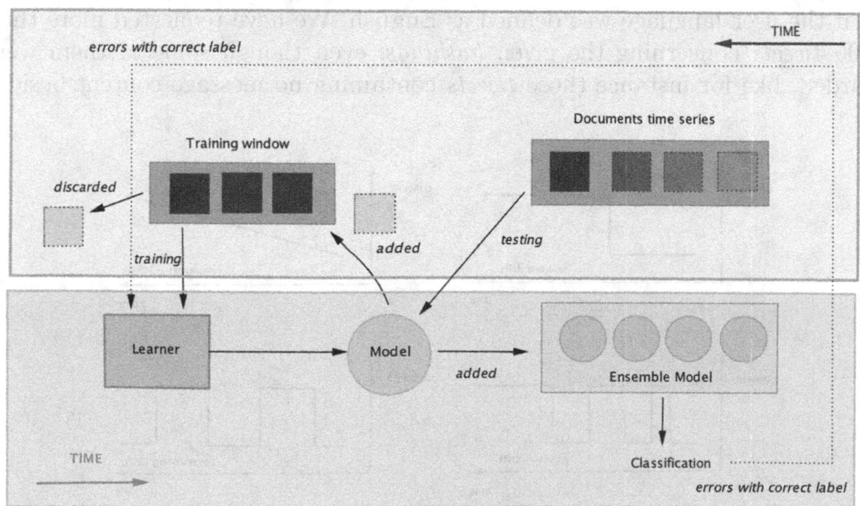

Fig. 1. Proposed models

difference between both models is using the outcome of the classification as a new incoming in the subsequent training phase.

4 Experimental Setup

In this section we will detail the dataset we propose to test and evaluate our approach. We then characterize the methodology for document representation and proceed dealing with the pre-processing methods. Finally, we conclude by introducing the performance metrics used to evaluate the proposed approach.

4.1 Dataset

The dataset we have defined to evaluate and validate our strategy was carried out by defining 10 different *hashtags* that would represent our drifts (see Fig. 2), based on the assumption that they would denote mutually exclusive concepts, like *#realmadrid* and *#android*. By trying to use mutually exclusive concepts we intend to avoid misleading a classifier, as two different *tweets* could represent the same concept, and that way introducing a new variable to our scenario that could mislead the possible obtained results. In order to achieve a considerable amount of *tweets*, and consequently diversity, we have chosen trending *hashtags* like *#syrisa* and *#airasia*. Table 1 shows the chosen *hashtags* and the corresponding drift they represent. This correspondence was done arbitrarily and do not correspond to any possible occurrence in the real *Twitter* scenario, since as stated above, no information is known about the occurrence of drifts in *Twitter*.

The *Twitter API* (https://dev.Twitter.com/) was then used to request public *tweets* that contain the defined *hashtags*. The requests have been cared of between 28 December 2014 and 21 January 2015 and *tweets* were only considered if the user language was defined as English. We have requested more than 75.000 *tweets* concerning the given *hashtags*, even though some of them were discarded, like for instance those *tweets* containing no message content besides

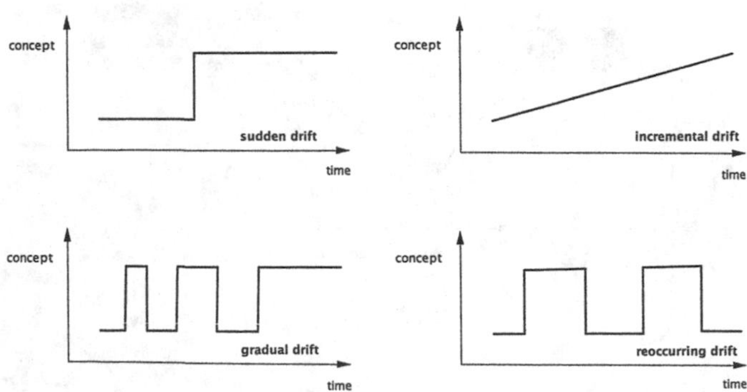

Fig. 2. Different types of drift

Table 1. Mapping between type of drift and hashtag.

Drift	Hashtag
Sudden #1	#syrisa
Sudden #2	#airasia
Gradual #1	#isis
Gradual #2	#bieber
Incremental #1	#android
Incremental #2	#ferrari
Reoccurring	#realmadrid
Normal #1	#jobs
Normal #2	#sex
Normal #3	#nfl

the *hashtag*. The *hashtag* was then removed from the message content in order to be exclusively used as the document label. The *tweets* matching this presumptions were considered labelled and suited for classification purposes, and were used by their appearing order in the public feed.

We have simulated the different types of drift by artificially defining timestamps to the previously gathered *tweets*. Drift Oriented Tool System (DOTS) is a drift oriented framework we have presented in [5]. DOTS was developed to dynamically create datasets with drift and is available for free download at http://dotspt.sourceforge.net/. DOTS is used to create the described dataset. It receives the *tweets* requested by the *Twitter API* and reproduces the defined artificial time-stamped time-windows. Time is represented as 100 continuous time windows, in which the frequency of each *hashtag* is altered in order to represent the defined drifts. Each *tweet* is then timestamped so it can belong to one of the time windows we have defined. For instance, Sudden #1 is represented by the appearance of 500 *tweets* with the *hashtag #syrisa* in each time windows from 25 to 32. It does not appear in any other time windows. Differently from *Sudden #1*, *Sudden #2* is represented with only 200 *tweets* with the *hashtag #airasia* in each time windows from 14 to 31. We tried to simulate a more soft occurring drift, but with a more long-standing appearance.

By making both concepts disappear, in time windows, 32 and 31, respectively, we also intended to simulate the opposite way of the [19] proposed sudden drift. Due to space constraints it is unbearable to present a table with the frequency of each *hashtag* in each time window, but it is important to state that *Incremental #2* and *Gradual #2* are represented by the same number of *tweets* in an equal number of time windows, but in a descent way than represented in *Incremental #1* and *Gradual #1.Normal #1*, *Normal #2* and *Normal #3* differ in the number of *tweets* that appear in a constant way in all time windows. Our final dataset contains 34.240 *tweets*.

4.2 Representation and Pre-processing

A *tweet* is represented as one of the most commonly used document representation, which is the vector space model, also known as *Bag of Words*. The collection of features is built as the dictionary of unique terms present in the documents collections. Each *tweet* of the document collection is indexed with the *bag* of the terms occurring in it, i.e., a vector with one element for each term occurring in the whole collection. The weighting scheme used to represent each term is the *term frequency - inverse document frequency*, also know as *tf-idf*.

High dimensional space can cause computational problems in text classification problems where a vector with one element for each occurring term in the whole connection is used to represent a document. Also, overfitting can easily occur which can prevent the classifier to generalize and thus the prediction ability becomes poor. In order to reduce feature space pre-processing methods were applied. These techniques aim at reducing the size of the document representation and prevent the mislead classification as some words, such as articles, prepositions and conjunctions, called *stopwords*, are non-informative words, and occur more frequently than informative ones. An english-based *stopword* dictionary was used, but *Twitter* related words like "*rt*" or "*http*" were also considered as they can be seen as *stopwords* in the *Twitter* context. *Stopword removal* was then applied, preventing those non informative words from misleading the classification.

Stemming method was also applied. This method consists in removing case and inflection information of each word, reducing it to the word stem. Stemming does not alter significantly the information included, but it does avoid feature expansion. Pre-processing methods were applied in DOTS.

4.3 Learning and Evaluation

The evaluation of our approach was done by the previously described dataset and using the Support Vector Machine (SVM) as stated above. This machine learning method was introduced by Vapnik [20], based on his Statistical Learning Theory and Structural Risk Minimization Principle. The idea behind the use of SVM for classification consists on finding the optimal separating hyperplane between the positive and negative examples. Once this hyperplane is found, new examples can be classified simply by determining which side of the hyperplane they are on. SVM constitute currently the best of breed kernel-based technique, exhibiting state-of-the-art performance in text classification problems [21–23]. SVM were used in our experiments to construct the proposed models. Based on [18] a 4 time-window size training window will be used.

In order to evaluate a binary decision task we first define a contingency matrix representing the possible outcomes of the classification, as shown in Table 2.

Several measures have been defined based on this contingency table, such as, error rate ($\frac{b+c}{a+b+c+d}$), recall ($R = \frac{a}{a+c}$), and precision ($P = \frac{a}{a+b}$), as well as

Table 2. Contingency table for binary classification.

	Class Positive	Class Negative
Assigned Positive	a	b
	(True Positives)	(False Positives)
Assigned Negative	c	d
	(False Negatives)	(True Negatives)

combined measures, such as, the van Rijsbergen F_β measure [24], which combines recall and precision in a single score:

$$F_\beta = \frac{(\beta^2 + 1)P \times R}{\beta^2 P + R}. \quad (1)$$

F_β is one of the best suited measures for text classification used with $\beta = 1$, i.e. F_1, an harmonic average between precision and recall (2), since it evaluates well unbalanced scenarios that usually occur in text classification settings and particularly in text classification in the *Twitter* environment.

$$F_1 = \frac{2 \times P \times R}{P + R}. \quad (2)$$

Considering the proposed approach and the fact that we are working with a time series and we use a one-against-all strategy, we will have a classifier for each batch of the time series that is composed by $|Y|$ binary classifiers, being $|Y|$ the collection of possible labels. To perceive the performance of the classification for each drift pattern, we will consider all the binary classifiers that were created in all the time series batches. To evaluate the performance obtained across time, we will average the obtained results. Two conventional methods are widely used, specially in multi-label scenarios, namely macro-averaging and micro-averaging. Macro-averaged performance scores are obtained by computing the scores for each learning model in each batch of the time series and then averaging these scores to obtain the global means. Differently, micro-averaged performance scores are computed by summing all the previously introduces contingency matrix values (a,b,c and d), and then use the sum of these values to compute a single micro-averaged performance score that represents the global score.

5 Experimental Results

In this section we evaluate the performance obtained on the *Twitter* data set using the two approaches described in Sect. 3, namely the baseline model approach and the reinforced model approach. In the reinforced model approach we obtained results by storing examples during 4 time-windows, represented as "$Reinforced_4$", and by storing examples *ad eternum*, named "$Reinforced_{forever}$". Table 3 summarises the performance results obtained by classifying the dataset, considering the micro-averaged F_1 measure.

Table 3. Micro-averaged F_1

Drift	Baseline	$Reinforced_4$	$Reinforced_{forever}$
Sudden #1	74,80 %	75,45 %	72,37 %
Sudden #2	87,80 %	88,06 %	86,55 %
Gradual #1	52,55 %	54,82 %	89,03 %
Gradual #2	62,21 %	63,43 %	89,21 %
Incremental #1	88,58 %	88,99 %	94,86 %
Incremental #2	77,21 %	79,26 %	74,28 %
Reoccurring	35,33 %	36,63 %	72,42 %
Normal #1	70,89 %	71,86 %	91,89 %
Normal #2	90,49 %	91,01 %	94,69 %
Normal #3	81,52 %	83,74 %	73,60 %
Average of micro-averaged F_1	78,27 %	79,33 %	83,75 %

Analysing the table we can observe that globally, and considering the average of the micro-averaged F_1, the storage of the priorly misclassified examples improves the overall classification. This is normal and expected as the learning models are trained with more informative examples and this leads to a better performance.

It is particularly important to note that model "$Reinforced_4$", which stores relevant examples for 4 time-windows, presents an improvement in the classification performance of all classes, regardless the type of drift they represent. This demonstrates the importance of the misclassified examples to improve the classification performance of the subsequent time-windows.

Nevertheless, when considering storing those examples for longer periods, specially *ad eternum*, one must notice that this improve is not straightforward. Most classes benefit from storing examples, and we have a significant improve in the average of the micro-averaged F_1, that increases from 78,27 % to 83,27 %, but some classes, namely *Sudden#1*, *Sudden#2*, *Incremental#2* and *Normal#3* have a worst classification performance. Firstly, both classes that represent a sudden drift have a performance decrease, *Sudden#1* from 74,80 % to 72,37 % and *Sudden#2* from 87,80 % to 86,55 %. We are confident that this decrease might be explained by the nature of the drift pattern.

A sudden drift is characterized by an abrupt increase of the frequency of a given class that occur during a period of time, followed by its disappearance. Storing examples that were misclassified, specially the positive ones that appeared firstly and remained misclassified until the classifier identified them as positive, will delude future classifiers, when the drift pattern is no longer represented. Secondly, we have identified a performance decrease in the classification of the class that represent *Incremental#2* drift. Similarly to what is happening with the sudden drift, the positive misclassified examples might be contributing to this decline, as the frequency of examples of this class is vanishing in time.

Finally, we have the *Normal#3* drift. Differently from the mentioned above, the frequency of this drift is not diminishing in the time serie. There is nothing in the drift pattern that allows us to infer what is happening, as it is exclusive to the *Normal#3* drift, and does not appear in the *Normal#1* or *Normal#2*, with the same nature. Although this is a supposition, that must be validated in future work, we do believe that it might be related to the class, that is the *hashtag* we have chosen to represent it. One of the possible problems that might arise from our approach is to store examples that are not representative of the class.

As we cannot guaranteed that a message is well-classified by its *hashtag*, we might be propagating errors by storing examples that were misclassified, but, differently from what we want, that is to store the most informative ones, we might be propagating the ones that do not represent the class at all. This is a problem that might arise in a dataset like ours, because it is impossible to validate that the *Twitter* user that wrote the *tweet*, is using the *hashtag* correctly.

6 Conclusions

In this paper we propose a method to determine the most relevant examples, by analysing their behaviour when defining separating planes or thresholds between classes. Those examples, deemed better than others, are kept for a longer time-window than the rest. The main idea is to boost the classification performance of learning models by providing additional and significant information.

We have used a *Twitter* case study to show that keeping those examples enhances the final classification performance. Since it is not known which types of drift occur in the context of social networks, and particularly in *Twitter*, we have also simulated different types of drift in an artificial dataset to evaluate and validate our strategy.

The results revealed the usefulness of our strategy, as the results improved by 5 % in comparing to the baseline approach, considering the average of the micro-averaged F_1.

It is also important to conclude that we have shown that retaining informative examples during the right amount of time can improve the learners' ability to identify a given class, independently from the drift pattern the class is representing. We do believe that it is problem dependent, even thought it is an important insight in dynamic models, as they are particularly difficult learning scenarios. A special attention must be given to classes that tend to disappear, as retaining examples, in this particular case, for long periods can lead to misclassications.

Our future work will include not only a more profound study about the longevity of those examples, i.e., for how long is it relevant to retain those examples. By understanding the suitable longevity of those examples, we can maximize the cost benefit relation between the storage computational complexity and the classification performance increase. Another effort should be done in minimizing the use of those examples. In our approach we retain the relevant examples and present them to all models that compose the ensemble, but in future work we want to understand if we could have a similar income if we retain examples in a model based strategy, instead of an ensemble based one. The question that arises is that an example can be relevant to a model but irrelevant to another, and thus we can retain the example and provide it just to the model that needs it, instead of all models that compose the ensemble.

Acknowledgments. This work is financed by the ERDF - European Regional Development Fund through the Operational Programme for Competitiveness and Internationalisation - COMPETE 2020 Programme within project "POCI-01-0145-FEDER-006961", and by National Funds through the FCT - Fundação para a Ciência e a Tecnologia (Portuguese Foundation for Science and Technology) as part of project UID/EEA/50014/2013.

This work was supported by national funds through the Portuguese Foundation for Science and Technology (FCT), and by the European Regional Development Fund (FEDER) through COMPETE 2020 – Operational Program for Competitiveness and Internationalization (POCI).

References

1. Costa, J., Silva, C., Antunes, M., Ribeiro, B.: Concept drift awareness in Twitter streams. In: Proceedings of the 13th International Conference on Machine Learning and Applications, pp. 294–299 (2014)
2. Mejri, D., Khanchel, R., Limam, M.: An ensemble method for concept drift in nonstationary environment. J. Stat. Comput. Simul. **83**(6), 1115–1128 (2013)
3. Ditzler, G., Polikar, R.: Incremental learning of concept drift from streaming imbalanced data. IEEE Trans. Knowl. Data Eng. **25**(10), 2283–2301 (2013)
4. Tsymbal, A.: The problem of concept drift: definitions and related work, Department of Computer Science, Trinity College Dublin. Technical report (2004)
5. Costa, J., Silva, C., Antunes, M., Ribeiro, B.: DOTS: drift oriented tool system. In: Arik, S., Huang, T., Lai, W.K., Liu, Q. (eds.) ICONIP 2015. LNCS, vol. 9492, pp. 615–623. Springer, Heidelberg (2015)
6. Widmer, G., Kubat, M.: Effective learning in dynamic environments by explicit context tracking. In: Proceedings of European Conference on Machine Learning, pp. 227–243 (1993)
7. Costa, J., Silva, C., Antunes, M., Ribeiro, B.: Defining semantic meta-hashtags for twitter classification. In: Tomassini, M., Antonioni, A., Daolio, F., Buesser, P. (eds.) ICANNGA 2013. LNCS, vol. 7824, pp. 226–235. Springer, Heidelberg (2013)
8. Kim, J., Bentley, P., Aickelin, U., Greensmith, J., Tedesco, G., Twycross, J.: Immune system approaches to intrusion detection - a review. Nat. Comput. **6**(4), 413–466 (2007)
9. Elwell, R., Polikar, R.: Incremental learning of concept drift in nonstationary environments. IEEE Trans. Netw. **22**, 1517–1531 (2011)
10. Kolter, J.Z., Maloof, M.A.: Dynamic weighted majority: a new ensemble method for tracking concept drift. In: Proceedings of the 3rd International Conference on Data Mining, pp. 123–130 (2003)
11. Huang, J., Thornton, K.M., Efthimiadis, E.N.: Conversational tagging in Twitter. In: Proceedings of the 21st ACM conference on Hypertext and hypermedia, pp. 173–178 (2010)
12. Merriam-webster's dictionary, October 2012

13. Zappavigna, M.: Ambient affiliation: a linguistic perspective on Twitter. New Media Soc. **13**(5), 788–806 (2011)
14. Johnson, S.: How Twitter will change the way we live. Time Mag. **173**, 23–32 (2009)
15. Tsur, O., Rappoport, A.: What's in a hashtag?: content based prediction of the spread of ideas in microblogging communities. In: Proceedings of the 5th International Conference on Web Search and Data Mining, pp. 643–652 (2012)
16. Yang, L., Sun, T., Zhang, M., Mei, Q.: We know what @you #tag: does the dual role affect hashtag adoption? In: Proceedings of the 21st International Conference on World Wide Web, pp. 261–270 (2012)
17. Chang, H.-C.: A new perspective on Twitter hashtag use: diffusion of innovation theory. In: Proceedings of the 73rd Annual Meeting on Navigating Streams in an Information Ecosystem, pp. 85:1–85:4 (2010)
18. Costa, J., Silva, C., Antunes, M., Ribeiro, B.: The impact of longstanding messages in micro-blogging classification. Int. Joint Conference on Neural Networks (IJCNN) **2015**, 1–8 (2015)
19. Zliobaite, I.: Learning under concept drift: an overview. Vilnius University, Faculty of Mathematics and Informatic, Technical report (2010)
20. Vapnik, V.: The Nature of Statistical Learning Theory. Springer, New York (1999)
21. Joachims, T.: Learning Text Classifiers with Support Vector Machines. Kluwer Academic Publishers, Dordrecht (2002)
22. Tong, S., Koller, D.: Support vector machine active learning with applications to text classification. J. Mach. Learn. Res. **2**, 45–66 (2002)
23. Costa, J., Silva, C., Antunes, M., Ribeiro, B.: On using crowdsourcing and active learning to improve classification performance. In: Proceeding of the 11th International Conference on Intelligent Systems Design and Applications, pp. 469–474 (2011)
24. van Rijsbergen, C.: Information Retrieval, 2nd edn. Butterworths, London (1979)

Semi-supervised Modeling

Semi-supervised Hybrid Modeling of Atmospheric Pollution in Urban Centers

Ilias Bougoudis, Konstantinos Demertzis, Lazaros Iliadis, Vardis-Dimitris Anezakis, and Antonios Papaleonidas(✉)

Democritus University of Thrace,
193 Pandazidou st., 68200 N Orestiada, Greece
ibougoudis@yahoo.gr,
{kdemertz, liliadis, danezaki}@fmenr.duth.gr,
antonis.pap@gmail.com

Abstract. Air pollution is directly linked with the development of technology and science, the progress of which besides significant benefits to mankind it also has adverse effects on the environment and hence on human health. The problem has begun to take worrying proportions especially in large urban centers, where 60,000 deaths are reported each year in Europe's towns and 3,000,000 worldwide, due to long-term air pollution exposure (exposure of the European Agency for the Environment http://www.eea.europa.eu/). In this paper we propose a novel and flexible hybrid machine learning system that combines Semi-Supervised Classification and Semi-Supervised Clustering, in order to realize prediction of air pollutants outliers and to study the conditions that favor their high concentration.

Keywords: Pollution of the atmosphere · Air quality · Semi-supervised learning · Semi-supervised clustering · Semi-supervised classification · Air pollution

1 Introduction

1.1 Contamination of the Atmosphere

Air pollution is the presence of air pollutants in quantity, concentration or duration, which can cause deterioration of the structure, composition and characteristics of the atmospheric air. The main sources of air pollution are associated with human activities and they are mainly located in urban areas. They are associated with the production of energy, transport, industry and the heating of buildings, engineering structures and households. Air pollution can cause serious health, environmental, social and economic problems. It is caused mainly from oxides, such as oxides of nitrogen, sulfur carbon and soot (unburnt carbon in air mixture gases). Nitrogen oxides cause photochemical smog, usually in cities or centers and the surrounding areas. Oxides of sulfur and carbon react with water vapor cloud creating acid rain, which affects forests, while the sulfuric acid (component of acid rain) attack the marble transforming them into plaster. Carbon dioxide and other gases produced by incomplete combustion, such as unburned

C. Jayne and L. Iliadis (Eds.): EANN 2016, CCIS 629, pp. 51–63, 2016.
DOI: 10.1007/978-3-319-44188-7_4

hydrocarbons, contribute to the greenhouse effect. There are many respiratory events and lung cancer cases in cities that are close to power plants that burn fossil fuels such as oil or lignite. The European Union has announced a strategy aimed at improving the legislation on air quality, in establishing maximum risk limits for various pollutants. The final target is the progressive drastic reduction of emissions, in order to achieve lower morbidity and mortality as a result air pollution. There are primary pollutants emitted directly into the air (e.g. CO, NO, NO_2, SO_2) and secondary formed by chemical reactions between primary ones (e.g. O_3). Although the atmosphere has physicochemical mechanisms that can remove air pollutants, pollution incidents are mainly due to "unfavorable" weather conditions that significantly limit this potential of the atmosphere and they act in a catalytic manner.

Sunshine helps catalyze the transformation of primary pollutants in secondary, speed and direction of the wind influence the dispersion and transport of the pollutants, the stability of the atmosphere due to excitation of the pressure gradient and temperature gradient of the atmosphere also affects the transport and dispersion of pollutants. Moisture creates the effect of atmospheric water vapor. Moreover, the combination of temperature and humidity (Discomfort Index) aggravates the consequences in people with respiratory or heart problems. To make a quantitative assessment of the impact, especially in densely populated urban areas, requires a detailed spatiotemporal analysis of the conditions that favor high pollutant concentrations, focusing on flexible and realistic modeling approaches. Passive monitoring is one of the traditional ways of coping with this phenomenon, without serious substantial forecasting or early intervention and prevention policies. Real-time monitoring and forecasting of pollutants' concentrations, based on advanced machine learning approaches is one of the most important issues of modern environmental science and research. This research proposes an innovative and effective hybrid forecasting system that does not require high computational power. It employs Semi Supervised Clustering and Classification in order to determine the most extreme air pollutants' values in urban areas.

1.2 Literature Review – Advantages of the Proposed System

In an earlier research of our team [1] we have made an effort to get a clear and comprehensive view of air quality in the center of the city of Athens and also in the wider Attica basin. This study was based on data that were selected from nine air pollution measuring stations during the temporal periods (2000–2004, 2005–2008 and 2009–2012). This method was based on the development of 117 partial ANN whose performance was averaged by using an ensemble learning approach. The system used also fuzzy logic in order to forecast more efficiently the concentration of each feature. The results showed that this approach outperforms the other five ensemble methods. Also, in a previous research effort, Iliadis et al. [2] applied Self Organizing Maps (SOM) in order to cluster air pollution concentrations in groups. The ultimate goal was to find the most isolated cluster where all of the extreme values of pollutants were gathered. This specific cluster would contain vital information about the hazardous pollutants and would also specify the meteorological and temporal conditions under which they occur. Moreover, they tried to evaluate the clustering outcome, using

Pattern Recognition. The inputs were related to 5 temporal parameters, 7 meteorological and 5 pollutants. Bougoudis et al. [3] present the EHF forecasting system which allows the prediction of extreme air pollutant values. EHF was introduced and tested with a vast volume of actual data records. Its main advantage is that though it takes no pollutants as inputs it manages to operate quite efficiently. Moreover, it used a small number of inputs (7), which comprised of 4 temporal features, air temperature, a station identification code (which was determined automatically by geolocation based services) and a cluster identification code. Four unsupervised learning algorithms were employed in EHF, namely: SOM, Neural Gas ANN, Fuzzy C-Means and a fully unsupervised SOM algorithm. Every algorithm, aimed in detecting the most extreme clusters, which contained the most hazardous pollutants' values. Thereafter, they gathered all the records from the extreme clusters, in order to create four datasets, one for each algorithm. These four datasets were used as inputs to the EHF model, which has given promising results in forecasting pollutants' concentrations.

There are other similar studies in the literature that are trying to forecast the air pollution values. However, they have certain limitations that do not guarantee their generalization ability. More specifically they train ANN models with data related to a narrow area (e.g. city center) and they consider this data sample as representative of a wider area that covers locations varying from a topographic, micro climate or population density point of view. However, such research efforts [4–7] are quite interesting and they offer motivation to scientists from diverse fields to employ artificial intelligence in air pollution modeling. Also there are important seasonal studies in the literature [8–13] that do not offer more generalized annual models. Finally, a very interesting approach with objective criteria has been proposed for the specific problem in China [14]. Also Vong et al. [15] have built a forecasting system based on Support vector machines (SVMs), Xiao et al. [16] proposes a novel hybrid model combining air mass trajectory analysis and wavelet transformation to improve the artificial neural network (ANN) forecast accuracy of daily average concentrations of $PM_{2.5}$ and Zabkar and Cemas [17] have applied methods of machine learning to the problem of ground level ozone forecasting. This requires the use of actual raw data and data calculated by the numerical weather prediction model or stations. On the other hand, Lopez-Rubio et al. [18] introduced Bregman divergences in self-organizing models, which are based on stochastic approximation principles, so that more general distortion measures can be employed. A procedure is derived to compare the performance of networks using different divergences. Moreover, a probabilistic interpretation of the model is provided, which enables its use as a Bayesian classifier. Experimental results show the advantages of these divergences with respect to the classical Euclidean distance. Also Menéndez et al. [19] proposed a new algorithm, named genetic graph-based clustering (GGC), which takes an evolutionary approach introducing a genetic algorithm (GA) to cluster the similarity graph. The experimental validation shows that GGC increases robustness of spectral clustering and has competitive performance in comparison with classical clustering methods. Donos et al. [20] have presented a study to provide a seizure detection algorithm that is relatively simple to implement on a microcontroller, so it can be used for an implantable closed loop stimulation device. The classification of the features is performed using a random forest classifier. Finally, Quirós et al. [21] have extended the traditional definitions of k-anonymity, l-diversity and t-closeness of

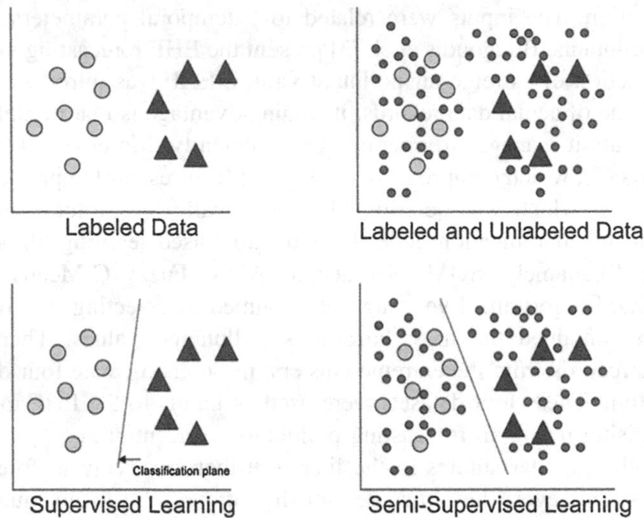

Fig. 1. Semi-supervised versus supervised learning

fuzzy sets as a way to improve the protection of privacy in microdata. The performance of these new approaches is checked in terms of the risk index. The methodology described herein is an extension of a previous research effort of our team [4]. More specifically, [4] describes a system that performs fast and reliable air pollution modeling in mobile devices with limited computational resources. The Semi-Supervised system described in this paper, manages to perform effective air pollution modeling (available to the public) with the minimum amount of data. Unlike paper [4] that uses data input from all the measuring stations in Attica, the proposed system herein uses only data from Athens city center "Athena station". The resulting model has shown high generalization ability for the whole city. Moreover, the system proposed in this research uses fewer features than the initial one presented in [4]. The main advantage of the approach discussed here is its portability. Due to its low requirements and to its generalization potential it can be used in many other cities with the same problem.

The main drawback of the classic classification methods with full supervision (Supervised approach) is that they require a vast number of labeled training examples to construct a predictive model with satisfactory accuracy. The classification of the training set is usually done manually by the instructor, which is a tedious and time consuming process. Instead, the key characteristic of training with partial supervision (Semi-Supervised method) is the production of the final model with the use of pre-classified along with unsorted examples. The Semi-Supervised Clustering approach operates on the condition that the input patterns with and without data tags, belong to the same marginal distribution, or they follow a common cluster structure. Generally, unclassified data provide useful information for the exploration of the overall dataset data structure, while respectively the sorted data are offering in the learning process. Overall, it should be stressed that the success of learning with partial supervision (Semi-Supervised Learning) depends on some basic assumptions imposed by each model or algorithm. This fact makes each case depending on these assumptions which are related to the logic of machine learning methods. Thus, even the most serious real world problems can be modeled effectively, based on the essential peculiarities that characterize them (Fig. 1).

This research effort proposes a Semi-Supervised Classification and Semi-Supervised Clustering Hybrid Air Quality system (SSC^2-HAQS). The system is capable of modeling air pollution in urban centers, after considering the actual positive or negative correlations between all of the involved features (meteorological or primary and secondary chemicals).

2 Data

The data used come from the "Athena" station. The station is located in the heart of Athens, so it provides a representative picture of the atmospheric pollution in modern cities. There were hourly data values available for CO, NO, NO_2, O_3 and SO_2, measured in $\mu g/m^3$ for the period 2000–2013. The model was built with data for 13 years (2000–2012) whereas the dataset of 2013 was used for testing the forecast framework with first time seen cases and to determine its validity.

Apart from the five pollutants, each record also consists of five calendar items namely: Year, Month, Day, D_ID (1 Monday, 2 for Tuesday and so on), Hour. Moreover, there are seven meteorological factors namely: Air Temperature (Temp), Relative Humidity (RH), Atmospheric Pressure (PR), Solar Radiation (except 2013) (SR), Percentage of Sunshine (till 2010) (SUN), Wind Speed (WS) Wind Direction (WD). The following tables present a typical statistical analysis of the whole available data and for the 2012 dataset which was chosen to be the pilot one for the determination of the classes (Tables 1 and 2).

Table 1. Statistical analysis for the period 2000–2013

2000–2013 (97201)	CO	NO	NO_2	O_3	SO_2
MAX	21.4	908	377	253	259
MIN	0.1	1	1	1	2
MODE	0.8	7	60	3	2
COUNT_MODE	5592	2651	1606	7137	10435
AVERAGE	1.79	57.88	61.86	33.16	9.40
STANDARD_DEV	1.45	88.29	26.98	28.47	9.06

3 Description of the SSC^2-HAQS Algorithm

We considered three situations for each entry: Tag (1) for the cases with extreme primary pollutants, (2) for records with extreme ozone values (secondary pollutant) and (0) for normal pollutant values. This was assumption was done in order to classify our data in three basic risk categories. Then from the set of available data for 2012, we have

Table 2. Statistical analysis for the year 2012

2012 (8644)	CO	NO	NO_2	O_3	SO_2
MAX	9.3	600	142	186	47
MIN	0.2	1	5	1	2
MODE	0.7	8	53	2	4
COUNT_MODE	672	299	206	436	1372
AVERAGE	1.29	42.36	51.11	38.29	6.88
STANDARD_DEV	0.91	59.67	19.05	29.33	3.28

chosen a small sample of approximately 10 %, which had records that could be clearly labeled as members of one of the three classes. This small sample was used as a pilot in order to classify the rest of the data by employing the Naïve Bayesian algorithm described below.

Algorithm 1. The semi supervised **Naive Bayesian** clustering

Inputs: Input data, clusters of the input data and testing data to which a label should be assigned

Step 1:

Identify the discrete number of clusters

For every cluster, create matrices with the mean and standard deviation of all their input data

Step 2:

For every cluster, recreate these matrices, based on the testing data

Calculate a variable, based on the formula below:

x =(1./(2*pi*ns.^2)).*exp(-((test-nm).^2)./(2.*sn.^2))

where *ns* is the new standard deviation matrix, *nm* is the new mean matrix and *test* is the testing data

Sum all these variables for each cluster

Step 3:

For every testing data, find the maximum value of the summary calculated before.

Once completing the clustering with the use of the Naive Bayesian algorithm, we have managed to obtain a clear view for the risk level of each record. The corresponding class was added as a new attribute to the final dataset. However, the values assigned to the "0" label were of no interest because the main target was the determination of the extreme cases, regardless the normal ones. The addition of this feature has ensured uniformity as to the classification of the cases and it has solved the following problem:

The concentrations of O_3 in many cases appear to be extremely high, whereas at the same time the relative concentrations of CO and No appear to be extremely low and vice versa. Thus, an overall risk index for both the primary and the secondary pollutants is not possible. The final version of the dataset includes as independent variables the time profile (Year, Month, Day, Day_Id, Hour), meteorological indications (Air-Temp, RH, PR, SR, WS, WD) and the value of the cluster determination to which each record belongs (Cluster). The five pollutants (CO, NO, NO_2, O_3, SO_2) were used as dependent variables.

Then, the Yatsi algorithm was used to classify the unlabeled data, using the classified 10 % as a pilot model. It should be mentioned that the Yatsi algorithm is semi-supervised and it applies the Weighted Nearest Neighbor approach.

Collective classification [22] is a combinatorial optimization problem, in which we are given a set of nodes, V = {V1,...,Vn} and a neighborhood function N, where Ni ⊆ V\{Vi}. Each node in V is a random variable that can take a value from an appropriate domain. V is further divided into two sets of nodes: X, the nodes for which we know the correct values (observed variables) and, Y, the nodes whose values need to be determined. The actual task is to label the nodes Yi ∈ Y with one of a small number of labels, L = {L1,...,Lq}; The lower case yi will be used to denote the label of node Yi.

Algorithm 2. High level pseudo code for the two-stage Yatsi algorithm [23]

```
Input: a set of labeled data Dl and a set of unlabeled data Du, an of-the-shelf
classifier C and a nearest neighbor number K; let N = |Dl| and M = |Du|
Step 1:
    Train the classifier C using Dl to produce the model Ml
    Use the model Ml to "pre-label" all the examples from Du
    Assign weights of 1.0 to every example in Dl
    and of F × N/M to all the examples in Du
    Merge the two sets Dl and Du into D
Step 2:
    For every example that needs a prediction:
    Find the K-nearest neighbors to the example from D to produce set NN
    For each class:
    Sum the weights of the examples from NN that belong to that class
    Predict the class with the largest sum of weights.
```

In this research effort, semi-supervised classification has been applied to isolate the potential extreme records. The reasoning of the method is based on the concept that performing classification for a robust subset of the available data (not less than 10 % of the whole) can provide a prototype for the effective classification throughout the dataset. The following specific steps were applied to achieve this task:

We have initially determined the actual three risk classes, working with the 2012 dataset (pilot data). This was chosen as the actual robust dataset, because it is an extensive one (951 vectors) with the vast majority of the selected values being valid. Also the range of the values for each involved feature was representative of the total potential fluctuation for each pollutant.

Thus it was determined that all 2012 vectors that had CO concentration higher than 3.2 mg/m^3 were labeled as class 1, whereas the ones that had $O_3 > 60$ $\mu g/m^3$ were tagged as class 2. All of the rest of the cases were assigned class 0. The above boundary values were selected to represent the extreme cases, based on the results emerging from a previous research effort of our team [3]. Also it is really important that in [2] we had shown that a record can be an outlier, either according to the concentration of primary pollutants (CO, NO, NO_2, SO_2) or based on the secondary O_3 concentrations but not for both types of features at the same time. The parameter CO was selected as representative of the extreme pollutant group 1, because according to [2] it played the most crucial role for its determination, with the extreme values of the rest of primary pollutants to "follow". So we adopted three risk classes for each record: The extreme one for the primary pollutants (1), the extreme in relation to ozone (2) (secondary pollutant) and the class of normal pollutants' values (0). Running the semi-supervised algorithm, we obtained a very effective classification for the whole available data records related to all of the years under study (Table 3).

Correctly Classified Instances 935 (98.3176 %), Incorrectly Classified Instances 16 (1.6824 %), Root Mean Squared Error 0.1024. Figure 2 presents a graph of the proposed method.

After the classification, a dataset with the extreme values of the pollutants was developed. Also the class attribute was added, having the corresponding values 0, 1, 2. This addition ensured uniformity as to the classification of the cases, which appear to have inverse effects over periods of time due to their physico-chemical composition. For example, in cases where there were O_3 outliers, the values of the primary pollutants CO and NO appeared to be extremely low and vice versa.

We have developed feed forward Artificial Neural Networks (ANN) in order to forecast the extreme values of pollutants. Specifically, for each pollutant an ANN has been developed. The input parameters are the following: YEAR, MONTH, DAY, HOUR, AIR_TEMPERATURE and finally the attribute produced by the SSC^2-HAQS, CLUSTER_ID. The network had 10 neurons in the hidden layer, it employed the tansig transfer function, the training function trainlm and the learngdm learning function. The Root Mean Square Error metric (RMSE) was used to evaluate the performance.

4 Results and Comparative Analysis

4.1 Results

Thus, after running the SSC^2-HAQS approach in order to obtain the extreme dataset and after having generated a neural network for each pollutant, the testing process considered the 2013 data vectors, originating from the "Athenas" station. The ANN were not fed with the desired output pollutants' values (targets). Using first time seen inputs, the models predicted some values which were compared to the actual ones (Table 4). The following Table 5 contains the ANN testing results.

4.2 Comparative Analysis

The application of the semi supervised algorithm gives reliable results, especially in the classification process. More specifically, the SSC^2-HAQS model outperforms the approaches that have been proposed by our research team in the literature [3].

Table 3. Confusion matrix for the assignment of the classes

Confusion matrices 951 instances (0 normal values) (1 extreme primary) (2 extreme O_3)		
A (0)	B (1)	C (2)
239	5	0
5	281	1
2	3	415

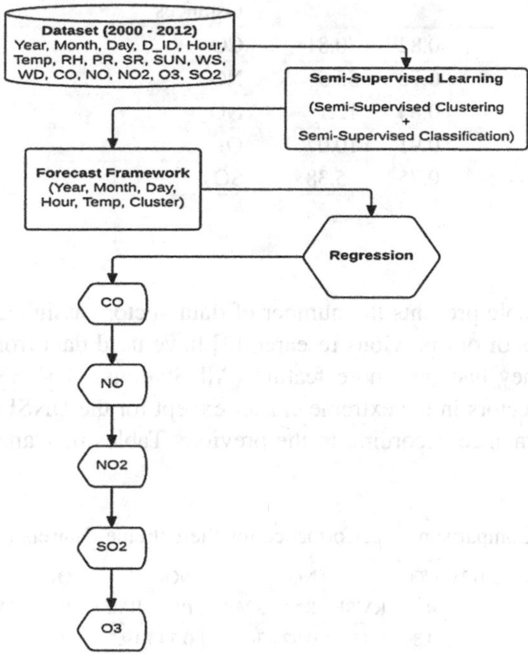

Fig. 2. Graph of the developed algorithm

The main advantage of our approach is that it runs only for one measuring station aiming to offer an overall classification for the whole area under study, much faster and in a simpler way. The same method can be applied for any other station. Also the Semi-Supervised Learning employed runs effectively by using only three classes whereas the fuzzy c-means required 5 classes and the SOM needed 9 classes in order to determine effectively the extreme pollutants' groups. The hypothesis that a pollution record is either harmless or dangerous for the public health, being related to high concentrations of primary or secondary pollutants is rather rational, flexible and moreover effective. The following Tables 6 and 7 present a comparison between the performance of the herein proposed method SSC^2-HAQS and the SOM, GAS, FUZZY and Unsupervised SOM that were applied in a previous research effort of our team [3]. The SSC^2-HAQS has better performance (for 3 out of 5 features). Specifically, it is more reliable for the NO_2, O_3 and SO_2 whereas it is equally reliable (though a little worse) for the CO and NO cases. However, it is a good compromise since it is much faster it models the whole area with the use of a single measuring station and it requires fewer classes in order to group the extreme values effectively.

Table 4. Training results

Training (2000–2012) 44601 instances	R^2	RMSE
CO	0.82	0.81
NO	0.78	55.5
NO_2	0.84	12.1
O_3	0.91	10.07
SO_2	0.75	5.38

Table 5. Testing results

Testing (2013) 5098 instances	R^2	RMSE
CO	0.78	0.59
NO	0.82	37.34
NO_2	0.53	12.88
O_3	0.70	19.94
SO_2	0.12	3.35

The following table presents the number of data vectors assigned the extreme tag. The four approaches of our previous research [3] have used data from four measuring stations and thus they had one more feature (All Stations). The SSC^2-HAQS incorporates more data vectors in the extreme cluster except for the UNSUPERSOM, which has very bad performance according to the previous Tables 6, 7 and 8.

Table 6. Comparison of performance for the extreme datasets (Training)

Training comparison (2000–2012)	CO		NO		NO_2		O_3		SO_2	
	R^2	RMSE	R^2	RMSE	R^2	RMSE	R^2	RMSE	R^2	RMSE
SOM	0.86	0.75	0.92	36	0.74	19.2	0.86	14	0.71	15.7
GAS	0.90	0.7	0.94	33	0.74	17.6	0.83	17.5	0.62	13.7
FUZZY	0.88	0.62	0.92	30.27	0.72	15.4	0.83	15.4	0.64	10.7
UNSUPER SOM	0.42	1.29	0.37	76.39	0.54	23.63	0.9	10.27	0.34	16.23
SEMI	0.82	0.81	0.78	55.5	0.84	12.1	0.91	10.07	0.75	5.38

5 Discussion–Conclusions

This work presents an innovative and effective method of analyzing high concentrations of air pollutants with a combined hybrid Semi-Supervised Learning system. The proposed approach was tested successfully, in classifying and also in forecasting the extreme primary and secondary pollutant values for the center of Athens. It uses a

Table 7. Comparison of performance for the extreme datasets (Testing)

Testing comparison (2013)	CO		NO		NO_2		O_3		SO_2	
	R^2	RMSE	R^2	RMSE	R^2	RMSE	R^2	RMSE	R^2	RMSE
SOM	0.77	0.53	0.83	40	0.48	17.9	0.71	33.3	0.13	6.88
GAS	0.76	0.62	0.9	30.1	0.49	16.2	0.4	36.9	0.14	6.69
FUZZY	0.76	0.57	0.85	40.6	0.53	14.5	0.69	19.5	0.1	6.51
UNSUPER SOM	0.19	0.98	0.38	58	0.25	25.1	0.27	35.4	0.03	7.13
SEMI	0.78	0.59	0.82	37.34	0.53	12.88	0.7	19.94	0.12	3.35

Table 8. Comparison of the extreme records' number (number of records)

Number of extreme records	Training (2000–2012)		Testing (2013)	
	All stations	Athinas	All stations	Athinas
SOM	30077	3383	14129	4378
GAS	53589	9354	13965	4343
FUZZY	91440	24834	14273	3987
UNSUPER SOM	213058	51304	19950	7757
SEMI	-	44601	-	5098

sophisticated technique of combined learning, which ensures fast, robust and effective forecasting and classification performance. Moreover, it is a general model which does not require specific characteristics of the area under study. All the above, add generalization ability to the methodology which is easily adjustable and applicable to other areas (cities) of research. The SSC^2-HAQS employs a Semi-Supervised Learning algorithm which is considered a realistic machine learning method that can model the most serious problems of the real world, based on the essential peculiarities that might characterize them. A main innovation introduced by the proposed scheme, concerns the data classification in homogeneous classes (distinction between primary and secondary pollutants). This process is done based on a sample of few pre-classified data vectors, something that incorporates the hidden knowledge and the correlations between the features. This hybrid system was tested effectively, with data that have specific particularities as they originate from a period of financial crisis for Greece, which has a significant effect on air quality in major urban centers.

Future work could involve testing of the system data in other urban centers with different climatic conditions and moreover it should consider climate change scenarios in these regions. Additionally, it would be very important to apply a new weights learning algorithm which will modify and adjust them based on specificity rates that are deemed necessary for the local climate. Thus the system could be made more flexible in achieving results in future evaluations and investigations of a region.

References

1. Bougoudis, I., Iliadis, L., Papaleonidas, A.: Fuzzy inference ANN ensembles for air pollutants modeling in a major urban area: the case of Athens. In: Mladenov, V., Jayne, C., Iliadis, L. (eds.) EANN 2014. CCIS, vol. 459, pp. 1–14. Springer, Heidelberg (2014)
2. Bougoudis, I., Iliadis, L., Spartalis, S.: Comparison of self organizing maps clustering with supervised classification for air pollution data sets. In: Iliadis, L., Maglogiannis, L., Papadopoulos, H. (eds.) AIAI 2014. IFIP AICT, vol. 436, pp. 424–435. Springer, Heidelberg (2014)
3. Bougoudis, I., Demertzis, K., Iliadis, L.: Fast and low cost prediction of extreme air pollution values with hybrid unsupervised learning. Integr. Comput.-Aided Eng. **23**(2), 115–127 (2016). doi:10.3233/ICA-150505. IOS Press
4. Bougoudis, I., Demertzis, K., Iliadis, L.: HISYCOL a hybrid computational intelligence system for combined machine learning: the case of air pollution modeling in Athens. EANN Neural Comput. Appl. 1–16 (2016). doi:10.1007/s00521-015-1927-7
5. Roy, S.: Prediction of particulate matter concentrations using artificial neural network. Resour. Environ. **2**(2), 30–36 (2012). doi:10.5923/j.re.20120202.05
6. Robles, L.A., Ortega, J.C., Fu, J.S., Reed, G.D., Chow, J.C., Watson, J.G., Moncada-Herrera, J.A.: A hybrid ARIMA and artificial neural networks model to forecast particulate matter in urban areas: the case of Temuco, Chile. Atmos. Environ. **42**(35), 8331–8340 (2008)
7. Ordieres Meré, J.B., Vergara González, E.P., Capuz, R.S., Salaza, R.E.: Neural network prediction model for fine particulate matter (PM). Environ. Model Softw. **20**, 547–559 (2005)
8. Wahab, A., Al-Alawi, S.M.: Assessment and prediction of tropospheric ozone concentration levels using artificial neural networks. Environ. Model. **17**, 219–228 (2002)
9. Paschalidou, A., Iliadis, L., Kassomenos, P., Bezirtzoglou, C.: Neural modeling of the tropospheric ozone concentrations in an urban site. In: Proceedings of 10th International Conference Engineering Applications of Neural Networks, pp. 436–445 (2007)
10. Ozcan, H.K., Bilgili, E., Sahin, U., Bayat, C.: Modeling of trophospheric ozone concentrations using genetically trained multi-level cellular neural networks. Advances in Atmospheric Sciences, vol. 24, pp. 907–914. Springer, Heidelberg (2007)
11. Ozdemir, H., Demir, G., Altay, G., Albayrak, S., Bayat, C.: Prediction of tropospheric ozone concentration by employing artificial neural networks. Environ. Eng. Sci. **25**(9), 1249–1254 (2008)
12. Inal, F.: Artificial neural network prediction of tropospheric ozone concentrations in Istanbul, Turkey. CLEAN – Soil Air Water **38**(10), 897–908 (2010)
13. Paoli, C.: A neural network model forecasting for prediction of hourly ozone concentration in Corsica. In: Proceedings IEEE of 10th International Conference on EEEIC (2011)
14. Kadri, C., Tian, F., Zhang, L., Dang, L., Li, G.: Neural network ensembles for online gas concentration estimation using an electronic nose. IJCS **10**(2), 1 (2013)
15. Vong, C.-M., Ip, W.-F., Wong, P.-K., Yang, J.-Y.: Short-term prediction of air pollution in Macau using support vector machines. J. Control Sci. Eng. **2012**, 4 (2012). Article ID 518032
16. Xiao, F., Li, Q., Zhu, Y., Hou, J., Jin, L., Wang, J.: Artificial neural networks forecasting of PM 2.5 pollution using air mass trajectory based geographic model and wavelet transformation. Atmos. Environ. **107**, 118–128 (2015). doi:10.1016/j.atmosenv.2015.02.030. Elsevier
17. Zabkar, R., Cemas, D.: Ground-level ozone forecast based on ML. AIR040051 (2004)

18. Lopez-Rubio, E., Palomo, E.J., Dominguez, E.: Bregman divergences for growing hierarchical self-organizing networks. Int. J. Neural Syst. **24**, 4 (2014). 1450016
19. Menendez, H., Barrero, D.F., Camacho, D.: A genetic graph-based approach to the partitional clustering. Int. J. Neural Syst. **24**, 3 (2014). 1430008
20. Donos, C., Duemoelmann, M., Schulze-Bonhage, A.: Early seizure detection algorithm based on intractable EEG and random forest classification. IJNS **25**, 5 (2015). 1550023
21. Quirós, P., Alonso, P., Díaz, I., Montes, S.: On the use of fuzzy partitions to protect data. Integr. Comput.-Aided Eng. **21**(4), 355–366 (2014)
22. Sen, P., Namata, G., Bilgic, M., Getoor, L., Galligher, B., Eliassi-Rad, T.: Collective classification in network data. Assoc. Adv. AI **29**(3), 93 (2015)
23. Driessens, K., Reutemann, P., Pfahringer, B., Leschi, C.: Using weighted nearest neighbor to benefit from unlabeled data. In: Ng, W.-K., Kitsuregawa, M., Li, J., Chang, K. (eds.) PAKDD 2006. LNCS (LNAI), vol. 3918, pp. 60–69. Springer, Heidelberg (2006)

Classification Applications

Predicting Abnormal Bank Stock Returns Using Textual Analysis of Annual Reports – a Neural Network Approach

Petr Hájek(✉) and Jana Boháčová

Institute of System Engineering and Informatics, Faculty of Economics and Administration, University of Pardubice, Pardubice, Czech Republic
petr.hajek@upce.cz, st30300@student.upce.cz

Abstract. This paper aims to extract both sentiment and bag-of-words information from the annual reports of U.S. banks. The sentiment analysis is based on two commonly used finance-specific dictionaries, while the bag-of-words are selected according to their *tf-idf*. We combine these features with financial indicators to predict abnormal bank stock returns using a neural network with dropout regularization and rectified linear units. We show that this method outperforms other machine learning algorithms (Naïve Bayes, Support Vector Machine, C4.5 decision tree, and *k*-nearest neighbour classifier) in predicting positive/negative abnormal stock returns. Thus, this neural network seems to be well suited for text classification tasks working with sparse high-dimensional data. We also show that the quality of the prediction significantly increased when using the combination of financial indicators and bigrams and trigrams, respectively.

Keywords: Stock return · Prediction · Text mining · Sentiment · Neural network

1 Introduction

In recent years, the importance and volume of firm-related textual information have steadily increased. Stakeholders make decisions based on a wide range of information, much of it subjective. It is therefore becoming increasingly difficult to ignore the contribution textual analysis may have in finance. The past decade has seen the rapid development of textual analysis of many financial problems, such as the modelling of abnormal stock returns [1–3], volatility modelling [4–6], fraud detection [6] and financial-distress prediction [7–9].

In the literature, two general approaches have been used to analyse firm-related text: (1) sentiment analysis and (2) machine learning. The former approach calculates overall sentiment based on the frequency of words chosen by financial experts, thus addressing the context-specific nature of financial vocabulary better than using general dictionaries like Harvard IV-4. Machine-learning approaches, on the other hand, automatically construct word lists and their weights based on a classification of texts. This approach may provide more accurate predictions, but it is problem-specific and

C. Jayne and L. Iliadis (Eds.): EANN 2016, CCIS 629, pp. 67–78, 2016.
DOI: 10.1007/978-3-319-44188-7_5

difficult to interpret. Both approaches have shown promising results in predicting the reactions of financial markets.

For example, Li [10] demonstrated that changes in sentiment about risk (uncertainty) in annual reports significantly affects future earnings and stock returns. Li [11] found some evidence that managers may hide adverse information from investors by using harder-to-read language in annual reports. Feldman et al. [12] also reported that market reactions (two days after the U.S. Securities and Exchange Commission filing date) are significantly associated with the tone (net positive) of the Management Discussion and Analysis (MD&A) section of the annual report.

Machine-learning approaches to the textual classification of annual reports have also been reported in the literature. For example, Balakrishnan and Srinivasan [13] found that significantly positive, size-adjusted returns can be achieved by using the predictions of a machine-learning model. More specifically, textual information was reported to affect investors' use of price momentum, which then became a key determinant of these excess returns.

In this study, we use a hybrid textual analysis approach combining sentiment analysis and machine learning. Here, the sentiment analysis is based on two commonly used finance-specific dictionaries developed respectively by [1, 6]. The use of finance-specific dictionaries in this approach has shown significantly higher prediction accuracy compared to the use of general dictionaries [6, 14]. Moreover, Loughran and McDonald [15] reported that general dictionaries were especially inappropriate for sentiment analysis of financial disclosures, causing a high percentage of sentiment misclassification. The dictionary by [6] has become particularly dominant in the literature for finance-related analysis. Loughran and McDonald [6] reported that event period excess returns are positively affected by a frequent use of litigious terms (but only in cases of proportional weights of terms), whereas other financial dictionaries (negative, positive, uncertainty, weak and strong modal) have negative effects for both proportional and *tf-idf* weights of terms. Negative, uncertainty, weak and strong modal word lists displayed statistically significant effects for both weighting schemes.

The aim of this paper is to predict abnormal stock returns using the analysis of text in the annual reports of U.S. banks. Most studies in the field tended to focus on either sentiment or machine learning approach, paying little attention to their synergistic effects. Here we use the combination of financial indicators, sentiment and bag-of-words (BoW) to increase prediction accuracy. First, adopting the approach of prior studies, we employ predefined dictionaries to show the effect of sentiment on abnormal stock returns in the banking industry. We show that the chosen sentiment categories displayed in the annual reports of banks negatively affects abnormal stock returns, with the exception of sentiment tone. Second, we use a BoW representation to detect the most relevant terms in the annual reports. To perform the prediction of abnormal stock returns, we employ a Neural Network (NN) with dropout regularization and rectified linear units [16] and compare it with four machine learning approaches commonly used in text classification [17], namely Naïve Bayes, Support Vector Machine (SVM), C4.5 decision tree, and *k*-nearest neighbour (*k*-NN) classifier. We demonstrate that the NN performs best using the combination of financial indicators and BoW approach.

The remainder of this paper is organised in the following way. Section 2 outlines finance-specific aspects of textual analysis. Section 3 presents the corpus of documents and the results of its pre-processing. The prediction of abnormal stock returns is performed in Sect. 4. In addition to textual information in annual reports, the financial indicators of banks are used for analysis, in line with previous literature. Section 5 discusses the obtained results and concludes the paper.

2 Textual Analysis in Finance – Literature Review

Kearney and Liu [18] classified the sources of textual information in the financial domain into three categories: corporation-expressed, media-expressed, and Internet-expressed. Corporation-expressed information is usually extracted from annual reports [6, 10] or from earnings press releases and conference calls [1]. MD&A sections of annual reports are widely considered to be the most important source of insider information, because they provide management's perspective on past performance, current financial positions and future prospects [12]. These sections may therefore be particularly important for the prediction of firm performance and stock prices. Researchers have shown increasing interest in the analysis of firm-related narratives partly due to the requirements of the U.S. Securities and Exchange Commission (SEC) for electronic filings. 10-K filings (forms) provide both audited financial statements and a comprehensive overview of the firm's business and financial condition. Therefore, they are the most widely used source of data. However, the information provided by management may be rather subjective and not entirely true, making analysis difficult.

Li [19] examined the MD&A sections of 10-K (and 10-Q) filings using a Naïve Bayes method, demonstrating that a positive tone in the documents indicates positive future earnings. General dictionaries, on the other hand, failed to predict future financial performance. Demers and Vega [20] examined the impact on future earnings of net optimism and uncertainty of managerial communications regarding a firm's quarterly earnings results, suggesting that net optimism is positively associated with future earnings, whereas uncertainty indicates a decrease in future earnings. Davis et al. [21] calculated net optimism in earnings press releases, finding that this measure (1) is positively associated with future return on assets and (2) generates a significant market response in a short window of time around the date of the earnings announcement.

In contrast to the abovementioned studies, which used a general dictionary, Loughran and McDonald [6] developed a finance-specific dictionary to measure the sentiment in company-related textual documents. They reported that general dictionaries misclassified many negative words, such as "taxes" or "liabilities", thus adding noise to prediction models. Moreover, other industry-specific words ("oil", "cancer") do not carry the generally negative connotation they do in general language. In addition to negative words, Loughran and McDonald [6] considered other effects by using five other word classifications (positive, uncertain, litigious, strong modal, and weak modal). Taken together, higher sentiment (across all word categories) in annual reports significantly and negatively affected future abnormal returns, whereas it significantly and positively impacted both abnormal volume and return volatility.

Meanwhile, media-expressed information is the information of outsiders contained in news stories and analyst reports [5]. Tetlock et al. [2] studied the effect of news stories on future earnings and stock returns, demonstrating that the fraction of negative words in firm-related news stories predicts both low earnings and low stock returns. Schumaker and Chen [22] examined a SVM approach for financial news articles analysis using several textual representations: BoW, Noun Phrases, and Named Entities. The majority of the sources used are major news websites such as The Wall Street Journal [4], Bloomberg [23] and Yahoo! Finance [24].

Internet-expressed sentiment is used to extract the information from small investors [4]. For example, in their stock price prediction model, Li et al. [25] combined news information with the information obtained from online financial discussion boards. Similarly, Yu et al. [26] have investigated content from the social media, including blogs, forums and Twitter. Their findings suggest that social media has a stronger impact on firm stock performance than conventional media.

Finally, several researchers have investigated a variety of firm-related textual documents. For example, Kothari et al. [27] examined corporate reports, analyst disclosures and briefings, and disclosures made in the general business press. Their results showed that favourable disclosures have a significantly negative effect on firm's perceived risk (as proxied by the cost of capital, stock return volatility, and analyst forecast dispersion).

3 Data and Research Methodology

Our study encompasses 180 U.S. banks listed on the New York Stock Exchange (NYSE) or Nasdaq and with a reported stock price of at least 3 USD before the 10-K filing date (usually within 90 days after the end of the firm's fiscal year). This limit was chosen to reduce the contribution of bid/ask bounce in reaction to 10-K filing [6]. We also required market capitalisation of at least 100 million USD to reduce the effect of risk factors for stocks [28]. We downloaded all 10-Ks for such banks from the EDGAR system for the period 2013. To control for variables that have shown significant impacts on abnormal stock returns in prior literature [1, 14], we collected corresponding data from the Marketwatch database for the following variables: (1) log of the market capitalisation (lnMC), (2) price-earnings ratio (P/E), (3) price to book value (P/B), (4) return on equity (ROE), (5) total debt to total assets (TD/TA), and (6) a dummy variable for NYSE versus Nasdaq listing. ε-SVR (Support Vector Regression) was used for the imputation (with average RMSE = 4.65). All attributes except the missing one were used to estimate the missing value. The completed data on financial indicators were used afterwards to predict abnormal stock returns.

Following previous studies [1], abnormal returns were calculated as accumulated returns in excess of the return on the CRSP (Center for Research in Security Prices) equal-weighted market portfolio. Consistent with related studies, we also adopted a three-day event window, from day t-1 to $t + 1$, where t represents the 10-K filing day. The U.S. banks were categorized into two classes, with positive (139 banks) and negative abnormal returns (41 banks), indicating an imbalanced dataset. Table 1 shows

Table 1. Descriptive statistics on financial indicators

Class	Positive		Negative	
Var.	Mean	Std. dev.	Mean	Std. dev.
lnMC	6.20	1.30	7.05	1.30
P/E	16.53	6.13	17.44	9.60
P/B	1.36	0.44	1.42	0.45
ROE [%]	7.89	4.16	9.61	8.51
TD/TA [%]	9.81	4.31	10.54	7.19

basic descriptive statistics of the sample. Nasdaq listings predominated in the data at 81.06 % of considered firms.

In accordance with prior studies [19], we extracted only the most important textual section from the downloaded 10-Ks, namely Item 7: Management's Discussion and Analysis of Financial Condition and Results of Operations (MD&A). This section provides managements' perspective on their firms' past, current and future financial performance [18].

To obtain their tone, we compared the extracted documents with several finance-specific word categorisations: those developed by [1, 6]. Henry [1] created two word categories, one positive and one negative, both containing 85 words. However, this approach has two limitations. First, the limited number of words contained in each category has been reported as insufficient in the domain of business communication. Second, other important word categories besides positive and negative are ignored, such as uncertainty, modality, and so on. Loughran and McDonald [6] have addressed these drawbacks, leading to extensive word lists of 354 positive and 2,329 negative words. In addition, word categories for uncertainty (291 words), litigious (871 words), modal strong (19 words) and modal weak (27 words) were created as part of their work.

The use of negative words seems unambiguous, whereas the use of positive words in a negative statement has been one of the main challenges addressed in the literature on sentiment analysis [15]. To handle the problem of negations, we followed the approach proposed by [6], performing a collocation analysis with positive words to detect one of six negation words (no, not, none, neither, never, nobody) occurring within three words preceding a positive word. The frequency of net positive words was then calculated as the positive term count minus the count for negation (positive terms are easily qualified or compromised). Although this procedure should provide a more accurate measurement of positive tone, previous studies have shown that positive word lists can generally locate only a little incremental information [5, 6].

Another issue to be addressed is the choice of an appropriate term-weighting scheme to evaluate how important a word is within a document in a corpus [29]. Using raw term frequency, all terms are considered equally important. However, this scheme assigns higher weights to terms that occur frequently in the text and it does not consider, moreover, the length of the document. Therefore, we used the most common term-weighting scheme, *tf-idf* (term frequency-inverse document frequency), in which weights w_{ij} are defined as follows:

$$w_{ij} = \begin{cases} (1 + \log(tf_{ij})) \log \frac{N}{df_i} & \text{if } tf_{ij} \geq 1 \\ 0 & \text{otherwise} \end{cases}, \quad (1)$$

where N represents the total number of documents in the corpus, df_i denotes the number of documents with at least one occurrence of the i-th term, and tf_{ij} is the frequency of the i-th term in the j-th document.

The weight for each word category was then calculated as the average frequency of the words in that category. In addition to the abovementioned word categories, we also calculated the overall tone, defined as the count of positive words minus the count of negative words, divided by the sum of both positive and negative word counts [1]. Table 2 shows that banks with a stronger sentiment orientation (in all word categories) performed worse, this is that sentiment was not taken positively by investors. On the other hand, the overall tones (Henry Tone and LM Tone) were higher for the banks with positive abnormal return.

Table 2. Descriptive statistics on sentiment indicators

Class	Positive		Negative	
Var.	Mean	Std. dev.	Mean	Std. dev.
Henry Pos.	0.27	0.15	0.32	0.17
Henry Neg.	0.26	0.14	0.34	0.19
Henry Tone	0.02	0.27	−0.03	0.24
LM Pos.	0.23	0.12	0.29	0.16
LM Neg.	0.19	0.09	0.27	0.18
LM Tone	0.09	0.21	0.04	0.20
LM Uncertainty	0.21	0.09	0.26	0.12
LM Modal Weak	0.23	0.19	0.31	0.22
LM Modal Strong	0.21	0.19	0.29	0.22
LM Litigious	0.16	0.10	0.31	0.24

Legend: LM denotes word categories developed by Loughran and McDonald [6].

To match the data from the EDGAR system and Marketwatch database, we used the ticker symbols of the banks (see Table 3 for a data sample).

To identify a set of useful N-grams, we first removed stop-words, performed stemming using the Snowball stemmer, and converted all word tokens to lower case letters. Finally, all unigrams, bigrams and trigrams were identified in the training data and ranked according to their *tf-idf*. For our experiments, we used the top 200, 500 and 1000 N-grams in a bag-of-words fashion. Table 4 presents the N-grams with the highest information gain, indicating overlaps and potential value provided by bigrams and trigrams (e.g., flow, cash flow, future cash flow).

Table 3. Data sample of financial and sentiment indicators

Ticker	lnMC	P/E	…	Henry pos.	Henry neg.	Henry tone	…	LM litig.	Class
ACNB	4.83	13.14	…	0.14	0.50	-0.57	…	0.26	neg
BBT	10.25	14.15	…	0.34	0.38	-0.05	…	0.26	pos
…	…	…	…	…	…	…	…	…	…
ZION	8.63	16.33	…	0.74	0.89	-0.09	…	0.44	pos

Table 4. Terms with the highest information gain

Category	Terms
Unigrams	flow, approximately, lending, acquisition, yield
Bigrams	cash flow, interest margin, fixed rate, credit losses, acquired loans
Trigrams	net interest margin, positive and negative, provision for loan, future cash flows, board of directors

4 Experimental Results

The survey on text mining for stock market prediction [29] concludes that SVM and Naïve Bayes are heavily favoured by researchers, whereas NNs are significantly under-researched in the field of stock market predictive text-mining at this stage, despite that NNs have shown promising potentials for textual classification and sentiment analysis. NNs equipped with advanced techniques such as rectified linear units, AdaGrad and dropout regularization have been reported to be particularly effective compared with state-of-the-art approaches to text classification [30].

In our experiments, we examined NN with dropout regularization and rectified linear units [16]. Dropout regularization [5] was utilized because fully connected NNs are prone to overfitting. This regularization randomly sets a given proportion of the activations to the fully connected layers to zero during training. Thus, hidden units that activate the same output are decoupled. This largely improves generalization ability and prevents overfitting. Rectified linear units have attracted increased attention because traditional sigmoidal units suffer from the vanishing gradient problem, which may cause slow optimization convergence to a poor local minimum [31]. The synergistic effects of combining rectified linear units with dropout regularization have been demonstrated by [32]. We trained this NN using gradient descent algorithm with the following parameters: input layer dropout rate = 0.2, hidden layer dropout rate = 0.5, number of units in the hidden layer = {10, 20, 50, 100, 200}, learning rate = {0.05, 0.10}, and the number of iterations = 1000. The structure and parameters of the NN learning were found using grid search procedure. The large number of neurons in the hidden layer was examined due to the high number of input features (more than 1000). However, adding too many neurons was not necessary because it would lead to modelling the noise in the training data, eventually causing poor generalization performance.

To demonstrate the effectiveness of this NN, we compared the results with four methods commonly used in text classification tasks, namely Naïve Bayes, SVM, C4.5 decision tree, and k-NN classifier.

Naïve Bayes is the most commonly used generative classifier in text classification. The posterior probabilities of classes are calculated based on the distribution of the words in the document. The main assumption of Naïve Bayes is that the words in the documents are conditionally independent given the class value.

Further, we used the SVMs trained by the sequential minimal optimization algorithm. Since SVMs are robust to high dimensionality, they are well suited for text classification because of the sparse high-dimensional nature of the text. The following parameters of the SVMs were examined: kernel functions = polynomial, the level of polynomial function = {1,2,3}, complexity parameter $C = \{2^0,2^1,2^2,\ldots,2^8\}$.

Error based pruning algorithm was used to train the C4.5 decision tree. This algorithm uses single-attribute splits at each node. The feature with the highest information gain is used for the purpose of the split. For this algorithm, confidence factor is used when pruning the tree. The following parameters of C4.5 were examined to obtain the best classification performance: confidence factor = {0.1,0.25,0.4}, minimum number of instances per leaf = {1,2, …,5} and number folds = 3.

Linear nearest neighbour search algorithm with Euclidean distance function was used for the *k*-NN classifier. The number of neighbours was set to 3. The main idea is that documents belonging to the same class are likely to be close to one another based on a similarity measure.

It was reported that the use of common classification performance criteria such as accuracy may yield misleading conclusion in the case of class imbalance [33]. More accurate measures such as ROC (receiver operating characteristic) curve have been predominantly used for imbalanced datasets. Therefore, we measured the quality of abnormal return prediction using the area under the ROC curve. To avoid overfitting, all experiments were performed using 10-fold cross-validation.

In the first set of experiments, we used the financial, sentiment and BoW features separately. Table 5 shows the classification performance on the abnormal bank stock returns dataset. We report the Average ± Std.Dev. values of ROC from the 10-fold cross-validation. The best performance of the algorithm is marked in bold.

Table 5. Comparison of ROC classification performance – single approaches

	NB	SVM	C4.5	*k*-NN	NN
Fin.	0.652 ± 0.201*	0.499 ± 0.005	0.496 ± 0.044	0.602 ± 0.162	0.670 ± 0.190*
Sent.	0.592 ± 0.172*	0.500 ± 0.000	**0.567 ± 0.135**	0.525 ± 0.172	0.598 ± 0.195*
BoW_200uni	0.651 ± 0.150	0.600 ± 0.143	0.562 ± 0.156	**0.636 ± 0.167**	0.692 ± 0.155*
BoW_200bi	0.667 ± 0.148	**0.650 ± 0.156**	0.550 ± 0.190	0.547 ± 0.169	0.736 ± 0.160*
BoW_200tri	0.672 ± 0.180	0.625 ± 0.139	0.554 ± 0.172	0.571 ± 0.166	0.757 ± 0.144*
BoW_500uni	0.689 ± 0.164*	0.558 ± 0.128	0.535 ± 0.157	0.533 ± 0.153	0.710 ± 0.148*
BoW_500bi	0.725 ± 0.153*	0.596 ± 0.148	0.528 ± 0.146	0.496 ± 0.148	0.755 ± 0.125*
BoW_500tri	0.694 ± 0.200	0.608 ± 0.114	0.532 ± 0.141	0.596 ± 0.166	0.756 ± 0.131*
BoW_1000uni	0.661 ± 0.169	0.582 ± 0.121	0.555 ± 0.173	0.498 ± 0.159	0.706 ± 0.179*
BoW_1000bi	**0.729 ± 0.161**	0.619 ± 0.144	0.547 ± 0.151	0.547 ± 0.166	0.762 ± 0.137*
BoW_1000tri	0.726 ± 0.159	0.585 ± 0.115	0.556 ± 0.171	0.557 ± 0.147	**0.778 ± 0.126***

* ROC significantly higher at $p = 0.05$.

SVM, C4.5 and *k*-NN algorithms performed generally better on the lower dimensional datasets, whereas NB and NN performed best for the BoW with 1000 features. In case of the NB, this suggests a high variance in the data that this probability based term weighting scheme better distinguishes documents in the minor category. Moreover, the quality of the prediction significantly increased when using bigrams and trigrams, respectively. We employed Student's paired *t*-test at $p = 0.05$ to test the differences in ROC. The results show that NN performed particularly well on the BoW datasets.

Table 6. Comparison of ROC (Accuracy) performance – combinations of approaches

	NB	SVM	C4.5	*k*-NN	NN
Fin. + Sent.					
ROC	0.603 ± 0.172	0.499 ± 0.007	0.533 ± 0.139	0.628 ± 0.153	0.652 ± 0.191*
Accuracy	51.18 ± 12.72#	77.10 ± 1.31	**73.91 ± 7.12**	**74.46 ± 8.77**	77.25 ± 0.66
Fin. + BoW					
ROC	0.729 ± 0.160	**0.651 ± 0.148**	0.558 ± 0.168	**0.661 ± 0.157**	**0.786 ± 0.121***
Accuracy	71.93 ± 11.88	76.99 ± 10.30	67.42 ± 11.98#	71.71 ± 10.60	**77.57 ± 8.45**
Sent. + BoW					
ROC	**0.737 ± 0.156**	0.646 ± 0.156	**0.581 ± 0.159**	0.628 ± 0.152	0.775 ± 0.125*
Accuracy	**72.54 ± 11.32**	**77.35 ± 10.66**	68.92 ± 11.12#	70.70 ± 10.51	77.26 ± 8.26
Fin. + Sent. + BoW					
ROC	**0.737 ± 0.155**	0.643 ± 0.149	0.574 ± 0.177	0.625 ± 0.147	0.776 ± 0.129*
Accuracy	72.46 ± 11.45	76.99 ± 10.54	69.02 ± 11.79#	68.52 ± 9.87#	77.49 ± 7.53

* significantly higher ROC at $p = 0.05$, # significantly lower accuracy at $p = 0.05$.

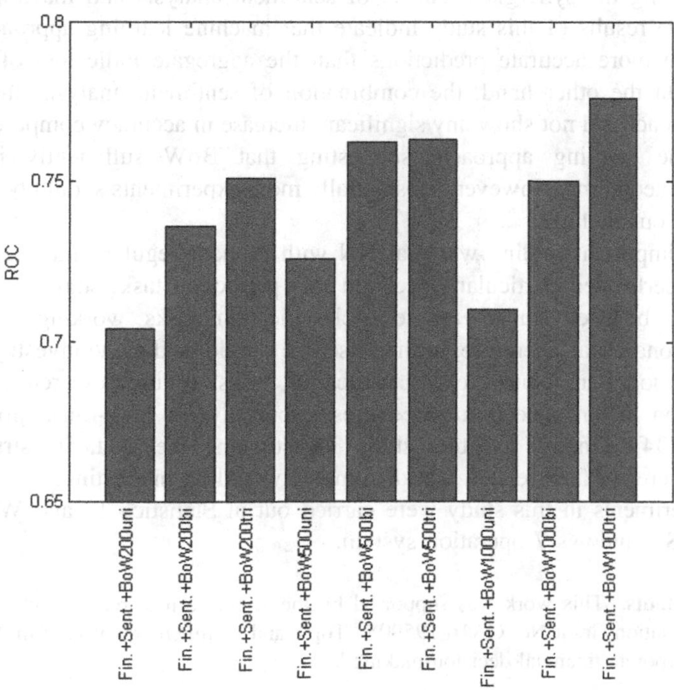

Fig. 1. Average area under the ROC curve for NN. We used the combinations of financial (Fin.), sentiment (Sent.) and bag-of-words (BoW) for 200, 500 a 1000 unigrams, bigrams and trigrams.

In the second set of experiments, we combined the categories of features to demonstrate the synergistic effect of financial, sentiment and BoW information. Specifically, we examined the following combinations: (1) financial and sentiment, (2) financial and BoW, (3) sentiment and BoW, and (4) financial, sentiment and BoW features. Table 6 shows that the classification performance of all algorithms increased compared with single approaches presented in Table 5. To save space, we show only the best results for the BoW combinations. In addition to ROC, we also show accuracy of the classifiers. Regarding precision and recall, we obtained *F*-measure of 0.338 ± 0.272 for the best NN model (Fin. + Bow).

For all algorithms, the performance was best when the financial or sentiment indicators were combined with the BoW approach. In terms of ROC, the NN method significantly outperformed the remaining methods in all four sets of experiments. Figure 1 shows the detailed results for the combination of the financial, sentiment and BoW features used by the NN method. Obviously, the classification performance increases with both the number and length of terms, providing the best performance for 1000 features and trigrams.

5 Conclusion

A strong relationship between textual information extracted from annual reports and abnormal stock return has been reported in the literature. This study set out with the aim of assessing the synergistic effects of sentiment analysis and machine learning approach. The results of this study indicate that machine learning approaches using BoW provide more accurate predictions than the aggregate indicators of sentiment categories. On the other hand, the combination of sentiment analysis and machine learning approach did not show any significant increase in accuracy compared with the pure machine learning approach, suggesting that BoW sufficiently incorporate sentiment-related terms. However, substantially more experiments should be conducted to generalize our findings.

Another important finding was that NN with dropout regularization and rectified linear units performed particularly well on this prediction task, suggesting that this method may be well suited for text classification tasks working with sparse high-dimensional data. Therefore, further research should be done to investigate the use of this NN model in related text classification tasks. Future research should also concentrate on feature selection procedures, especially for high-dimensional imbalanced data [34]. Finally, a future study investigating the syntactic structure and semantic features of firm-related text documents would be interesting.

The experiments in this study were carried out in Statistica 12 and Weka 3.7.13 using the MS Windows 7 operation system.

Acknowledgments. This work was supported by the scientific research project of the Czech Sciences Foundation Grant No: GA16-19590S "Topic and sentiment analysis of multiple textual sources for corporate financial decision-making".

References

1. Henry, E.: Are investors influenced by how earnings press releases are written? J. Bus. Commun. **45**(4), 363–407 (2008)
2. Tetlock, P.C., Saar-Tsechansky, M., Macskassy, S.: More than words: quantifying language to measure firms' fundamentals. J. Financ. **63**(3), 1437–1467 (2008)
3. Doran, J.S., Peterson, D.R., Price, S.M.: Earnings conference call content and stock price: the case of REITs. J. Real Estate Financ. Econ. **45**(2), 402–434 (2012)
4. Antweiler, W., Frank, M.Z.: Is all that talk just noise? the information content of internet stock message boards. J. Financ. **59**(3), 1259–1294 (2004)
5. Tetlock, P.C.: Giving content to investor sentiment: the role of media in the stock market. J. Financ. **62**, 1139–1168 (2007)
6. Loughran, T., McDonald, B.: When is a liability not a liability? textual analysis, dictionaries, and 10-Ks. J. Financ. **66**(1), 35–65 (2011)
7. Hájek, P., Olej, V.: Evaluating sentiment in annual reports for financial distress prediction using neural networks and support vector machines. In: Iliadis, L., Papadopoulos, H., Jayne, C. (eds.) EANN 2013, Part II. CCIS, vol. 384, pp. 1–10. Springer, Heidelberg (2013)
8. Hajek, P., Olej, V., Myskova, R.: Forecasting corporate financial performance using sentiment in annual reports for stakeholders' decision-making. Technol. Econ. Dev. Econ. **20**(4), 721–738 (2014)
9. Hájek, P., Olej, V.: Intuitionistic fuzzy neural network: the case of credit scoring using text information. In: Iliadis, L., et al. (eds.) EANN 2015. CCIS, vol. 517, pp. 337–346. Springer, Heidelberg (2015). doi:10.1007/978-3-319-23983-5_31
10. Li, F.: Do Stock Market Investors Understand the Risk Sentiment of Corporate Annual Reports? (2006). SSRN 898181
11. Li, F.: Annual report readability, current earnings, and earnings persistence. J. Account. Econ. **45**(2), 221–247 (2008)
12. Feldman, R., Govindaraj, S., Livnat, J., Segal, B.: Management's tone change, post earnings announcement drift and accruals. Rev. Account. Stud. **15**(4), 915–953 (2010)
13. Balakrishnan, R., Qiu, X.Y., Srinivasan, P.: On the predictive ability of narrative disclosures in annual reports. Eur. J. Oper. Res. **202**(3), 789–801 (2010)
14. Price, S.M., Doran, J.S., Peterson, D.R., Bliss, B.A.: Earnings conference calls and stock returns: the incremental informativeness of textual tone. J. Bank. Financ. **36**(4), 992–1011 (2012)
15. Loughran, T., McDonald, B.: The use of word lists in textual analysis. J. Behav. Financ. **16** (1), 1–11 (2015)
16. Hinton, G.E., Srivastava, N., Krizhevsky, A., Sutskever, I., Salakhutdinov, R.R.: Improving neural networks by preventing co-adaptation of feature detectors (2012). arXiv preprint arXiv:1207.0580
17. Baharudin, B., Lee, L.H., Khan, K.: A review of machine learning algorithms for text-documents classification. J. Adv. Inf. Technol. **1**(1), 4–20 (2010)
18. Kearney, C., Liu, S.: Textual sentiment in finance: a survey of methods and models. Int. Rev. Finan. Anal. **23**(33), 171–185 (2014)
19. Li, F.: The information content of forward-looking statements in corporate filings - a naïve Bayesian machine learning approach. J. Account. Res. **48**(5), 1049–1102 (2010)
20. Demers, E.A., Vega, C.: Soft Information in Earnings Announcements: News or Noise? Working paper. In: INSEAD (2010)
21. Davis, A.K., Piger, J.M., Sedor, L.M.: Beyond the numbers: measuring the information content of earnings press release language. Contemp. Account. Res. **29**(3), 845–868 (2012)

22. Schumaker, R.P., Chen, H.: Textual analysis of stock market prediction using breaking financial news: the AZFin text system. ACM Trans. Inf. Syst. (TOIS) **27**(2), 12 (2009)
23. Jiang, S., Pang, G., Wu, M., Kuang, L.: An improved K-nearest-neighbor algorithm for text categorization. Expert Syst. Appl. **39**(1), 1503–1509 (2012)
24. Schumaker, R.P., Zhang, Y., Huang, C.N., Chen, H.: Evaluating sentiment in financial news articles. Decis. Support Syst. **53**(3), 458–464 (2012)
25. Li, Q., Wang, T., Gong, Q., Chen, Y., Lin, Z., Song, S.K.: Media-aware quantitative trading based on public web information. Decis. Support Syst. **61**, 93–105 (2014)
26. Yu, Y., Duan, W., Cao, Q.: The impact of social and conventional media on firm equity value: a sentiment analysis approach. Decis. Support Syst. **55**(4), 919–926 (2013)
27. Kothari, S.P., Li, X., Short, J.E.: The effect of disclosures by management, analysts, and business press on cost of capital, return volatility, and analyst forecasts: a study using content analysis. Account. Rev. **84**(5), 1639–1670 (2009)
28. Fama, E.F., French, K.R.: Common risk factors in the returns on stocks and bonds. J. Finan. Econ. **33**(1), 3–56 (1993)
29. Nassirtoussi, A.K., Aghabozorgi, S., Wah, T.Y., Ngo, D.C.L.: Text mining for market prediction: a systematic review. Expert Syst. Appl. **41**(16), 7653–7670 (2014)
30. Nam, J., Kim, J., Mencía, E.L., Gurevych, I., Fürnkranz, J.: Large-scale multi-label text classification - revisiting neural networks. In: Calders, T., Esposito, F., Hullermeier, E., Meo, R. (eds.) Machine Learning and Knowledge Discovery in Databases, pp. 437–452. Springer, Heidelberg (2014)
31. Maas, A.L., Hannun, A.Y., Ng, A.Y.: Rectifier nonlinearities improve neural network acoustic models. In: Proceedings of the 30th International Conference on Machine Learning (ICML), vol. 30, pp. 1–6. Atlanta, Georgia (2013)
32. Jaitly, N., Hinton, G.: Learning a better representation of speech soundwaves using restricted boltzmann machines. In: IEEE International Conference on Acoustics, Speech and Signal Processing (ICASSP), pp. 5884–5887 (2011)
33. Chawla, N.V., Japkowicz, N., Kotcz, A.: Editorial: special issue on learning from imbalanced data sets. ACM Sigkdd Explor. Newsl. **6**(1), 1–6 (2004)
34. Yin, L., Ge, Y., Xiao, K., Wang, X., Quan, X.: Feature selection for high-dimensional imbalanced data. Neurocomputing **105**, 3–11 (2013)

Emotion Recognition Using Facial Expression Images for a Robotic Companion

Ariel Ruiz-Garcia(✉), Mark Elshaw, Abdulrahman Altahhan, and Vasile Palade

Faculty of Engineering, Environment and Computing,
School of Computing, Electronics and Mathematics,
Coventry University, Priory Street, Coventry CV1 5FB, UK
ariel.ruizgarcia@coventry.ac.uk

Abstract. Social robots are gradually becoming part of society. However, social robots lack the ability to adequately interact with users in a natural manner and are in need of more human-like abilities. In this paper we present experimental results on emotion recognition through the use of facial expression images obtained from the KDEF database, a fundamental first step towards the development of an empathic social robot. We compare the performance of Support Vector Machines (SVM) and a Multilayer Perceptron Network (MLP) on facial expression classification. We employ Gabor filters as an image pre-processing step before classification. Our SVM model achieves an accuracy rate of 97.08 %, whereas our MLP achieves 93.5 %. These experiments serve as benchmark for our current research project in the area of social robotics.

Keywords: Emotion recognition · Support Vector Machine · Gabor filter · Image classification · Neural networks · Social robots

1 Introduction

Robotic machines are gradually becoming present in a diverse number of fields, such as national and international security, transport, social media, industry, education [1], and health care [2], amongst others. Social robots, in particular, are progressively becoming more intelligent and capable of interacting with human beings. Despite this perpetual development and albeit the unceasing advancements in technology, social robots are still far from being able to effectively interact in a human-like manner. As a result, robotic researchers are focusing their research on developing robots that go beyond a simple mechanical machine; great efforts are being made to create human-like social robots with the ability to interact in a manner that a human would [3]. The first step towards achieving human-like interaction skills is for the robot to adequately recognise the user's emotional state and adjust its responses according to this state, thus mimicking human empathic behaviours. Consequently, in this paper we discuss

C. Jayne and L. Iliadis (Eds.): EANN 2016, CCIS 629, pp. 79–93, 2016.
DOI: 10.1007/978-3-319-44188-7_6

experimental results on emotion recognition through the use of facial expression images. These experiments serve as the base concept of our research project: the development of an empathic robot, a robot with the ability to: (i) recognise human emotions through facial expressions, (ii) illustrate emotional states itself, and (iii) automatically and autonomously produce and associate responses to specific emotional states. The following section discusses existing background material on human emotions, social robots, and machine learning approaches for emotion recognition and empathic robots. Section 3 explores the dataset we use and our experiments. Section 4 presents results and discussion. The last section focuses on future work followed by a list of references.

2 Background and Literature Review

Research has previously suggested that when interacting with non-human beings our responses depend upon the number of human characteristics we attribute to the object or animal we are interacting with. This process of attributing human characteristics to objects is known as anthropomorphism [4]. Anthropomorphising can facilitate the process of empathising with robots, more precisely with humanoid robots or robots that can show some sort of facial expressions: since humanoid robots already possess human-like characteristics such as a mouth and a pair of eyes, the need to recreate a human image from scratch is eliminated. This highlights the importance and the key role of facial expressions in human-robot interactions, hence our research focuses on the development of a social robot that can accurately recognise these.

2.1 Social Robots

The benefits that social robots can offer to society are numerous. Social robots can be used in nurseries and homes to act as companions for the elderly [5,6], in hospitals to help with the recovery of patients with specific conditions such as cancer, in clinics specialised on therapy for children with autism [8], amongst others. A good example of the latest trends in social robotics is Paro, a therapeutic seal robot developed by the National Institute of Advanced Industrial Science and Technology in Japan. This robot is used all around the world for therapeutic purposes. Aside from its abilities to act as a live animal by demonstrating proactive, reactive, and physiological behaviour, Paro has the ability to learn and adjust its behaviour based on these. Paro places positive values on preferred stimulations and negative ones on undesired stimulations for long term memory [5]. Another socially interactive robot is GeriJoy, a virtual care companion developed by Massachusetts Institute of Technology researchers. This virtual companion offers wellness coaching, therapeutic programs, reminders, safety supervision, companionship and care for the elderly [6]. The KSERA project is another example of existing research in the field of Sociorobotics. This project aims to provide a social robot that can assist the elderly with their daily activities and care needs. Moreover, this project is specially designed to assist the elderly with conditions such as Chronic Obstructive Pulmonary Disease [7].

Another added benefit provided by social robots is their assistance in educational institutions to serve as tutors for children [1]. The success of social robots in the education sector are also heavily dependable upon their ability to illustrate empathic states. Castellano et al. [1] have introduced EMOTE, a project aimed to capture some of the empathic and human elements that characterise a traditional teacher. The authors have identified a number of crucial points for the success of empathic robotic tutors including the development of a set of cues that should create social bonding despite the fact that not all features will be anthropomorphic [1]. We speculate that if we can develop a robot capable of illustrating such behaviours, it will significantly facilitate the process of anthropomorphising and thus, bring social robots a step closer to (i) be able to illustrate signs of intelligence, (ii) create closer and more intimate relationships with humans by empathising with them, and (iii) be fully accepted by society.

One of the remarking characteristics of social robots such as Paro [5] is their ability to adjust their behaviour in order to suffice the user's needs. However, these machines lack a number of abilities due to the limitations imposed by technology. Our aim is to provide robots with the ability to illustrate signs of intelligence and be able to effectively interact with a user in order to provide users with affordance and increase user satisfaction. We take a biologically inspired approach in an attempt to diminish the gap between natural and artificial systems. Just like Castellano et al. [1], we use empathy as a base concept of our research.

2.2 Biological Basis of Empathy

In order to establish meaningful relationships with humans, social robots must interact with people in natural ways and employ social mechanisms such as empathy in the same manner as humans do when interacting with other humans [9]. Empathy, at an abstract level, can be defined as our human ability to understand and share the feelings of other beings [10]. Research has previously suggested that there exist two separate systems for empathy: an emotional systems supporting our ability to empathise emotionally and a cognitive system involving cognitive understanding of the other's perspective [11]. Studies have shown the activation of a number of regions including: the inferior front gyrus, ifnerior parietal lobule, anterior cingulated, anterior insula, ventromedial prefrontal cortex, dorsemdeial preforontal cortes, temporoparietal junction, and medial temporal lobe, during empathic states in human [11]. Some of these regios are activated when empathy for pain is expressed and others during emotional contagion.

Many researchers in the field of social neuroscience have attributed the ability to empathise emotionally to mirror neurons [12]. Mirror neurons are a class of neuron that modulate their activity both when an individual executes a specific motor act and when they observe the same or similar act performed by another individual [13]. These neurons were first observed in the macaque monkeys by Rizzolatti et al. in the 1990s. Rizzolatti et al. [14] observed the existence of neurons, in the F5 area of the macaque premotor cortex, that discharge when the macaque performs an action and when it observes a similar action being done by another monkey or experimenter. The contributions made by Rizzolatti et al. [14] served as a base concept for a number of studies attempting to replicate the same results in the human brain. However, most of the studies confirming the existing of mirror neurons in the human brain have employed neuroimaging techniques to monitor brain activity during the observation of and execution of specific task such as seeing someone in pain and responding to it. Singer et al. [15] investigated pain related empathy using functional magnetic resonance imaging techniques (fMRI). The authors observed and assessed brain activity on females when pain was induced to their partners right hand through an electrode attached to the back of the hand. After being discovered, the existence of mirror neurons seemed to suggest that their function is hardwired and thus that if these exist in the human brain then we are predetermined to resonate with the emotions of others because of our mirror neurons [16]. These controversial claims created a dispute in the scientific community given that all the evidence confirming the existence of mirror neurons in the human brain seem to be based on the assumptions that they exist in the first place. Nonetheless, the empathic characteristics of mirror neurons make them a viable path to create empathic behaviours in social robots. A social robot with the ability to create such behaviours would not only be able to create appropriate automatic responses but create closer and even intimate relations with the user.

The empathy system involving cognitive understanding of the other's perspective is directly linked to theory of mind (ToM) [17]. Theory of mind refers to the ability to understand the intentions of others and covers two concepts: (i) the knowledge that other animals have mental states such as beliefs, goals, intentions or emotions, which may be the same or different to our own, and (ii) the ability to infer what these states may be [17]. If humans can infer similar states to social robots and if a social robot can adequately interpret the meaning of human actions and behaviours, and produce autonomous and automatic responses to these actions, the interaction process between humans and robots would be facilitated. Moreover, the action of anthropomorphising for the human would require less effort. One of the main challenges in the development of such intelligent machines is the fact that the sources of inspiration come from what exists around us [18]. Building machines that can be as intelligent and versatile as humans and with the ability to socialise and interact as if they were humans themselves, requires employing the human frame of reference to a certain extent [18]. At present, existing state of the art machines and frameworks such as Paro and EMOTE rely on machine learning algorithms to illustrate intelligent behaviours.

2.3 Machine Learning Approaches for Emotional Face Recognition

Although existing social robots such as Paro are good examples of the current technological advancements, these robotic machines still lack the ability of adequately illustrating signs of intelligence in a natural way. Consequently, in this paper we present experimental results on emotion recognition through the use of facial expression images, a first step towards an emphatic robot. The most common approaches to emotion recognition from facial expression images make use of machine learning algorithms such as artificial neural networks [19,20], self-organised maps [21,22] or support vector machines [23,24] to classify facial expression images as a specific emotion. In this work we employ and compare the performance of Support Vector Machines (SVM), and Multilayer Perceptron Neural Networks (MLP) to classify the following emotions: Happy, Sad, Angry, Surprise, Fear, Disgust, and Neutral.

Sohail and Bhattacharya [24] make use of facial feature point localisation to reconstruct a neutral face to use as reference. When combined with an MLP, the authors obtain a recognition rate of 92 %. Hewahi and Baraka also use a similar approach in which they utilise ethnic background information as an input to the neural network [25]. Khashman also employs an MLP for classification and Global Pattern Averaging as an image pre-processing step to achieve 87.78 % accuracy [26]. Ouellet [27] uses Convolutional Neural Networks combined with Support Vector Machine to obtain 94.4 % accuracy rate. Burkert et al. [28] also use Convolutional Neural Networks and set a state of the art benchmark with 99.6 % accuracy. Levi and Hassner [29] have proposed an approach which uses Local Binary Patter (LBP) features as input to Convolutional Neural Network models. The LBP codes are produced by applying thresholds on pixel intensity values in small neighbourhoods using the intensity of each neighbourhoods central pixel as the threshold. The resulting pattern of 0s or 1s is used as the representation [29]. Their pre-processing of images reduces variability due to illumination changes. In addition to this, the authors use network ensembles trained using different image representations as well as different network architectures in order to boost recognition performance [29].

Given that the performance of classifier algorithms heavily relies on the quality of the feature vector representing the image, and thus the emotional state, it is essential that the optimum image pre-processing method is applied to the images used for training. Gabor filter is one of the most popular methods in image processing due to its ability to detect edges. This process resembles the mechanism in the human visual system [30] and is characterised by multi-resolution and multi-orientation properties. Essentially a Gabor filter is a Gaussian kernel modulated by a sinusoidal plane [30]. The authors of [30] successfully applied Local Transitional Pattern and Gabor filters to classify facial expression images with Support Vector Machines, obtaining an accuracy rate of 95 %. A similar approach was done by Chelali and Dejardi [31].

While the results obtained by researchers offer a good degree of accuracy, there exist controversy with regards to which facial features play a bigger role in the recognition of specific emotions. Beaudry et al. [32] conducted a study aimed to provide a scientific answer to this paradox. Beaudry et al. [32] deducted that the eyes and eyebrows play a bigger role for the recognition of sadness, whereas the mouth is more influential to recognise happiness. In contrast, the authors determined that a holistic processing could be called upon fear but could not determine the best approach to recognise other emotions. Given that the primary objective of our research is the development of an empathic robot with the ability to (i) recognise human emotions through facial expressions and

(ii) show emotions itself, we plan to exploit the findings by Beaudry et. al. [32] by emphasising on the features that are relevant for identifying a specific emotional state. Consequently, in this paper we employ Gabor filters as an image pre-processing technique due to their ability to highlight these features. In the next section we describe the dataset used in our experiments and the selection of features.

3 Experimental Setup and Methodology

3.1 Emotional Facial Expression Corpus

In this paper we use the Karolinska directed Emotional faces database (KDEF) [33]. It contains a set with 70 individuals: 35 males and 35 females, all between 20 and 30 years old, each displaying seven different emotional expressions in five different angles. All images were taken under a controlled environment: subjects wore uniform T-Shirt colours, faces were centred with a grid, and eyes and mouths were positioned in fixed image coordinates [33]. In our experiments we only use front angle images: a subset containing 140 front angle images per emotion, thus a total of 980 images. The first step towards obtaining a feature vector was to locate the face and crop out the irrelevant spatial features such as background, hair, and ears. Face images where then converted to grayscale in order to reduce their dimensionality. Given that the resulting face images varied in size and they were resized to a standard 120 × 110 size to maintain the aspect ration of the resulting face image. Figure 1 illustrates sample face images obtained from the KDEF database.

Fig. 1. Sample extracted face images from the KDEF database [33]. Subject F06 displaying seven emotions: angry, disgust, fear, happy, neutral, sad, surprise.

In order to avoid overfitting due to the reduced amount of front angle images contained in the KDEF dataset, after applying the Gabor filter with five scales and eight orientations over an image I, we split the resulting feature vector into four smaller vectors and treat each one as an input sample. Applying this method quadruples the number of sample inputs from 980 to 3920 input samples, giving us more training data. Essentially, each individual input contains an image convolved with 10 filters expanded over 2 dimensions and 8 orientations.

3.2 Gabor Filter

Taking into consideration that the aim of our research is to create a system that bridges the gap between natural and artificial mechanisms, we employed

Gabor filter as an image pre-processing technique due to it resemblance to the perception in the human visual system [30]. Gabor filters have a unique ability to highlight salient facial features such as the mouth, eyes, and eyebrows. Studies have previously demonstrated that the mouth, eyes, and eyebrows play a key role in the recognition of specific emotions [32]. After applying this filter to our dataset we obtain image representations that highlight similar facial features as illustrated in Fig. 2. The commonly used Gabor filters in face recognition are defined as follows [34]:

$$G_{\mu,v}(Z) = \frac{||k_{\mu,v}||^2}{\sigma^2} \exp\left(\frac{-||k_{\mu,v}||^2||Z||^2}{2\sigma^2}\right)\left[e^{ik_{\mu,v}z} - e^{-\sigma^2/2}\right]. \quad (1)$$

$Z(x, y)$ is the point of coordinates (x, y) in image space, μ and v define the orientation and scale of the Gabor filters, $k_{\mu,v}$ is the wave vector [34]. We experimented with a Gabor filter with eight orientations and five frequencies. The Gabor filter applied also down samples the original image producing a reduced feature vector. Once the Gabor filters were applied the feature vector values were normalised to obtain values in range zero to one. Moreover, as done by Chelali and Djeradi [31] we only used the magnitude information given that it highlights areas of interest such as the mouth and eyes, and discards the effect of noises.

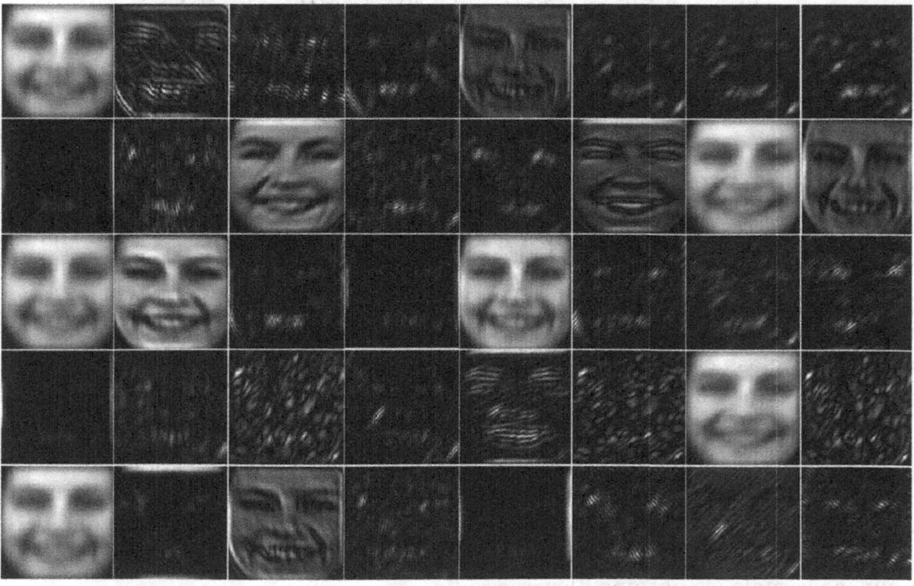

Fig. 2. Magnitude response of Gabor filter with 8 orientations and 5 dimensions. Image corresponding to subject F06 from KDEF database [33].

Our Gabor filter is essentially a sinusoidal modulated by a Gaussian kernel function [31] in which orthogonal directions are represented by real and imaginary components. We tried using a combination of the two as a complex component, however we obtained lower classification results as compared to when using the real component only. Let λ represent the frequency of the sinusoidal, θn represents the orientation, and σ represents the standard deviation of the Gaussian over x and y dimensions of the sinusoidal plane, the real component of the Gabor filter applied to an image with dimensions the x and y is defined as follows:

$$G_{\lambda,\theta}(x,y) = \exp\left[-\frac{1}{2}\left\{\frac{x_{\theta n}^2}{\sigma_x^2} + \frac{y_{\theta n}^2}{\sigma_y^2}\right\}\right]\cos(2\pi * \theta n * \lambda). \tag{2}$$

where

$$x_{\theta n} = x(\sin\theta n) + y(\cos\theta n)$$
$$y_{\theta n} = x(\cos\theta n) + y(\sin\theta n)$$

After trying a number of parameters we concluded that initialising the Gabor filter with the following parameters results in the best magnitude response vector for emotion classification: $\theta = 2pi/3$, $\lambda = 6$, $\gamma = 0.5$, and $\sigma = 4$. Figure 1 illustrates the resulting magnitude response after convolving the Gabor filter with these parameters over an image. This response vector is given by:

$$||G_{\lambda,\theta}(x,y)|| = \sqrt{\Re^2\{G_{\lambda,\theta}(x,y)\} + \Im^2\{G_{\lambda,\theta}(x,y)\}}. \tag{3}$$

Where $\Re\{G_{\lambda,\theta}(x,y)$ represents the real part of the filter and $\Im\{G_{\lambda,\theta}(x,y)\}$ represents the imaginary part.

3.3 Facial Expression Classification

Support Vector Machines are non-probabilistic binary classifiers well known for performing notably well in image classification problems. We follow similar approaches to [30, 31] to classify facial expression images into seven emotions: angry, disgust, fear, happy, neutral, sad, and surprised. There exist two methods for multiclass classification problems: *one-versus-all* and *one-versus-one*. If we have c unique classes in the training set, the one versus all approach builds c binary classifiers in which the classifier with the highest output determines the classification in a *winner-takes-all* approach. The one versus one approach creates $c * (c1)/2$ classifiers in which the selected class is the one which most classifiers predict. We tested both approaches and concluded that the one versus one approach produces better results with our training set. Moreover, when nonlinear kernels we obtained lower results.

Let b represent the bias, K be a linear kernel function, our facial expression classification is determined by:

$$f(x) = \text{sgn}\left(\sum_{i+l}^{l} y_i\alpha_i|K(x_i,x) + b\right). \tag{4}$$

Table 1. Left: Gabor filter + SVM confusion matrix; Right: Gabor filter + MLP confusion matrix. A: angry; D: disgust; F: fear; H: happy; N: neutral; Sa: sad; Su: surprised. Displayed overall results are rounded.

	A	*D*	*F*	*H*	*N*	*Sa*	*Su*	
A	74	0	1	2	0	0	4	91.36
D	1	96	0	0	0	0	0	98.97
F	0	0	89	0	2	0	0	97.80
H	0	0	0	73	0	0	0	100
N	0	0	0	0	77	0	0	100
Sa	1	0	1	0	1	82	0	96.47
Su	4	0	0	0	0	0	80	95.24

	A	*D*	*F*	*H*	*N*	*Sa*	*Su*	
A	91	0	0	0	0	0	3	96.81
D	0	61	2	0	0	3	1	91.05
F	1	0	77	1	2	1	0	93.90
H	0	2	0	105	0	1	1	96.34
N	0	0	1	0	73	4	1	92.41
Sa	0	0	0	0	2	70	2	94.60
Su	1	2	1	2	4	0	73	88.00

where x_i is the training vector, x is the testing vector with $\alpha_i > 0$, y_i represents Lagrange multipliers of dual optimization problem [35] and α and b are solutions of a quadratic programing problem. We randomly selected 85 % of the input vector as our training and validation set and 15 % for testing. This model produced an accuracy rate of 100 % on training and validation set and 97.08 % testing set after training with a 10-fold-crossvalidation approach. Table 1 illustrates the confusion matrix produced by this model.

Given the popularity of feed forward neural networks for classification problems we decided to compare the performance of Support Vector Machine against that of a Multilayer Perceptron Network. We experimented with different network topologies and obtained best results with the following Multilayer Perceptron Network configuration: one input layer with 8400 neurons taking 10 28×30 filtered images as input, one hidden layer with 93 neurons, and one output layer

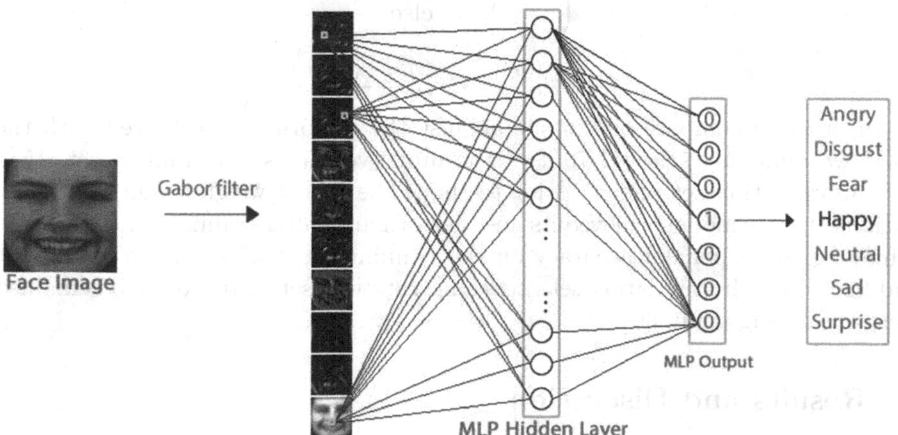

Fig. 3. Proposed fully connected Multilayer Perceptron Network taking Gabor filter magnitude response as input. Only sample connections are illustrated. Face image extracted from KDEF database [33].

with 7 neurons. The target values contained a one in the place of the target class and zero for the rest. See Fig. 3 for a visualisation. We use the sigmoid activation function defined as:

$$S(t) = \frac{1}{1 + e^{-t}}. \tag{5}$$

Though, sigmoid activation functions often create an increased risk of falling into a local minimum when used in conjunction with steepest decent to train MLPs; since the gradient can have very small magnitude and cause small changes in the weights and biases, the network may never reach global minimum if the weights and bias are far from their optimal values. We employed Resilient Backpropagation (Rprop) as our learning algorithm to avoid such side effects. Resilient backpropagation tackles these side effects by performing a direct adaptation of the weight step based on local gradient information [36]. Let Δ_{ij} represent the individual update-value which determines the size of the weight-update, then the evolution of the adaptive update-value during learning is based on the error function E according to [36]:

$$\Delta_{ij}^{(t)} = \begin{cases} n^{+} * \Delta_{ij}^{(t-1)}, & \text{if } \frac{\partial E}{\partial w_{ij}}^{(t-1)} * \frac{\partial E}{\partial w_{ij}}^{(t)} > 0 \\ n^{-} * \Delta_{ij}^{(t-1)}, & \text{if } \frac{\partial E}{\partial w_{ij}}^{(t-1)} * \frac{\partial E}{\partial w_{ij}}^{(t)} < 0 \\ \Delta_{ij}^{(t-1)}, & \text{else} \end{cases} \cdot \tag{6}$$

$$\text{where } 0 < n^{-} < 1 < n^{+}$$

Then the weight-update is determined according to the following rule [36]:

$$\Delta_{ij}^{(t)} = \begin{cases} -\Delta w_{ij}^{(t)}, & \text{if } \frac{\partial E}{\partial w_{ij}}^{(t)} > 0 \\ +\Delta w_{ij}^{(t)}, & \text{if } \frac{\partial E}{\partial w_{ij}}^{(t)} < 0 \\ 0, & \text{else} \end{cases} \cdot \tag{7}$$

$$w_{ij}^{(t+1)} = w_{ij}^{(t)} + \Delta w_{ij}^{(t)}$$

In order to compare one-to-one against the performance achieved with the SVM we randomly selected 70 % of the input vector as our training set, 15 % for our validation set and 15 % for testing. The initial weights were randomly initialised. This model achieved its best performance after training for 175 epochs achieving 100 % percent accuracy on the training set, 93.5 % on the testing set and 92.9 % on the validation set. Learning rate was set to 0.0001 and remained constant during training.

4 Results and Discussion

In this paper we make progress towards developing an empathic robot. The architectures proposed provide the robot with the ability to accurately recognise emotions in people. Humans have this unique ability to autonomously recognise emotional states in other people by simply observing their body language and facial expressions when expressing emotions. Robotic machines on the contrary, have to be provided with a way to obtain this information. This can conveniently be achieved through the use of emotion facial expression images. The robot then needs a way to analyse this data in a timely manner. Machine learning approaches have proved to be an efficient way to learn patterns and classify facial expression images as particular emotion. In this work we have employed a multiclass Support Vector Machine and a Multilayer Perceptron Network to classify facial expression images, obtained from the KDEF database, as a particular emotion. We compare classifier performances in order to determine their suitability to be used by a social robot to identify emotional states in humans.

Given that facial expression images contain a large amount of emotion-irrelevant features, we utilise Gabor filters to highlight the emotion-determinant features such as the mouth, eyes, and eyebrows. After applying Gabor filters to our images and obtaining a reduced feature vector we feed this to a multiclass Support Vector Machine and obtain a classification accuracy rate of 97.08 %. Moreover, as it can be observed on Table 1, this classifier learns to classify disgust, happy, and neutral emotions with an accuracy rate of 100 % and confuses images label as angry the most, followed by surprised. Our second classifier, Multilayer Perceptron Network, produces marginally lower results with an overall accuracy rate of 93.5 % on the testing set. This classifier confuses images labelled as surprised the most, followed by images labelled as disgust. It's best classification performance is on images labelled as angry and happy.

Although both classifiers obtain high rates on happy expressions there exist some discrepancy on angry emotions; SVM confuses angry faces the most whereas MLP achieves its best performance on this emotion. Likewise, SVM recognises 100 % of disgust faces whereas the MLP has its second worse performance on this emotion. The correlation on happy facial expressions can be explained by the clear difference on the shape of facial features of this emotion as compared to the rest as observed in Table 1. We speculate that the inconsistency on angry and disgust faces is due to the number of angry and disgust images present in each training set. Given that training and testing sets are randomly selected the number of instances for each individual emotion varies. A more accurate comparison would be best done by using the same number of class instances for all emotions. In effect, this would also reduce the risk of overfitting for some emotions, which could have happened in the case of the SVM; this classifier used a higher number of angry faces during training compared to the MLP, possibly leading to overfitting problems.

The results we obtained with SVM outperform those achieved by the MLP and RBF architectures proposed by Chelali and Djeradi [31]. The authors test their architectures on the Computer Vision (CV) and the Olivetti-Oracle Research (ORL) databases. The results achieved with our MLP also outperform the results obtained by the authors with the ORL database and are slightly outperformed by the results they obtain with the CV database. Nonetheless, our models need to be tested with other databases to obtain a more accurate comparison.

Support Vector Machines seem to be the most popular approach for facial expression classification due to their high performance rate. However, this often depends on the image pre-processing techniques used to process the data before using it for training and testing. Based on our experimental results we have concluded that Gabor Filters are a powerful pre-processing method that increases accuracy. However, in the future we plan to adopt deep learning techniques to allow the robot system to extract the relevant features for emotion recognition autonomously.

5 Future Work and Discussion

Despite the good results achieved in these experiments we have to take into account the fact that the training and testing sets contain images of similar quality. These images were taken under a controlled environment and this could impact the results produced by the classifier models. Furthermore, given that our main objective is to incorporate this model within a social robot and allow it to interact with users in real time, it may not always be possible to obtain good quality images. The distance from where a picture is taken will most likely vary, producing very different quality images and potentially leading to overfitting problems. Moreover, the ability of the robot to show and recognise emotions will be constrained by its ability to demonstrate facial expressions and its ability to obtain a good quality image with enough emotion-related information. Taking into account that one of the main challenges in computer vision is the computationally intensive tasks required to process images, we will need to explore possible ways to further reduce the size of our input feature vector and thus speed up the training process. We will explore the approach proposed by Altahhan [37]; obtain a random set of patches from input images to learn a set of feature maps and utilise these as input for our learning architecture. These concepts will be tested with our humanoid robot called NAO. Designed by Aldebaran, NAO is 58 cm in height and has a number of sensors and abilities such as moving, feeling, seeing, speaking, hearing, and thinking [38]. See Fig. 4 for image of our robot.

Fig. 4. NAO, our humanoid robot.

Our goal is to bring in some of the biological inspiration explored in Sect. 2.2 as the basis for an empathic robot. Even though mirror neurons have lost popularity, there is still further support for their role in empathy in primates and humans [16], some of the existing theories related to them do provide an interesting basis for neural architectures and learning approaches to aid the development of human-robot interactions. We speculate that if we can develop an artificial neural architecture with some of the same observed properties of mirror neurons, we will be able to develop a robotic companion capable of recognising a persons emotional state by simply observing the person's behaviour. This will lead the robot to produce responses itself and illustrate a state of empathy. However, we will need to take into consideration the ability of our robot to display emotions; due to the high costs and limitations of existing technology, NAO is constrained by its ability to demonstrate facial expressions. Therefore, we need to take into consideration the fact that we are far off from having a mechanical machine that can accurately imitate human facial expressions. However, we hope that appropriate behaviour can be a good start to begin introducing social robots to society.

Another concern that we have is the fact that empathy is sensitive to deeply rooted parochialism and ingroup bias [16]. This implies that people will empathise better with people they feel a closer connection with, often being people with the same background, ethnicity, beliefs, etc. This raises a new issue to consider: how will different groups react to a social robot? As pointed by Hewahi and Baraka [25], people from different ethnic groups do not only have different appearance characteristics they also express emotional states in different ways. Finally, given that our main objective is to incorporate this model with a social robot and allow it to interact with users in real time in real environments such as a user's home, we need to take into consideration issues such as data protection, security, safety, among others. We hope that our future architecture will compensate for light and slight angle changes. Our ideal neural architecture would aim to take inspiration from the processing of the human brain by activating a specific region when an emotion is seen as opposed to the entire architecture. We also plan to use reinforcement learning to allow the robot adjust its responses when interacting with a user. Our plan is to continue looking at existing architectures such as SVM and benchmark against their performance. In addition, we will analyse the use of other deep learning techniques to obtain feature vectors without the need to apply filters manually.

References

1. Castellano, G., Paiva, A., Kappas, A., Aylett, R., Hastie, H., Barendregt, W., Nabais, F., Bull, S.: Towards empathic virtual and robotic tutors. In: Lane, H.C., Yacef, K., Mostow, J., Pavlik, P. (eds.) AIED 2013. LNCS, vol. 7926, pp. 733–736. Springer, Heidelberg (2013)
2. Dautenhahn, K., Campbell, A., Syrdal, D.: Does anyone want to talk to me? Reflections on the use of assistance and companion robots in care homes. In: 4th International Symposium on New Frontiers in Human-Robot Interaction (2015)

3. Cameron, D., Fernando, S., Collins, E., Millings, A., Moore, R., Sharkey, A., Evers, V.: Presence of life-like robot expressions influences childrens enjoyment of human-robot interactions in the field. In: 4th International Symposium on New Frontiers in Human-Robot Interaction (2015)
4. Darling, K.: Who's Johnny? Anthropomorphic framing in human-robot interaction, integration, and policy. SSRN Electron. J. (2015)
5. Wada, K., Asada, T., Musha, T.: Robot therapy for prevention of dementia at home results of preliminary experiment. J. Robot. Mechatron. **19**, 691–697 (2007)
6. GeriJoy: Care and Companionship for Seniors-GeriJoy (2016). http://www.gerijoy.com/
7. KSERA: Knowledgeable Service Robots for Aging (2016). http://www.aat.tuwien.ac.at/index_en.html/
8. Wainer, J., Robins, B., Amirabdollahian, F., Dautenhahn, K.: Using the humanoid robot KASPAR to autonomously play triadic games and facilitate collaborative play among children with autism. IEEE Trans. Auton. Ment. Dev. **6**, 183–199 (2014)
9. Leite, I., Pereira, A., Mascarenhas, S., Martinho, C., Prada, R., Paiva, A.: The influence of empathy in human robot relations. Int. J. Hum. Comput. Stud. **71**, 250–260 (2013)
10. Lamm, C., Silani, G.: The neural underpinnings of empathy and their relevance for collective emotions. In: Scheve, C., Salmella, M. (eds.) Collective Emotions. Oxford University Press, Oxford (2014)
11. Shamay-Tsoory, S.: The neural bases for empathy. Neuroscientist **17**(1), 18–24 (2010)
12. Bernhardt, B., Singer, T.: The neural basis of empathy. Annu. Rev. Neurosci. **35**, 1–23 (2012)
13. Kilner, J., Lemon, R.: What we know currently about mirror neurons. Curr. Biol. **23**, R1057–R1062 (2013)
14. Rizzolatti, G., Fadiga, L., Gallese, V., Fogassi, L.: Premotor cortex and the recognition of motor actions. Cogn. Brain Res. **3**, 131–141 (1996)
15. Singer, T., Seymour, B., Doherty, J., Kaube, H., Dolan, R., Frith, C.: Empathy for pain involves the affective but not sensory components of pain. Science **303**, 1157–1162 (2004)
16. Lamm, C., Majdand, J.: The role of shared neural activations, mirror neurons, and morality in empathy A critical comment. Neurosci. Res. **90**, 15–24 (2015)
17. Agnew, Z., Bhakoo, K., Puri, B.: The human mirror system: a motor resonance theory of mind-reading. Brain Res. Rev. **54**, 286–293 (2007)
18. Duffy, B.R.: Fundamental issues in social robotics. Int. Rev. Inf. Ethics **6**, 31 (2006)
19. Boughrara, H., Chtourou, M., Ben Amar, C., Chen, L.: Facial expres-sion recognition based on a MLP neural network using constructive training algorithm. Multimed. Tools Appl. **75**, 709–731 (2014)
20. Kahou, S., Michalski, V., Konda, K., Memisevic, R., Pal, C.: Recurrent neural networks for emotion recognition in video. In: Proceedings of the 2015 ACM on International Conference on Multimodal Interaction (ICMI 2015), pp. 467–474 (2015)
21. Lawrence, S., Giles, C., Tsoi, A.C., Back, A.: Face recognition: a convolutional neural network approach. IEEE Trans. Neural Netw. **8**, 98–113 (1997)
22. Gupta, A., Garg, M.: A human emotion recognition system using supervised self-organising maps. In: 2014 International Conference on Computing for Sustainable Global Development (INDIACom), pp. 654–659 (2016)

23. Sarnarawickrame, K., Mindya, S.: Facial expression recognition using active shape models and support vector machines. In: 2013 International Conference on Advances in ICT for Emerging Regions (ICTer), pp. 51–55 (2013)
24. Sohail, A., Bhattacharya, P.: Classifying facial expressions using level set method based lip contour detection and multi-class support vector machines. Int. J. Pattern Recogn. Artif. Intell. **25**, 835–862 (2011)
25. Hewahi, N., Baraka, A.: Impact of ethnic group on human emotion recognition using backpropagation neural network. Broad Res. Artif. Intell. Neurosci. **2**, 20 (2011)
26. Khashman, A.: Application of an emotional neural network to facial recognition. Neural Comput. Applic. **18**, 309–320 (2008)
27. Ouellet, S.: Realtime emotion recognition for gaming using deep convolutional network features (2014)
28. Burkert, P., Trier, F., Afzal, M., Dengel, A., Liwicki, M.: DeXpression: deep convolutional neural network for expression recognition (2015)
29. Levi, G., Hassner, T.: Emotion recognition in the wild via convolutional neural networks and mapped binary patterns. In: Proceedings of the 2015 ACM on International Conference on Multimodal Interaction (ICMI 2015), pp. 503–510 (2015)
30. Ahsan, T., Jabid, T., Chong, U.: Facial expression recognition using local transitional pattern on gabor filtered facial images. IETE Tech. Rev. **30**, 47 (2013)
31. Chelali, F., Djeradi, A.: Face recognition using MLP and RBF neural network with Gabor and discrete wavelet transform characterization: a comparative study. Math. Probl. Eng. **2015**, 116 (2015)
32. Beaudry, O., Roy-Charland, A., Perron, M., Cormier, I., Tapp, R.: Featural processing in recognition of emotional facial expressions. Cogn. Emot. **28**, 416–432 (2013)
33. Lundqvist, D., Flykt, A., Öhman, A.: The Karolinska Directed Emotional Faces - KDEF. CD ROM from Department of Clinical Neuroscience, Psychology section, Karolinska Institutet (1998). ISBN: 91-630-7164-9
34. Wang, J., Cheng, J.: Face recognition based on fusion of Gabor and 2DPCA features. In: International Symposium on Intelligent Signal Processing and Communication Systems (2010)
35. Kuhn, H., Tucker, A.: Nonlinear programming. In: Proceedings of the Second Berkeley Symposium on Mathematical Statistics and Probability, pp. 481–492 (1951)
36. Riedmiller, M., Braun, H.: A direct adaptive method for faster backpropagation learning: the RPROP algorithm. Neural Netw. **1**, 586–591 (1993)
37. Altahhan, A.: Navigating a robot through big visual sensory data. Procedia Comput. Sci. **53**, 478–485 (2015)
38. Aldebaran.: Who is NAO? https://www.aldebaran.com/en/cool-robots/nao

Application of Artificial Neural Networks for Analyses of EEG Record with Semi-Automated Etalons Extraction: A Pilot Study

Hana Schaabova[1(✉)], Vladimir Krajca[1,2,3], Vaclava Sedlmajerova[1,3], Olena Bukhtaieva[1], Lenka Lhotska[4,6], Jitka Mohylova[5], and Svojmil Petranek[2,3]

[1] Faculty of Biomedical Engineering, Czech Technical University in Prague, Prague, Czech Republic
{hana.schaabova,krajcvla,vaclava.sedlmajerova, olena.bukhtaieva}@fbmi.cvut.cz

[2] Department of Neurology, Hospital Na Bulovce, Prague, Czech Republic
{vladimir.krajca,svojmil.petranek}@bulovka.cz

[3] National Institute of Mental Health, Prague, Czech Republic
{vladimir.krajca,vaclava.sedlmajerova,svojmil.petranek}@nudz.cz

[4] Faculty of Electrical Engineering, Czech Technical University in Prague, Prague, Czech Republic
lhotska@fel.cvut.cz

[5] Faculty of Electrical Engineering and Computer Science, VSB-Technical University of Ostrava, Ostrava, Czech Republic
jitka.mohylova@vsb.cz

[6] Czech Institute of Informatics, Robotics and Cybernetics, Czech Technical University in Prague, Prague, Czech Republic

Abstract. Application of artificial neural network (ANN) classification – multilayer perceptron (MLP) with simulated annealing for initialization and genetic algorithm for weight optimization on multi-channel EEG record is presented here. The novelty of the approach lies in the semi-automated etalon extraction. The etalons are suggested by the k-means algorithm and verified/edited by an expert. The whole process of EEG record consists of multichannel adaptive segmentation, feature extraction from segments, semi-automatic process of etalons extraction by the k-means cluster analysis leading to color segment identification and continuing with manual choice of segments for etalons by the expert and feature extraction of chosen etalons. Subsequent classification by ANN leads to unique color identification of segments in the EEG record and additionally in temporal profile. Our goal is to help the physician by mimetic software because the examination of long multichannel EEG is a tedious work.

Keywords: EEG classification · ANN · MLP · Semi-automated analysis · Adaptive segmentation · Segment identification by color · Supervised learning · Artifacts · Epileptic EEG · Genetic algorithm · Simulated annealing

C. Jayne and L. Iliadis (Eds.): EANN 2016, CCIS 629, pp. 94–107, 2016.
DOI: 10.1007/978-3-319-44188-7_7

1 Introduction

A supervised learning method of artificial neural networks (ANN) — multilayer perceptron (MLP) trained by genetic algorithm and initialized by simulated annealing is used for a classification of EEG segments. Supervised methods require etalons for learning. However, obtaining suitable etalons by an expert (physician) is a time-consuming work. The novelty in this process is the preparation of etalons by the k-means clustering method [2] and subsequent verification and editing by the expert. This approach helps to create larger groups of etalons more effectively.

The whole process of EEG analysis (see scheme in Fig. 1) consists of a signal preprocessing, a multichannel adaptive segmentation, a feature extraction from segments, a semi-automatic process of etalons extraction by the k-means cluster analysis leading to a color segment identification and continuing with a manual choice of segments for etalons by the expert and a feature extraction of chosen etalons. The subsequent classification by ANN leads to a unique color identification of the segments in the default multichannel EEG record and additionally in temporal profile and statistical summary sheets.

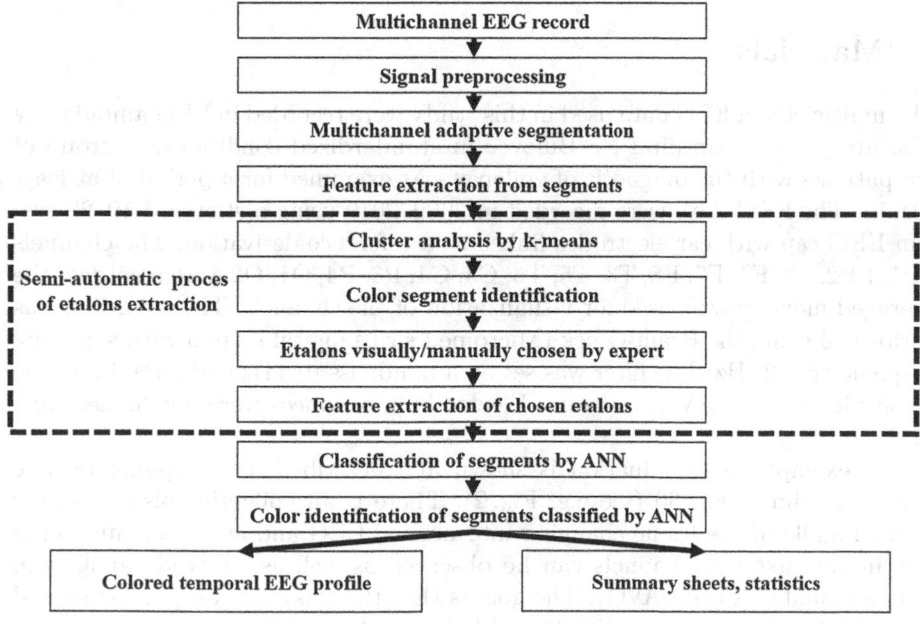

Fig. 1. Scheme of EEG process

The k-means algorithm is usually the main classification algorithm for EEG records. It is an unsupervised classification method, therefore no etalons are needed. However, we propose to use the k-means algorithm only for etalons preparation because we noticed that the k-means algorithm gives many falsely classified segments in the classes (see previous papers, where we discuss using the fuzzy k-means instead of the k-means [18] and the k-NN instead of the k-means [17]).

In the literature, there are other research groups [14,16,19,20] classifying EEG records by artificial neural networks — multilayer perceptron. However, they classify only the sharp EEG transients and they omit the other EEG segments, so they use around 4 classes at maximum. In this paper, our goal is to be as mimetic (meaning: close to what would the physician do) as possible, therefore we classify all of the EEG segments and not only the sharp transients. We have 7 classes based on the experience of our experts.

Another difference of our paper compared to the papers mentioned above is the adaptive segmentation method that we use for segmenting the whole EEG record in each channel. For example, in [20] the authors use averaging of the signal with subsequent threshold to find and segment sharp transients/EEG graphoelements out of the background that is eliminated. We do not eliminate the background activity, but we classify it as well.

The main highlight in our approach is the color identification of the EEG segments in the default multi-channel EEG record which is intuitive and more suitable for the physicians.

2 Materials

The multichannel EEG data used in this study were recorded in EEG ambulatory laboratory in the Hospital Na Bulovce in standardized conditions. A group of ten patients with the diagnosis of epilepsy was examined for a period of at least 20 min. The EEG data were recorded in 2009–2010 using a standard 10–20 system EEG cap with ear electrodes linked as a reference derivation. The channels FP1, FP2, F3, F4, F7, F8, T3, T5, T6, C3, C4, P3, P4, O1, O2 were recorded; the averaged montage was used for visualization of the channels. The recording was performed using the Brain-Quick (Micromed s.r.l.) digital system with sampling frequency of 128 Hz. The filter was set on a bandpass (0.4 Hz and 70 Hz) and the sensitivity was 100 μV per 10 mm. Electrode impedances were not higher than 5 kΩ.

An example of examination is shown and described in this paper on one patient at time 00:22:30 (see e.g. Fig. 2). There is an epileptic episode ending in the middle of the frame changing into normal EEG and an eye opening artifact in the first two channels can be observed as well as electrode artifact in channel number 15 (F4-AVG). The goal is that the classification process should distinguish between these artifacts and between these artifacts and the epileptic activity and normal EEG activity.

3 Methods

The WaveFinder software (WF [11]) was used as an EEG browser and also for signal preprocessing, adaptive/fixed segmentation, the k-means classification and colored visualization of results in default EEG signal. The main part of the WaveFinder software was created by one of the author (V.K.) and new separate module for classification by artificial neural networks was written in C++ by the authors. This ANN module reads data from WF such as segment boundaries, features of the segments, features of the chosen etalons and gives classified segments for visualizations of classified clusters back into WF.

3.1 Preprocessing Methods

The preprocessing methods of EEG signal usually include filtering of the signal and noise removal methods. In this paper, only the *mean removal* procedure was used and no other additional filtering was performed on the data (additional — meaning that the recording system Brain-Quick filters the signal during the recording, see note about bandpass filter above in Sect. 2).

3.2 Multichannel Adaptive Segmentation

The EEG signal must be segmented for the classification process. There are two approaches for EEG records: fixed segmentation and adaptive segmentation (see the difference in Fig. 2).

In the fixed segmentation, a window of a fixed length is used for obtaining boundaries of the segments. This approach is not suitable for the classification due to the fact that it separates EEG complexes that should be morphologically together (e.g. spike-and-wave complex). Our goal is to provide a mimetic algorithm to the physician. The fixed segmentation is not a mimetic approach because it does not take into account the non-stationary behavior of the EEG signal compared to the following method of adaptive segmentation that can provide piece-wise stationary segments.

We use the adaptive segmentation method based on A. Värri [21] due to its simplicity and computational efficiency compared to the newest developed adaptive segmentation methods using fractal dimensions [3,7] or wavelet transformation [6].

The Värri's adaptive segmentation method is based on two joint windows sliding along the signal and detecting local maxima in the total difference measure (see Fig. 3). The total difference measure is calculated from the amplitude and the frequency difference (eqs. in [15,21]). Further information about the principle of the adaptive segmentation can be also found in [4,5,10] and a simple scheme of the principle can be found in [15].

The Värri's adaptive segmentation is processed for all the channels simultaneously as it is shown in the Fig. 2. Parameters used for achieving this segmentation were chosen from the experience: the window length (WL) of 128 samples, the window length for local maxima identification (GWL) of 15 samples, the moving step of the two connected windows of 1 sample, minimum segment length of 70 samples. The threshold (THR) for elimination of fluctuations in the total difference measure is 100.

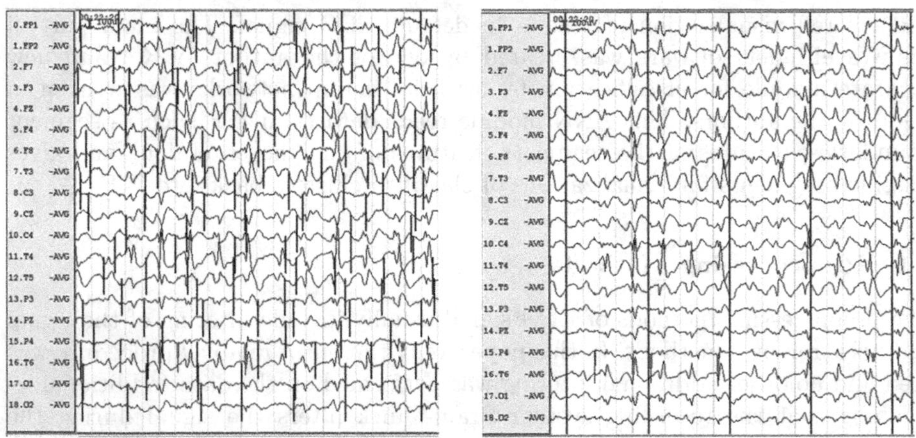

Fig. 2. Comparison of fixed and adaptive segmentation at time 00:22:30. Notice the fact that adaptive segmentation is a mimetic method that imitates the work of the physician.

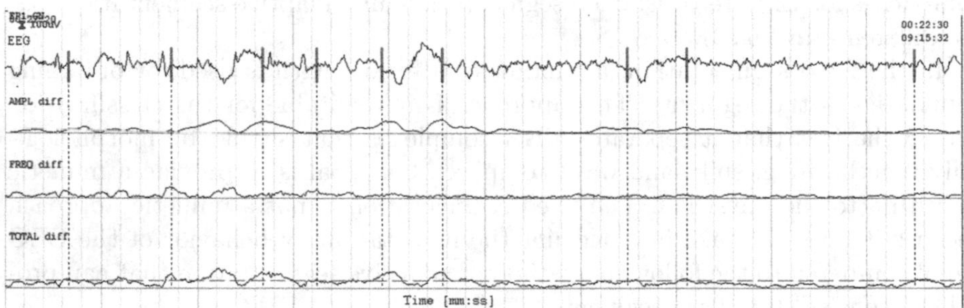

Fig. 3. The principle of Värri's adaptive segmentation shown on one channel of EEG (upper signal) in µV with time on the horizontal axis. The corresponding total difference measure (bottom signal) is calculated from the amplitude difference and the frequency difference (shown between the signal and the total difference measure) that are calculated from the original EEG signal. The red dashed line in the bottom plot is the threshold for elimination of fluctuation in the total difference measure. (Color figure online)

3.3 Feature Extraction from the Segments

The next step of the classification process is the feature extraction from the obtained segments. In this paper, we use 24 features (see Table 1) that evaluate each EEG segment (graphoelement) by its amplitude level, frequency, shape, variability of the signal and also spectral measures (further information about features can be seen in [9,10]).

Extracted features must be normalized for the classification process, therefore we use the min-max normalization, as we did in the previous paper [9].

Table 1. Features and their description.

Feature	Description
SIGM	variability of signal
APOS	maximal positive value in the signal segment
ANEG	maximal negative value in the signal segment
DELT1	FFT value in delta frequency band (0.5 Hz–1.5 Hz)
DELT2	FFT value in delta frequency band (2.0 Hz–3.5 Hz)
THET1	FFT value in theta frequency band (4.0 Hz–5.5 Hz)
THET2	FFT value in theta frequency band (6.0 Hz–7.5 Hz)
ALPH1	FFT value in alfa frequency band (8.0 Hz–10.0 Hz)
ALPH2	FFT value in alfa frequency band (10.5 Hz–12.5 Hz)
SIGMA	FFT value in frequency band (18.0 Hz–29.0 Hz)
BETA	FFT value in frequency band (13.5 Hz–29.0 Hz)
MAX1D	maximal value of the first derivation
MAX2D	maximal value of the second derivation
mf	the mean frequency
MD1	mean of the first derivation
MD2	mean of the second derivation
mob	Hjorth parameter, mobility
comp	Hjorth parameter, complexity
act	Hjorth parameter, activity
LOfC	length of the curve
NLinE	nonlinear energy [1]
ZC	zero crossing
Peaks	frequency value of the peak in spectrum
Infle	inflection point

3.4 Semi-automatic Process of Etalons Extraction

The semi-automatic extraction of the etalons (prototypes) consists of an automatic phase of cluster analysis that prepares possible groups of the etalons and subsequent manual phase where an expert revises the suggested groups and removes/adds segments in the groups of the etalons.

The first step of the etalons extraction is the cluster analysis, namely the k-means algorithm [2] coded according to the review of unsupervised clustering in [8]. We use 7 classes for the k-means algorithm — based on the experience of our experts.

After obtaining classes by the k-means algorithm, the EEG segments have numbers of their class membership and they are colored in the default EEG

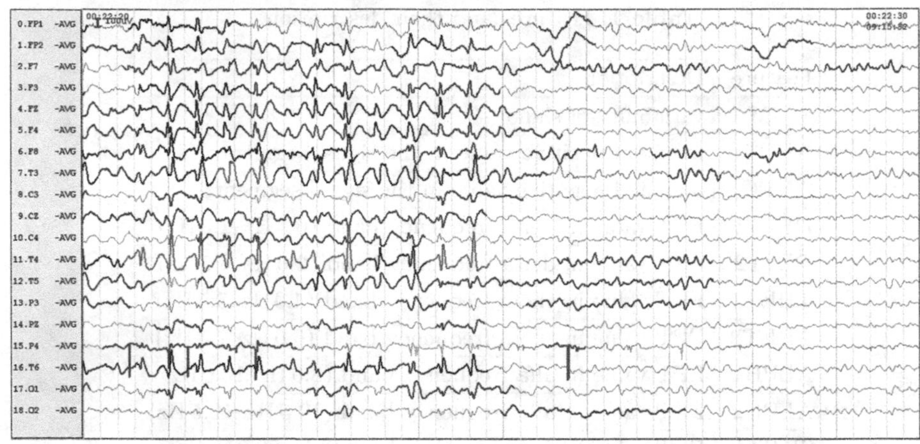

Fig. 4. Color segment identification of the segments in default EEG record according to the cluster analysis (k-means). This color identification helps the physician to snip etalons from the EEG record. Note that the eye artifact in the first two channels (FP1-AVG and FP2-AVG) are misclassified as smaller epileptic spikes. (Color figure online)

signal according to the class membership (see Fig. 4). This phase is called color identification based on the k-means classification of EEG segments.

Each class of segments that was created by the k-means algorithm will be now a suggested group of etalons for further classification by the ANN. However, these etalon groups made from clusters are not homogeneous (that is the reason why the ANN are studied), therefore a help of an expert (physician) is needed. The expert goes through the etalons in the groups and removes EEG segments (graphoelements) that do not belong to the class (etalon group). The expert can alternatively add segments snipped from the original EEG to the etalon groups.

The last part of this process is the feature extraction of those chosen etalons that uses the same features as for the EEG segments (see Sect. 3.3 above).

A special snipping tool was developed in the WaveFinder for the experts, in order to help the experts to add the chosen EEG segments from the EEG record directly into appropriate etalon group (see Fig. 5). To make this process easier, the EEG segments are identified by color based on the k-means algorithm processed before. This helps the expert to make this process quicker, easier and smoother.

Due to the combination of the k-means algorithm and the expert, who chooses the segments visually and manually based on the previous the k-means clustering, this method is a semi-automatic process of the etalons extraction. The approach of preparing the base group of the etalons by the k-means and verifying them by the experts gives larger groups of etalons in shorter time. The k-means clustering for choosing the etalons (training set) can be done on a shorter EEG

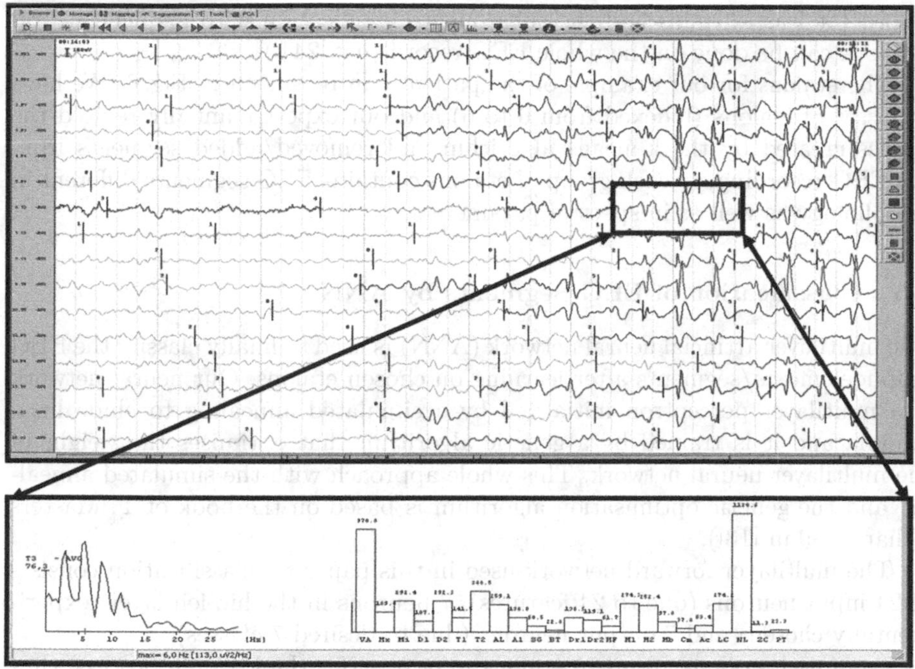

Fig. 5. The snipping tool for choosing etalons by expert with statistics inside. After the segment is chosen a new small window appears with list of classes, so the expert can choose the appropriate class for examined segment.

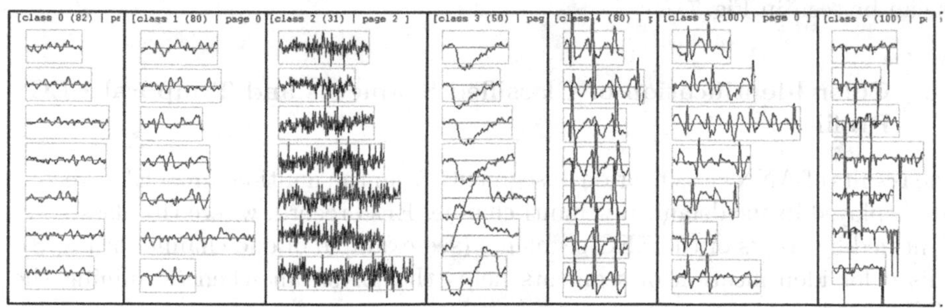

Fig. 6. Examples of etalon groups for 7 classes. The first class (0) contains normal EEG activity, the second class (1) contains higher EEG activity, the third class (2) contains EMG artifacts and noise artifacts, the fourth class (3) contains slow artifacts from eye movement etc., the fifth class (4) contains epileptic activity with higher amplitude compared to the sixth class (5) that contains epileptic activity with lower amplitude and the seventh class (6) contains the electrode artifacts that should never be misclassified as epileptic activity.

record (at least 20 min) and the subsequent classification with artificial neural network can be done on the whole EEG record (e.g. 24 h).

The etalons for our example of the patient can be shown in Fig. 6. We have 7 classes of etalons (indexed from 0 to 6) and our expert manually revised the classes created by the k-means algorithm and removed/added segments from the EEG recording, so each group of etalons contains EEG segments different in morphology as well as in spectral parameters.

3.5 Classification of EEG Segments by ANN

The multilayer artificial neural network (ANN) is used to finally classify the EEG graphoelements/segments after learning on chosen etalons. Our neural network is a multilayer perceptron initialized by a simulated annealing to elude local minima and it is trained by a genetic algorithm that optimises the weight of the multilayer neural network. This whole approach with the simulated annealing and the genetic optimisation algorithm is based on the book of T. Masters (Chaps. 6–9 in [13]).

The multilayer forward network used in this paper for classification consists of 24 input neurons (due to 24 features), 7 neurons in the hidden layer (experimentally chosen) and 7 output neurons (due to desired 7 classes).

From each etalons group 25 etalons were used for the learning process, thus the training set contains 175 etalons for all 7 classes. The whole patient record contains a total number of 31,179 EEG segments. The etalons were excluded from the total sum when testing the classification, so the training set is around 0,6 % of the whole set of segments.

The ANN module is capable of plotting feature space — 2 features at a time as can be seen in Fig. 7.

3.6 Color Identification of Classified Segments and Temporal EEG Profile

The results of ANN classification of segments from the multi-channel EEG record are displayed in the default raw multi-channel EEG record, where the classes are identified by colors of the EEG segments (see example of one channel in Fig. 8). This color identification of segments helps the physician when examining the EEG record.

The temporal profile is a method of visualization of classified EEG, it shows class membership plotted with time on the horizontal axis (see the principle in Fig. 8). The temporal profile can reveal hidden macrostructures in long time frames.

4 Results

The result of the whole process of EEG record analysis described above is shown in Fig. 9. The homogeneity of resulting classes is examined and it seems to

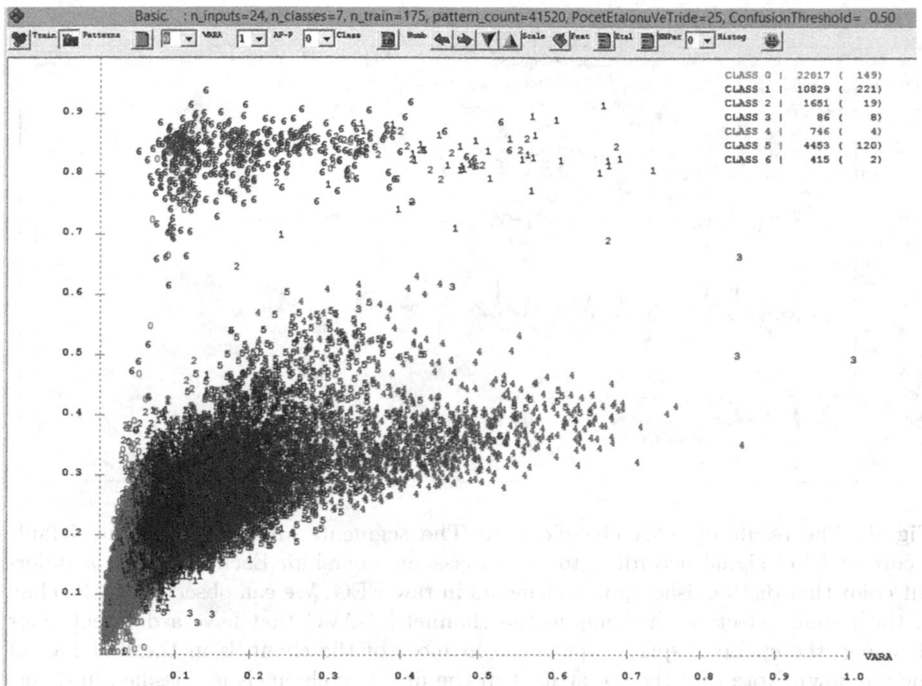

Fig. 7. Example of classified feature space projection. It is a result of multilayer perceptron classification with the simulated annealing as initialization method and the genetic algorithm as weight optimization. All clusters/classes are projected from 24D space in 2D. In this plot, there is an example of the projection of one of the first two features into 2D space. Notice the blue cluster (number six) of the electrode impulse artifacts and dark yellow cluster (number four) of EMG activity and noise artifacts. The normal EEG activity in grey color (number zero) is in the left bottom part of the plot in this projection example. (Color figure online)

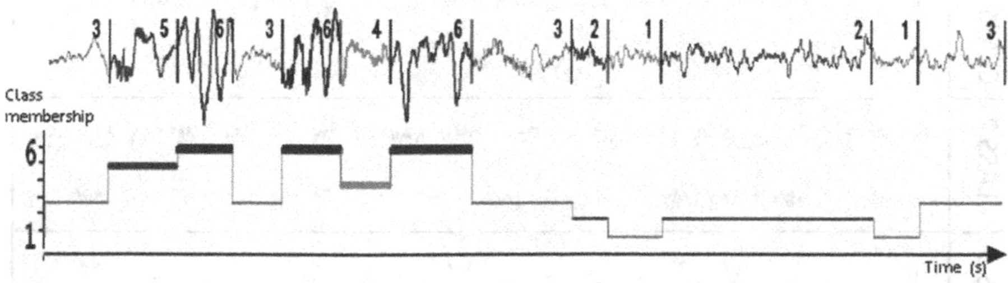

Fig. 8. Temporal profile visualization principle [12,18]. There is an example of one EEG channel with segment boundaries and numbers of class membership with time on the horizontal axis in the upper part. There is a corresponding temporal profile in the lower part — plot of class membership with time on the horizontal axis. (Color figure online)

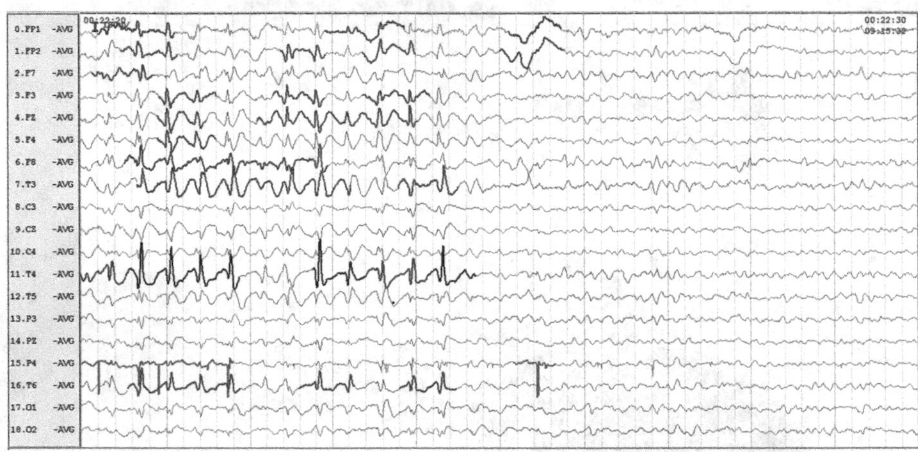

Fig. 9. The result of ANN classification: The segments are colored in the default recorded EEG signal according to their class membership. Each class has a different color that distinguishes graphoelements in raw EEG. We can observe e.g. the class of the impulse electrode artifacts in the channel F4-AVG that have a different color than e.g. the epileptic spikes (appearing in most of the channels in the first half of the window). Note that the eye artifact in the first two channels is classified in different class than the small epileptic spikes in comparison with the k-means classification in Fig. 4.

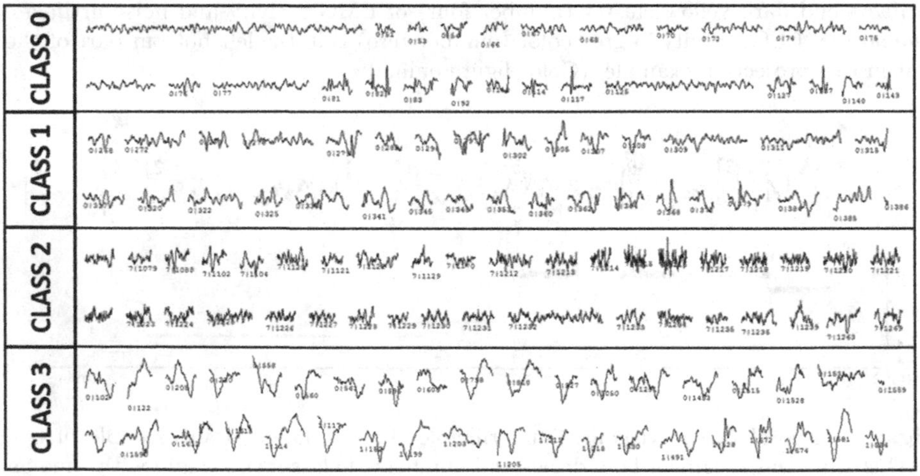

CLASS 0

CLASS 1

CLASS 2

CLASS 3

Fig. 10. The examples of two rows from the first four classes after ANN classification.

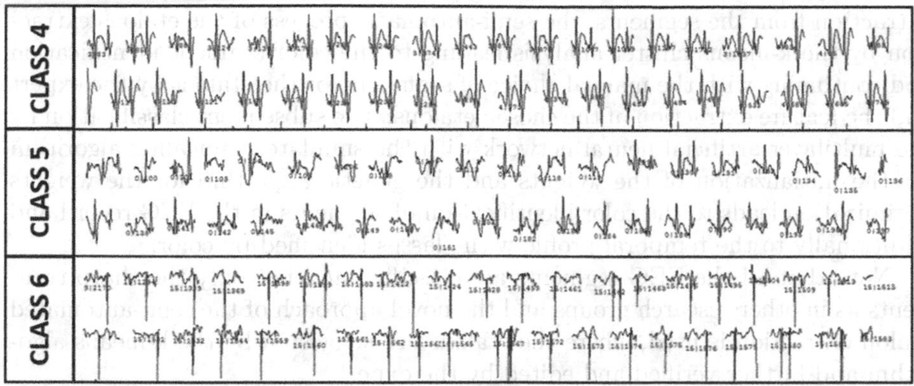

Fig. 11. The examples of two rows from the remaining three classes after ANN classification.

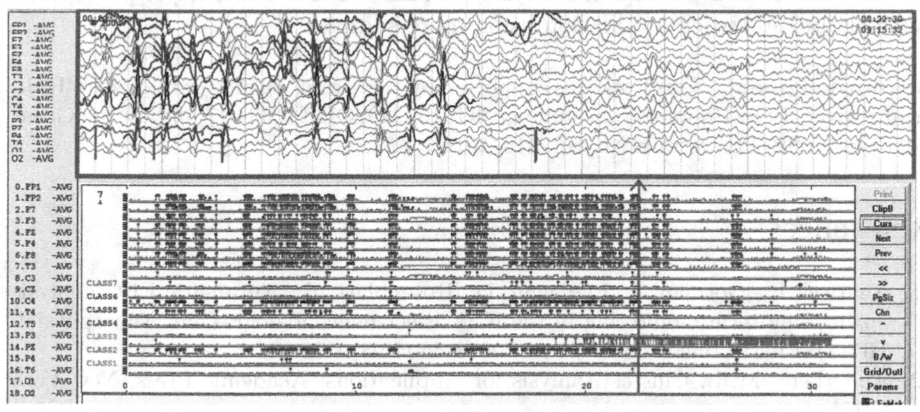

Fig. 12. The colored EEG signal (upper part) with the temporal profile of the whole recording (lower part) after the ANN classification. Each class has a different color that distinguishes the graphoelements in the raw EEG record (upper part) and all classes are plotted with time on the horizontal axis in the temporal profile. The red cursor in the temporal profile shows the position in the whole recording of the upper part of EEG. (Color figure online)

improve in comparison with the k-means classification. The homogeneity of the classes is examined visually (see Figs. 10 and 11).

The temporal profile of the whole EEG recording is shown in Fig. 12.

5 Conclusion

The multichannel ambulatory EEG record is analyzed by a process mimeting the work of the physician. The whole process of the EEG analysis consists of the signal preprocessing, the multichannel adaptive segmentation, the feature extraction from the segments, the semi-automatic process of the etalons extraction by the k-means cluster analysis leading to the color segment identification and continuing with the manual choice of segments for the etalons by the expert and the feature extraction of the chosen etalons. The subsequent classification by the multilayer artificial neural network with the simulated annealing algorithm for the initialization of the weights and the genetic algorithm for the weights optimization leads to the color identification of segments in the EEG record and additionally to the temporal profile with classes identified by color.

Note that all the EEG segments are classified and not only the sharp transients as in other research groups and the novel approach of the semi-automated etalon extraction is used, where the etalons are suggested by the k-means algorithm and further verified and edited by the expert.

The main goal is to help the physician with the tedious examination of the long multi-channel EEG records by offering segments identified by color in the default EEG record.

The future work will aim at studying parameters of the multilayer perceptron network and statistical evaluation using more samples.

Acknowledgment. This work was supported by the Grant Agency of the Czech Technical University in Prague, grant number SGS15/229/OHK4/3T/17 and by the Grant Agency of VSB-TU Ostrava No. SP2016/151.

References

1. Agarwal, R., Gotman, J., Flanagan, D., Rosenblatt, B.: Automatic EEG analysis during long-term monitoring in the ICU. Electroencephalogr. Clin. Neurophysiol. **107**(1), 44–58 (1998). doi:10.1016/S0013-4694(98)00009-1
2. Anderberg, M.R.: Cluster Analysis for Applications. Academic Press, New York (1974)
3. Anisheh, S.M., Hassanpour, H.: Adaptive segmentation with optimal window length scheme using fractal dimension and wavelet transform. Int. J. Eng. **22**, 257–268 (2009)
4. Barlow, J.S.: Methods of analysis of nonstationary EEGs with emphasis on segmentation techniques: a comparative review. J. Clin. Neurophys. **2**, 251–265 (1985)
5. Bodenstein, G., Praetorius, H.M.: Feature extraction from the electroencephalogram by adaptive segmentation. Proc. IEEE **65**(5), 642–652 (1977)
6. Hassanpour, H., Shahiri, M.: Adaptive segmentation using wavelet transform. In: International Conference on Electrical Engineering, pp. 1–5, April 2007
7. Hassanpour, H., Anisheh, S.M.: An improved adaptive signal segmentation method using fractal dimension. Ghaemshahr Branch, Islamic Azad University, Tehran, Iran (2010)
8. Jain, A.K., Dubes, R.C.: Algorithms for Clustering Data. Prentice Hall College Div, Englewood Cliffs (1988)
9. Krajca, V., Hozman, J., Mohylova, J., Petranek, S.: Pattern recognition of epileptic EEG graphoelements with adaptive segmentation, supervised and unsupervised learning algorithms. Biomedizinische Technik/Biomedical engineering, BMT (2012). Abstract of the Conference, Jena, Germany

10. Krajča, V., Petránek, S., Patáková, I., Värri, A.: Automatic identification of significant graphoelements in multichannel EEG recordings by adaptive segmentation and fuzzy clustering. Int. J. BioMed. Comput. **28**(1–2), 71–89 (1991)
11. Krajča, V., Petránek, S., Pietilä, T., Frey, H.: Wave-finder: a new system for an automatic processing of long-term EEG recordings. In: Rother, M., Zwiener, U. (eds.) Quantitative EEG Analysis-Clinical Utility and New Methods, pp. 103–106. Universittsverlag Jena (1993)
12. Krajča, V., Petránek, S., Mohylová, J., Paul, K., Gerla, V., Lhotská, L.: Neonatal EEG sleep stages modelling by temporal profiles. In: Moreno Díaz, R., Pichler, F., Quesada Arencibia, A. (eds.) EUROCAST 2007. LNCS, vol. 4739, pp. 195–201. Springer, Heidelberg (2007). doi:10.1007/978-3-540-75867-9_25
13. Masters, T.: Practical Neural Network Recipes in C++. Academic Press, New York (1993). ISBN:9780124790407
14. Park, H.S., Lee, Y.H., Lee, D.S., Kim, S.I.: Detection of epileptiform activity using wavelet and neural network. In: Proceedings of the 19th Annual International Conference of the IEEE Engineering in Medicine and Biology Society. 'Magnificent Milestones and Emerging Opportunities in Medical Engineering' (Cat. No. 97CH36136), pp. 1194–1197 (1997)
15. Paul, K., Krajča, V., Roth, Z., Melichar, J., Petránek, S.: Comparison of quantitative EEG characteristics of quiet and active sleep in new-borns. Sleep Med. **4**(6), 543–552 (2003)
16. Pang, C.C.C., Upton, A.R.M., Shine, G., Kamath, M.V.: A comparison of algorithms for detection of spikes in the electroencephalogram. IEEE Trans. Biomed. Eng. **50**(4), 521–526 (2003)
17. Schaabova, H., Krajca, V., Sedlmajerova, V., Bukhtaieva, O., Petranek, S.: Supervised learning used in automatic EEG graphoelements classification. In: E-Health and Bioengineering Conference (EHB), pp. 1-4. IEEE EHB Proceedings (2015). doi:10.1109/EHB.2015.7391470
18. Sedlmajerova, V., Schaabova, H., Krajca, V., Mohylova, J., Petranek, S.: Improving the homogeneity of classes of EEG patterns by fuzzy C-means algorithm, adaptive segmentation. In: YBERC 2014. FEI STU, pp. 99–103 (2014)
19. Tarassenko, L., Holt, M.R.G., Khan, Y.U.: Identification of inter-ictal spikes in the EEG using neural network analysis. IEE Proc. Sci. Meas. Technol. **145**(6), 270–278 (1998)
20. Tzallas, A.T., Karvelis, P.S., Katsis, C.D., Fotiadis, D.I., Giannopoulos, S., Konitsiotis, S.: A method for classification of transient events in EEG recordings: application to epilepsy diagnosis. Methods Inf. Med. **45**(6), 610–621 (2004)
21. Värri, A.: Algorithms and systems for the analysis of long-term physiological signals. Ph.D. Thesis, UTA (1992)

Clustering Applications

Economies Clustering Using SOM-Based Dissimilarity

Adam Chudziak(✉)

Warsaw School of Economics, Niepodległości 128, 02-554 Warsaw, Poland
ac53605@sgh.waw.pl

Abstract. Clustering of countries and economies has been done for a long time by expert comparing to ideal theoretical entities. More recently a data driven approach has been taken, including, so called, black box methods. In this paper a SOM-based dissimilarity measure is presented and used for agglomerative hierarchical clustering of economies. It turns out that the results differ significantly from those obtained via a more traditional Euclidean distance based approach.

Keywords: Self-Organising Maps · Hierarchical clustering · Dissimilarity · Political economy · Country clusters

1 Introduction

Policy makers compare countries and economies on the daily basis. It is necessary for proper evaluation of decisions and future projects. The comparisons are also made for assessment of past policies and theoretical models of social science. Within this context emerges the need for creating supporting tools which allow comparing countries between each other and clustering them. The natural choice seems to take a data-driven approach, since characteristics of such entities as countries are quantified nowadays on an unprecedent scale. It is important that the tools should not be too resource demanding and could be applied in a reasonable time.

There are three main ways of approaching this task, that is expert judgement clustering, mixture model-based clustering and so-called "black box" data-analytic approach. Expert judgement clustering has been performed by political economists for a long time now (e.g. [4,6]). It is usually conducted by comparing some quantitative and qualitative statistics of economies with an ideal model. Mixture model-based clustering assumes that the observed data is generated by a mixture of a finite family of probability distributions (e.g. [1]). In the "black box" approach (see e.g. [11]), no distributional assumptions restrict the analysis and the patterns in it are sought for with clustering algorithms such as hierarchical clustering.

Clustering algorithms are of many kinds, but many of them require some notion of dissimilarity (or similarity) between clustered objects. As with the choice of algorithm, the choice of the *right* or even *good enough* dissimilarity

C. Jayne and L. Iliadis (Eds.): EANN 2016, CCIS 629, pp. 111–122, 2016.
DOI: 10.1007/978-3-319-44188-7_8

notion is debatable. Usually they are created by some type of distance arbitrarily imposed on the feature space of the model. In this article we propose a measure of dissimilarity which comes from application of the Self-Organising Map (SOM) framework developed by Kohonen in [9]. The aim of this measure is to amplify the relationship between closest neighbours, instead of the actual distance between objects in the feature space. Using developed dissimilarity we perform a case study of clustering of economies using fiscal, geographical and socioeconomic data retrieved from the World Bank database.

The rest of the paper is organised as follows. In Sect. 2 we present a brief review of relevant research in political economy and data science. The following Sect. 3 contains a short characteristic of Self-Organising Maps and their properties. It also describes the data used for analysis. The similarity measure used is presented in Sect. 4 and the results of the analysis are discussed in Sect. 5. The conclusions can be found in the final Sect. 6 along with ideas for future research.

2 Related Work

Clustering of countries, economies and societies is of interest for many science disciplines spanning from political economy and sociology to data science. Many notable classifications were focused on classifying the desired group of countries into a small number of meaningful clusters [4,6,15]. Clusters were supposed to be meaningful in the sense that countries from the same group were to represent some ideal theoretic type of economy, such as e.g. Liberal Market Economy (according to the typology of the "Varieties of Capitalism" project, [6]). The classification was often performed by expert judgement on multidimensional data. This method raised questions about completeness of the typology or assigning the hard-to-cluster countries to clear-cut categories. It turned out later, that not all theoretical claims and classifications are supported by results of statistical data analysis (see e.g. [1]).

The second way of obtaining clusters for economies is data centric and proposes more general methods relying on data analysis. The data driven approach comes in a number of varieties. Model based clustering, (e.g. [1,2,7]) assumes that data is generated from an underlying distribution, which is a mixture of a family of probability distributions. This allows to transform the problem into a model selection task for which there is a well developed statistical theory. Then, there are black box data analytic methods (e.g. [11,14]), which do not assume an underlying statistical model, but have an exploratory rather then inferential nature.

There is a number of reasons why economies clustering problems are studied. Algorithmic tools can be used for development and testing of theoretical typologies. Social scientists often group objects and claim meaningfulness of such a grouping and there is a need for mechanisms providing sanity checks for expert clusterings. Another application possibility is explored i.a. in [14]. That is the idea of building early warning systems using machine learning toolkit. The goal here is to discover early patterns which may imply an unwanted event in the future, e.g. an economic crisis. Events from not so distant past, such as the 2008 crisis or 1997 South East Asia crisis exhibit the need for this type of systems. Finally, it is also important to remember, that governments and other policy makers may put ideas in context by comparing them with existing solutions. But to do it properly they have to gain information about similarity between economies or countries. Tools which allow it could be applied to enhance the decision process in national and international governing bodies. This type of research can be seen e.g. in [8,18].

3 Methodology and Data

In this paper we want to explore similarities between economies characterised by multidimensional data. We take a hierarchical clustering approach, as it allows for flexible granularity of clustering. This has an advantage of not limiting oneself to a specified number of clusters at the beginning of analysis. Hierarchical clustering requires some notion of dissimilarity and we will construct one using SOM framework.

3.1 Self-Organising Maps

Self-Organising Maps developed by Kohonen in a seminal paper [9] have a number of properties which are useful for our analysis. SOMs are essentially an unsupervised dimension reduction technique, which performs a non-linear projection of data on a one or two dimensional grid. A finite set of data points is represented by a set of nodes (neurons) which are organised into a lattice. This lattice has usually a rectangular or hexagonal topology. In the classic SOM the number of nodes and topology are predefined. While there are extensions and refinements to this technique, we use the classic, static version. In each node a weight vector is stored. It can be perceived as a prototype observation in the feature space. Weight vectors are the objects inputs are compared to.

During training of a SOM, data points are presented to the network one after another in a random order. For each input, a winner-takes-all competition between the nodes commences and is won by the node whose weight vector is the most resemblant to the input vector. Resemblance can be measured in a variety of ways. We will use a standard way, that is the Euclidean distance. After the winning neuron is selected, the weights in the neighbourhood (on the lattice) of the winning node are adjusted according to the formula

$$w_i(t+1) = w_i(t) + h_{i^*,i}(t)\,(x(t) - w_i(t)), \tag{1}$$

where $w_i(t)$ is the weight vector in the node i after input t, i^* is the number of the winning node, $x(t)$ is the t-th input and $h_{j,k}(t)$ is a neighbourhood function, which encodes the strength of excitement of neuron k, when j is the winner. The neighbourhood function is defined to be decaying with time to ensure convergence. After the map is trained, the data points are once again fed into the map and the winner neurons are selected. This way we get a list of data points associated with each node.

The common situation in real life applications is that the number of clusters equals the number of neurons in a SOM and all data points assigned to a node constitute a cluster. This approach, however, has its drawbacks, as one can get different clusters in different reruns of the algorithm. This is caused by the random order of inputting data into the algorithm. This can be solved by running the algorithm a number of times and then selecting a clustering using some kind of minimal intercluster variance criterion. We chose a different approach, which is explained in Sect. 4.

Let us note that the Self-Organising Maps have a number of properties (see [9,10]) which make them a good choice for representing high dimensional data. Because during the training many neurons of the map are moved in every step, a SOM is able to approximate the topology of the underlying dataset. The weight vectors from the nodes can be perceived as a discrete approximation of the data. Furthermore, the projections are non-linear and therefore may outperform many other dimension reductions techniques such as PCA. Also, SOM gives us an extra piece of structure — the rectangular or hexagonal lattice fitted to the data. Despite the advantages there are also drawbacks. Using a static SOM means that decision on the size and topology of the lattice has to be done before analysis. Usually it is done by experiments using auxiliary measures or simply expert judgement.

3.2 Used Data

The analysis uses data from the World Bank's World Development Indicators dataset (WDI). WDI contains more then 1300 time series for over 200 economies, some tracing 50 years back. We conduct a snapshot analysis and we decided to choose indicators from the year 2012, for their recency and relative completeness. Note that some indicators take longer to be computed and published. Of all indicators we chose 49 variables concerning 53 countries. The variables were chosen to constitute a group of statistics indicative of the structure and condition of economy. We chose a mixture of fiscal (e.g. tax revenue as a percentage of GDP), geographical (e.g. agricultural land as a percentage of total land area) and socioeconomic (e.g. employment to population ratio for people 15+ years old) indices.

4 Dissimilarity Measure

The clustering method we employ consists of two main steps. In the first step we use Self-Organising Maps to create a dissimilarity matrix. The matrix is then used as an input for the second step, when the actual clustering takes place. In this step we use agglomerative clustering.

There are a lot of ways one can measure how similar (dissimilar) two entities are. Many types of distances can be computed and used, such as e.g. Euclidean or Manhattan distances. Many of them have been successfully used in research. Taking a closer look at the Euclidean distance points to a realisation that it relies heavily on the scale of the features. It is also "absolute" in a sense, that it does not take into account relative position of points (objects) in the feature space against each other.

For these reasons we decided to choose a different way of defining (dis)similarity It is computed in a three step process.

1. First, we construct n Self-Organising Maps on the dataset. We call the number n the *averaging parameter.*
2. Then, let $\sigma(A, B)$ be the number of times that A and B were assigned to the same node. Similarity between objects A and B is defined to be the ratio

$$\mathrm{sim}_n(A, B) = \frac{\sigma(A, B)}{n}. \tag{2}$$

3. Finally, we define dissimilarity of two objects A and B to be

$$\mathrm{dsim}_n(A, B) = 1 - \mathrm{sim}_n(A, B). \tag{3}$$

Idea for measures (2) and (3) is that a SOM is able of recreating the inherent, high-dimensional topology of the data and hence, data points which are finally assigned to a node bear some similarity. This type of definition emphasises relative closeness of objects. It is important to note again that a SOM which is rerun on the same set of data with the same set of parameters can give different results. This depends on the order in which data points are fed into the algorithm. Thus the weight vectors in the nodes of a SOM are placed slightly differently in the feature space in different reruns of the algorithm. Hence the sets of objects associated with a node can differ. The ratio (2) gives then a good idea of a similarity of two objects represented by data points.

This of course leads to questions how the metric given by (2) is dependent on parameter n and do the similarity matrices created in different runs of the algorithm for a particular n differ. We address this questions experimentally. Table 1 presents experimental values of the Cauchy criterion. That is, for every element of an incrementing sequence $n(i)$ a similarity matrix $\mathrm{SM}_{n(i)}$ was computed. Then, for each $n(i)$ the value of

$$\max\left\{|\mathrm{SM}_{n(i)} - \mathrm{SM}_{n(j)}| \ : \ i < j\right\} \tag{4}$$

was calculated. The simulation results are presented in Table 1. They suggest that the values given by formula (4) indeed decay to 0 as n goes to infinity.

Secondly, we checked that for a particular n the similarity matrices created in reruns of the algorithm do not differ too much. Table 2 presents the maximum distance (using Frobenius norm) between similarity matrices computed 40 times for each n. The results suggests that the variability of similarity matrices decreases with growth of n.

Table 1. Maximum distance between similarity matrix created using sim_n and similarity matrix created using sim_m for $m > n$

n	Distance
5	4.501
10	2.734
50	1.338
100	0.906
200	0.57
300	0.405
500	0.411
700	0.451
900	0.378
1100	0.296
1300	0.346
1500	0.276
1700	0.233

Table 2. Maximum distance between similarity matrices created using sim_n for 40 reruns for a particular n

n	Maximum distance
5	6.946
10	5
50	2.965
100	1.714
200	1.277
300	0.916
500	0.765

5 Economy Clusters, Case Study

We used the algorithm described in Sect. 4 on a subset of the WDI. The dissimilarity matrix was created using a rectangular, two-dimensional Self-Organising Map of dimensions 5×4. The averaging parameter has been set to $n = 500$, as a reasonable compromise between convergence and computation time.

The choice of SOM dimensions is dependent on the problem and there is no perfect one-fits-all size. So is with the map topology. We wanted our SOM to be big enough to allow better approximation of the shape of data. In the same time, the map could not be too large. If it were, many nodes would have a unique country assigned to them. This would be harmful for our similarity measure. During preliminary experiments we decided on a 5×4 SOM with rectangular

Table 3. Clusters obtained using dsim and hclust method "ward.D2" for a chosen number of eight clusters

Cluster	Countries
1	Armenia Guatemala India Morocco Moldova Mozambique Paraguay El Salvador
2	Austria Belgium Germany Denmark Ireland Netherlands
3	Bosnia and Herzegovina Colombia Jordan Macedonia, FYR Peru Serbia Tunisia Ukraine South Africa
4	Brazil Cyprus Spain Finland Iceland Luxembourg Norway Sweden Thailand United States
5	Chile Croatia Hungary Latvia Poland
6	Czech Republic Estonia Greece Portugal Russian Federation Slovak Republic Slovenia
7	France United Kingdom Italy Japan
8	Mauritius Romania Turkey Uruguay

topology. For training we used the Self-Organising Maps implementation for R in the kohonen package [17]. The clustering was performed using built in hierarchical clustering function hclust in R using Ward criterion (method "Ward.D2" in R, [13]). Agglomerative hierarchical clustering procedures build clusters in a bottom-up fashion. Initially they create one cluster for each data point. In each of the following steps, the algorithm merges two clusters according to a predetermined linkage criterion with respect to selected dissimilarity measure. As a result we get a sequence of linkages between clusters, which can be neatly presented in a form of a dendrogram. Figure 1 presents a dendrogram for chosen economies and dissimilarity measure dsim.

After the dendrogram is computed the observations are partitioned into clusters. Table 3 presents the result of clustering economies into eight groups.

The results obtained are promising and some patterns can be observed. Clusters are relatively homogeneous with respect to World Banks classification of countries into income groups. Indeed, clusters 2, 4–7 consist predominantly of high income countries, clusters 3 and 8 consist mainly of countries classified as upper middle income and most countries assigned to cluster 1 are ranked by the World Bank as lower middle income.

On the other hand, there is a tendency to group countries from the same regions together — that is no surprise, as the economies in the same region are often similar. For example, almost all economies from post-communist Central and Eastern Europe can be found in clusters 3, 5 and 6. Also, well developed Western European countries such as France or Italy are grouped together. It is worth noting, that geographical criterion does not apply to Latin American countries, which are scattered across five of eight clusters. This can suggest more complex differences and divisions in the region going beyond the statism or free

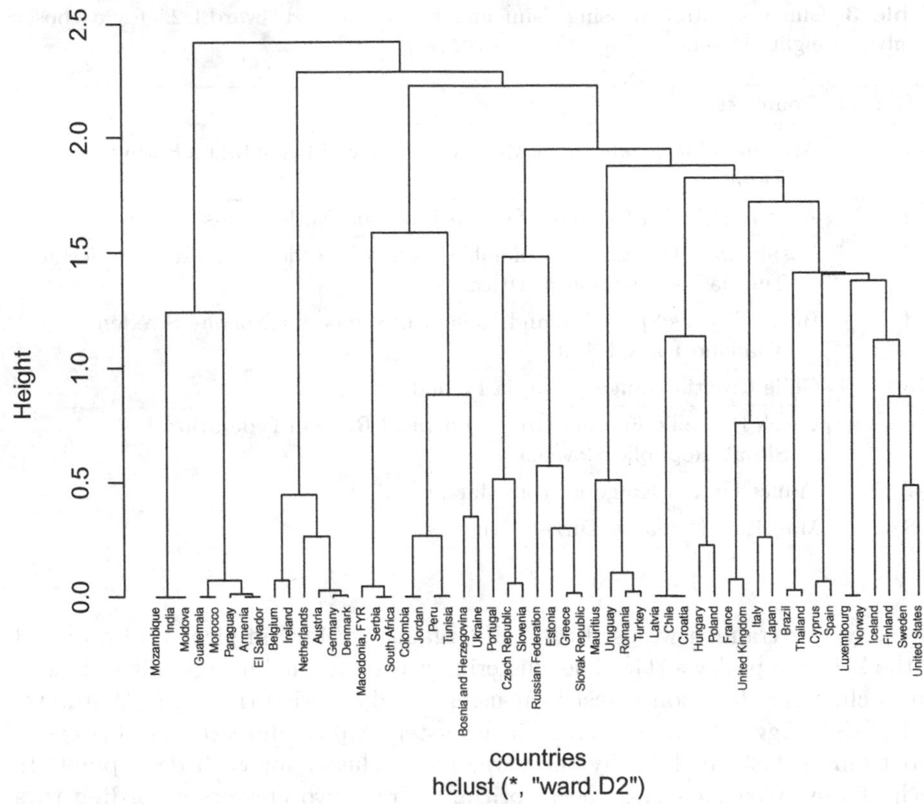

Fig. 1. Dendrogram for hierarchical clustering using Ward criterion and dsim dissimilarity measure

market dualism. Irregularities of such type can be among the biggest gains from this kind of clustering analysis, as data does not conform to expectations and point to potentially interesting or troubling issues for further investigation.

For comparison let us see the dendrogram and clustering prepared using Euclidean distance as a dissimilarity measure and the same hierarchical clustering method (function hclust in R using Ward criterion "Ward.D2", [13]). Figure 2 and Table 4 present the results. When using hierarchical clustering in social sciences it is common to use this combination of methods, i.e. Euclidean distance (or squared Euclidean distance) and Ward criterion (see e.g. [12, 15]).

Note that the dendrogram shown in Fig. 1 differs significantly from the clustering performed with Euclidean distance as a measure of dissimilarity (Fig. 2). Change of structure of a dendrogram should result in non-identical clusterings and indeed, clusters presented in Table 4 are noticeably different then the results presented in Table 3.

Table 4. Clusters using euclidean distance and hclust method "ward.D2" for a chosen number of eight clusters

Cluster	Countries
1	Armenia Bosnia and Herzegovina Colombia Guatemala India Jordan Morocco Moldova Macedonia, FYR Mozambique Peru Paraguay El Salvador Serbia Tunisia Ukraine South Africa
2	Austria Belgium Germany Denmark Finland Ireland Netherlands Sweden United States
3	Brazil Chile Croatia Hungary Latvia Mauritius Poland Romania Thailand Turkey Uruguay
4	Cyprus Spain France United Kingdom Italy Japan
5	Czech Republic Estonia Greece Portugal Russian Federation Slovak Republic Slovenia
6	Iceland
7	Luxembourg
8	Norway

There is an extensive literature on various measures for comparing clusterings (see [3,5,16]). Such a comparison is necessary to see whether the differences are significant or not. Table 5 presents a non-exhaustive list of four popular indices computed for clusterings in Tables 3 and 4. The Adjusted Rand Index has an expected value for random clusterings equal to 0 and the maximal value 1 (when two partitions agree perfectly). The Fowlkes-Mallows index admits value from 0 to 1, higher values imply greater similarity of clusterings. Baker's Gamma is measure of similarity of trees of hierarchical clustering. The values are in range -1 to 1, the values close to zero imply no statistical similarities of the trees. Variation of Information is a function that measures the distance between partitions of a dataset. It has an upper bound dependent on the size of the dataset and for our analysis it is approximately 4. The indices used suggest a significant difference between clusterings, but on the other hand do show that there are some similarities between them.

As hierarchical clustering is by its nature more of an exploratory technique and it is easy to overinterpret the obtained partition we do not call one clustering better than the other, as each can amplify different characteristics hidden in data. Let us also note that we do not imply any kind of a mutual causal relationship between the clustered countries. We argue that the economies grouped in the same clusters share some common qualities and exploring these similarities may be of use for policy makers or comparative policy analysis.

The idea behind using dsim as a dissimilarity measure was to amplify relative closeness between economies in favour of absolute distances in the feature space, so that a small number of significantly different observations do not distort the results. This goal was achieved. Notice that when using Euclidean distance three clusters, 6–8, consisted of only one country each (resp. Iceland,

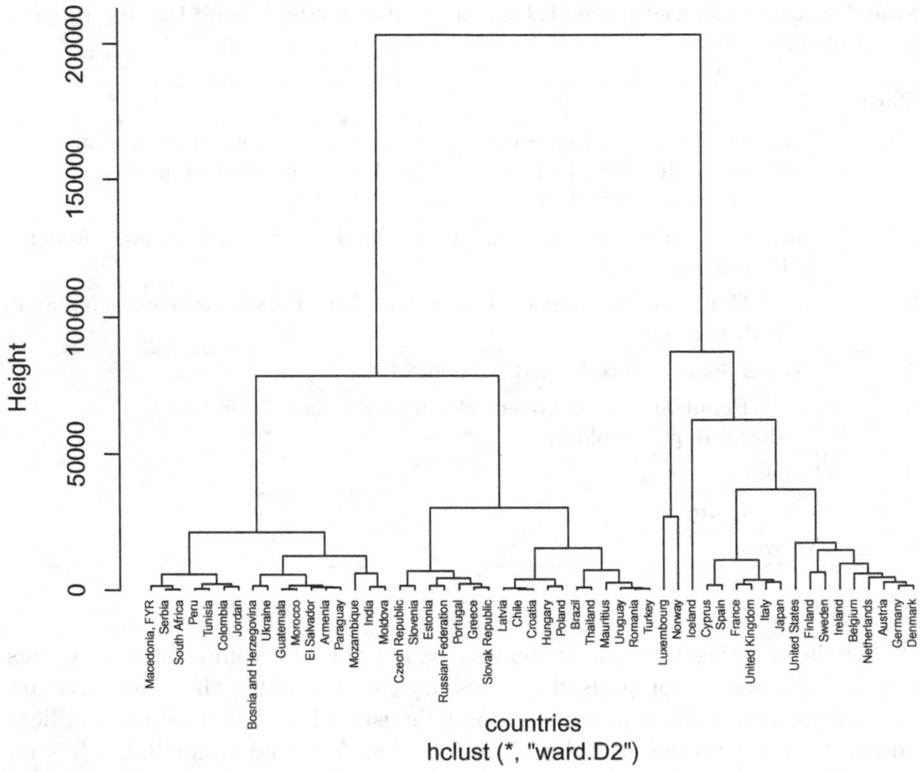

Fig. 2. Dendrogram for hierarchical clustering using Ward criterion and Euclidean distance as dissimilarity measure

Luxembourg and Norway). Using dsim, the algorithm grouped those “outlying” countries in one cluster and allowed a more even clustering. This is beneficial when we want to avoid clusters of size one, e.g. when clustering is performed in search for benchmarks for a particular economy.

Further research is necessary to discuss the mathematical properties of the dsim dissimilarity measure, but from the observation of the dendrograms we suppose that this measure supports grouping observations in small clusters early in the clustering process and indeed helps to group points which are far from each other but relatively similar. To illustrate this imagine a cloud of data in a two dimensional space which consists of two clusters easily distinguishable by hand-one dense (A) and the other one rather sparse (B). Using Euclidean distance as a dissimilarity criterion can result in putting the points from the sparse group (B) into many different clusters, as they are further away from each other, than points from (A). On the other hand, dsim as a dissimilarity measure encourages finding linkages between data points which are relatively close to each other.

Table 5. Clustering differences

Method	Value
Adjusted Rand Index	0.519
Fowlkes-Mallows Index	0.606
Baker's Gamma correlation coefficient	0.234
Variation of Information	1.352

The reason for that is that every data point has to be assigned to a node in a SOM and a vector of weights in a neuron can be the best match for data points which are distant according to Euclidean distance.

Whether such a property is desirable is a problem-specific discussion, the answer depending on the question posed by the researcher. It could be of use in situations when one does not want a number of outliers to overshadow the inherent data structure and does not want to get rid of them from the dataset.

6 Conclusions

We have presented a hierarchical clustering algorithm using a SOM-based dissimilarity matrix. The dissimilarity measure dsim was derived from a similarity measure sim, obtained by averaging neighbourhood relationship in the Self-Organising Map. Results of experiments implying convergence of sim_n for large averaging parameter n were presented. Using the dissimilarity matrix we performed a hierarchical clustering on the World Development Indicators. The analysis was performed on data for 53 countries and 49 variables in year 2012. Inspection of obtained clustering showed some regularities such as clustering countries from the same regions into same clusters. The clusters turned out also to be quite homogeneous with respect to the World Bank income group ratings. The results were confronted with hierarchical clustering using Euclidean distance as a dissimilarity measure. The clusterings in the two cases turned out to be significantly different, using dsim allowed to eliminate clusters consisting of one observation.

The directions for future research are twofold. The properties of dissimilarity measure dsim could be an object of further investigation, especially its ability to group objects according to relative distance from others. On the other hand, the we plan to explore in greater depth the economic applications of clustering mechanisms and their relation to economic theory.

Acknowledgments. The author would like to thank Prof. Michal Ramsza for critical reading of the manuscript.

References

1. Ahlquist, J., Breunig, C.: Country clustering in comparative political economy. MPIfG Discussion Paper 09/5, Max Planck Institute for the Study of Societies (2009)
2. Ahlquist, J., Breunig, C.: Model-based clustering and typologies in the social sciences. Polit. Anal. **20**(1), 92–112 (2012)
3. Baker, F.B.: Stability of two hierarchical grouping techniques case 1: sensitivity to data errors. J. Am. Stat. Assoc. **69**, 440–445 (1974)
4. Esping-Andersen, G.: The Three Worlds of Welfare Capitalism. Princeton University Press, New Jersey (1990)
5. Fowlkes, E.B., Mallows, C.L.: A method for comparing two hierarchical clusterings. J. Am. Stat. Assoc. **78**(383), 553–569 (1983)
6. Hall, P., Soskice, D.: Varieties of Capitalism: The Institutional Foundations of Comparative Advantage. Oxford University Press, Oxford (2001)
7. Jang, J., Hitchcock, D.: Model-based cluster analysis of democracies. J. Data Sci. **10**(2), 297–319 (2012)
8. Kiehlborn, T., Mietzner, M.: EU financial integration: is there a core europe evidence from a cluster-based approach. J. Econ. Asymmetries **2**(1), 81–104 (2005)
9. Kohonen, T.: Self-organized formation of topologically correct feature maps. Biol. Cybern. **43**(1), 59–69 (1982)
10. Kohonen, T.: Self-Organizing Maps, 3rd edn. Springer, Heidelberg (2001)
11. Koutsoukis, N.S.: Global political economy clusters: the world as perceived through black-box data analysis of proxy country rankings and indicators. Procedia Econ. Finan. **33**, 18–45 (2015)
12. Obinger, H., Wagschal, U.: Families of nations and public policy. W. Eur. Politics **24**, 99–114 (2001)
13. R Core Team: R: A Language and environment for statistical computing. R Foundation for Statistical Computing, Vienna (2016). https://www.R-project.org/
14. Resta, M.: Early warning systems: an approach via self organizing maps with applications to emergent markets. In: New Directions in Neural Networks: 18th Italian Workshop on Neural Networks, WIRN 2008 (2009)
15. Saint-Arnaud, S., Bernard, P.: Convergence or resilience? a hierarchical cluster analysis of the welfare regimes in advanced countries. Curr. Sociol. **55**, 499–527 (2003)
16. Wagner, S., Wagner, D.: Comparing Clusterings: An Overview (2007)
17. Wehrens, R., Buydens, L.: Self- and super-organising maps in R: the kohonen package. J. Stat. Softw. **21**(5), 1–19 (2007). http://www.jstatsoft.org/v21/i05
18. Yasar, E., Acikalin, S., Gezer, M.: Testing IDP hypothesis by cluster analysis: which countries in which stage? Procedia Econ. Finan. **23**, 1201–1209 (2015). 2nd Global Conference on Business, Economics, Management and Tourism

Elastic Net Application: Case Study to Find Solutions for the TSP in a Beowulf Cluster Architecture

Marcos Lévano[1(✉)] and Andrea Albornoz[2]

[1] Escuela Ingeniería Informática, Universidad Católica de Temuco, Av. Manuel Montt 56, Casilla 15-D, Temuco, Chile
mlevano@inf.uct.cl

[2] Departamento de Lenguas, Universidad Católica de Temuco, Av. Manuel Montt 56, Casilla 15-D, Temuco, Chile
aalbornoz2010@alu.uct.cl

Abstract. This study aims to apply the Durbin-Willshaw elastic net using parallel algorithms in order to solve the Traveling Salesman Problem (TSP) through a Beowulf cluster architecture for High-Performance Computing. The solutions for the TSP for the different number of cities are achieved by the minimization of the internal energy and by the maximization of the entropy in the information system. In this way, approximate solutions to the TSP can be determined. This work proposes a framework to implement a parallel algorithm to the Beowulf cluster. In order to find solutions for the TSP, we worked with 5000 cities with a net of 12500 nodes up to 10000 cities with 25000 nodes.

Keywords: Elastic net · Parallel strategy · Internal energy · TSP · Algorithm

1 Introduction

The Traveling Salesman Problem (TSP) is a problem of combinatorial optimization of the class NP-complete. The objective of the TSP is to solve a Hamiltonian [1], that means that we need to find a route that contains all the cities in a plane just once, in order to minimize the cost of the route of the salesman. No algorithms are known to ensure a polynomial time execution, but there are some heuristic methods like Genetic Algorithms, Simulated Annealing and Tabu Search that, starting from viable solutions, can determine approximate solutions to real ones; they all try to give solutions to the TSP [2,3]. A lot of methods have a high computational cost in time and memory; and they lead to exponential efforts when the number of cities increases [4]. The artificial neural networks also try to solve this problem. Some of the works that address this issue are "An analogue approach to the TSP using an elastic net method", which was proposed in 1985 for the same set of problems and implements the TSP in a uniprocessor, improving neural Hopfield and Tank network [5]; also

C. Jayne and L. Iliadis (Eds.): EANN 2016, CCIS 629, pp. 123–133, 2016.
DOI: 10.1007/978-3-319-44188-7_9

there is the work "An analysis of the elastic net approach to the travelling salesman problem" in which the functioning of the elastic net for the solution of the TSP is analyzed [7]. In 1998 a study in which the performance of the neural nets Elastic Net y Guilty Net in the TSP with 10, 30, 65 and 101 cities was carried out, concluding that the Elastic net is the most efficient one [5].

The class NP-complete includes a big amount of practical problems that can be found in the business and industrial areas. To demonstrate that a problem is NP-complete, also demonstrates that it is not in P class, thus, it does not have a deterministic solution in a polynomial time [8]. Some of the applications of the Durbin-Willshaw method are optimization problems, computer vision, pattern recognition and clustering [9,10].

In the solution proposed on this paper, the sequential Durbin-Willshaw (RE-DW) elastic net is analyzed and parallelized in a cluster Beowulf architecture, in order to approach minimal internal energy and maximum entropy in function of T (temperature) that informs the optimum convergence of the route of the seller. The tests were made for 5000 to 10000 cities with different elastic nets in the order n = 2.5 m (n = amount of nodes and m = amount of cities) with a rapid convergence and efficiency of the parallelized method.

In Sect. 2 of the paper, method, Durbin-Willshaw elastic net is explained, in Sect. 3 the methodology of the strategy of the parallel algorithm is described; in Sect. 4 the cluster Beowulf architecture is explained; in Sect. 5 the results are shown; in Sect. 6 discussion, results and conclusions are described, and finally the references are presented.

2 Method

2.1 Durbin-Willshaw Elastic Net

The elastic net is based on the principles proposed by Durbin-Willshaw (1987). The method tries to solve the TSP sequentially [7]. It consists on a unidimensional net with p nodes displayed in a linear or closed structure (like a circle) made of neurons or nodes. This structure is settled in a data system (cities) and the nodes are subjected to two types of forces, Ej, $j = 1..M$, which is the internal force that leads to a minimization of the size of the net; and then there is $C_{ij}, i = 1..N; j = 1..M$, which is the external force that corresponds to the attraction of the city i on the node j (See Fig. 1).

It is defined that the coordinates of city i are denoted by vector S_i and the nodes of the net are denoted by R_j. The force of attraction of the city on a node j in the net is proportional to the distance between the city and the node, having a lineal direction that goes through the city and the node, therefore: $C_{ij} = \Delta_{ij}(S_i - R_j)$, where Δ_{ij} is a parameter of proportionality, C_{ij} is the attraction of the city i on the node j of the net.

The two elastic forces that act in a node j of the net are proportional to the near nodes $j - 1$ and $j + l$, therefore, $E_j = E_{j+1} + E_{j-1} = K(R_{j+1} - R_j) + K(R_{j-1} - R_j) = K(R_{j+1} - 2R_j + R_{j-1})$, where K is a parameter of control of

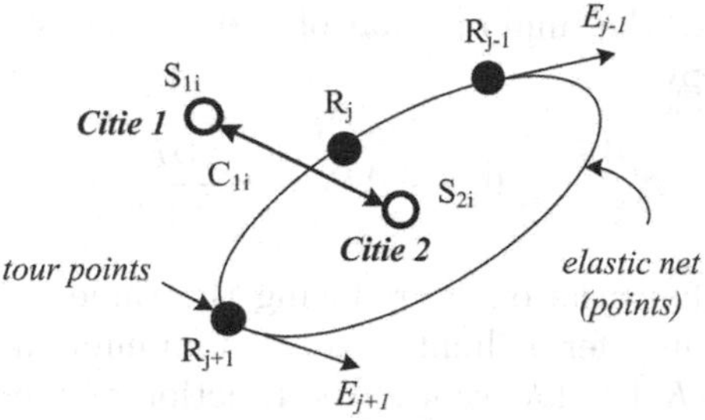

Fig. 1. Elastic net.

the algorithm. The summation of the types of forces leads to the resulting rate acting on the node j.

The change of ΔR_j of the node j in the elastic net is assumed to be proportional to this result,

$$\Delta R_j = \alpha_c \sum_i (S_i - R_j) + \alpha_t K(R_{j+1} - 2R_j + R_{j-1}) \tag{1}$$

where α_c and α_t are control coefficients that allow the convergence of the method, which is a key factor for the proper functioning of the net. The K parameter is a changing factor that decreases after several iterations. When K decreases allows the network to approach the cities because internal forces are small relative to the external forces. w_{ij} is the attraction that the city S_i exerts on R_j of the elastic net, ant it is calculated trough the power function,

$$w_{ij} = \frac{\Phi(|S_i - R_j|, K)}{\sum_{k=1}^{M} \phi(|S_i - R_k|, K)} \tag{2}$$

where ϕ is the power of function given by Yuille, (1990) [11] where all cities and nodes on the elastic net are described by a Boltzmann statistic distribution with probability of independent connection for each one,

$$\phi(|S_i - R_j|, K) = e^{-(S_i - R_j)^2/2K^2}.$$

If K has a high value, the power function will have a high influence in all the nodes of the net. If K decreases, the power function establishes a selective influence of a city on a near node in the net.

An energy functional $E\{R\}$ (optimization equation) decreases the attraction of the cities on the net nodes as the algorithm iterates:

$$E\{R_j\} = \frac{\alpha_t}{2} \sum_{j=1}^{M} |R_{j+1} - R_j|^2 - \alpha_c K \sum_{i=1}^{N} \ln\left(\sum_{j=1}^{M} \phi(d, K)\right). \tag{3}$$

The strategy of the elastic net is analogy to statistic mechanics [6,12,13]. In this analogy, K parameter is identical to temperature T by means of the relation $T = K^2$ and the Boltzmann constant is 1 [11].

The method allows the minimization of the internal energy and the maximization of the entropy,

$$\frac{\partial E}{\partial R_l} = 0 \longrightarrow \Delta R_l = -\frac{\partial E}{\partial R_l K}. \tag{4}$$

Equation (4) allows the variation R_j reducing the value of the Eq. (3) and due to E energy possess an inferior limit, a local minimum can be achieved when $K \longrightarrow 0$ When there K has low values, the function of energy contains a lot of local minima corresponding to possible routes around the cities, and the most optimal minimum is the shortest route [6].

3 Framework for the Parallel Strategy of the Elastic Network

3.1 Generation, Decomposition y Parallelization

In Fig. 2 the stages of the framework for the implementation of elastic net are shown. In short, the framework consists of decomposing from an amount of cities and a number of cities, so the energy is calculated at various slave computers. The result of the total energy is made in the master computer, which controls the other computers. Below, the phase of preparation of the cities and nodes and the parallelization phase are explained.

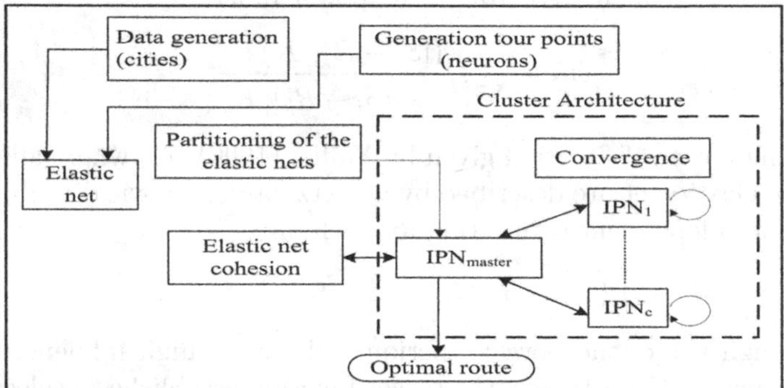

Fig. 2. Framework for the parallel elastic net in the cluster architecture.

Phase 1. Preparation of the cities and nodes. For the parallel test, the location of all the cities is randomly generated between −1 and 1, in the order n = 5000; n = 6000; n = 7000; n = 8000; n = 9000 and n = 10000 cities in a xy Cartesian plane. The m stops of the tour (nodes) start randomly in order of m = 2.5n [9], depending the amount of cities n, where for 1000 cities 2500 stops of the tour or nodes are generated, and so on with the rest of the cities.

The values of the control parameter used in the parallelization of this test where $\alpha_t = 0.01$ and $\alpha_c = 0.5$ [9], which gives comparable values of the two energies that are shown in the relation of the Eq. 3.

Phase 2. Parallelization. Equation 1 is divided for the different slave computers,

$$\Delta R_j = \alpha_c \sum_{i=1}^{N} \frac{\phi(S_i - R_j)(S_i - R_j)}{\sum_{k=1}^{M} \phi(S_i - R_j)} + \alpha_t K(R_{j+1} - 2R_j + R_{j-1}), \quad (5)$$

For the parallel strategy we used for computers, N is the number of cities and $M \approx 2.5N$ is the number of nodes.

Step 1. Calculation of the summation $X_i = \sum_{k=1}^{M} \phi(S_i - R_k)$ that appers in the denominator of the Eq. 5. The distribution of X_i on the processors in the following way:

$$X_i^l = \sum_{kl} (S_i - R_{kl}),$$

for $l = 1, 2, 3, 4, ..., n - 1$, where $K1$, $K2$, $K3$ and $K4$ consist of each group of cities.

$K1 = 1, ..., \frac{M}{4}$, $K2 = \frac{M}{4} + 1, ..., \frac{M}{2}$, $K3 = \frac{M}{2} + 1, ..., \frac{3M}{4}$, $K4 = \frac{3M}{4} + 1, ..., M$ This process is done for each city $i = 1, ..., N$. Therefore, it is required that each processor store all the coordinates of the cities S_i, $i = 1, ..., N$; and the fourth part of the coordinates of the tour R_{Kl} , $l = 1, ..., 4$.

Step 2. In the master computer $X_i = X_i^1 + X_i^2 + X_i^3 + X_i^4$ is calculated.

Step 3. The summation is distributed over the cities of the relation 5 (Eq. 5) in the same manner as in step 1 with the four computers.

$$\Delta R_{jl} = \alpha_c \sum_{i=1}^{N} \frac{\phi(S_i - R_{jl})(S_i - R_{jl})}{X_i} + \alpha_t K(R_{jl+1} - 2R_{jl} + R_{jl-1}), \quad (6)$$

with $l = 1, 2, 3, 4$. The storage is the same as in case X_i, that means that all the cities have to be stored in all the computers and the corresponding fourth part of the parameters R_{jl}.

Step 4. The summation $\Delta R_j = \sum_{l=1}^{4} \Delta R_{jl}$, is calculated in the master computer.

Phase 3. Convergence. The convergence of the parallelized elastic net is made by an annealing process in function of temperature T. The free energy functional (Eq. 5) that depends of R_j allows the minimization of the internal energy and the maximization of the entropy until the system is in equilibrium.

4 Cluster Beowulf Architecture

Beowulf is a design of cluster built in architectures of low price personal computers, and it is built to achieve parallel processing of high performance. The operating system used is GNU/Linux, distribution Fedora Core 5 with the installation package OSCAR (Open Source Cluster Application Resources).

OSCAR is a set of tools that allows the easy installation of a Beowulf cluster. It has everything necessary to manage a cluster HPC (High Performance Computing). The structure of hardware in the cluster currently has one master server node and 4 slave servers. The master node has 1 XEON processor of de 2 Ghz, 2 Gbytes of RAM and 800 Gbytes of Hard Disk in RAID 5. The 4 slave nodes are the equal among them (IPN = information processing node), with 1 XEON processor of 2.0 Ghz, 1 Gbyte of RAM, and 80 Gbytes of hard disk each one. Figure 3 shows the Cluster Beowulf architecture,

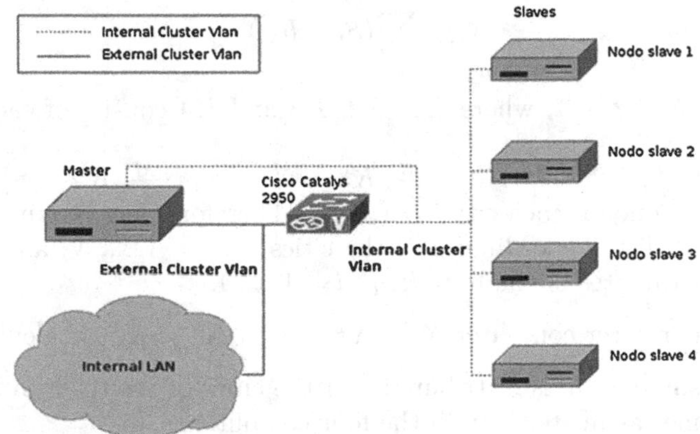

Fig. 3. Cluster Architecture.

In the next part, speedup and efficiency are defined: The gaining of velocity or acceleration **(speedup)** in a system N of computers is defined as $S(N) = \frac{t(1)}{t(N)}$, where $t(1)$ is the time to run just one computer and $t(N)$ is the time used to execute it in the parallel system with N computers. In conditions $t(N) = \frac{1}{N}$ the gaining of velocity for the equation $\frac{S(N)}{t(N)}$ will be: $S(N)_{ideal} = \frac{t(1)}{t(N)} = \frac{1}{(1/N)} = N$. One way to measure system performance, will be to compare the gaining of velocity of the system with the gaining of velocity given by $S(N)_{ideal}$, to this measure is called **efficiency** and is given by $E(N) = \frac{S(N)}{S(N)_{ideal}} = \frac{S(N)}{N} = \frac{(t(1)/t(N))}{N} = \frac{t(1)}{N_t(N)}$. Indicating the extent to which resources are used.

5 Results

The application of the method Durbin elastic net with the new parallel application for the TSP provided some results, having into account the amount of

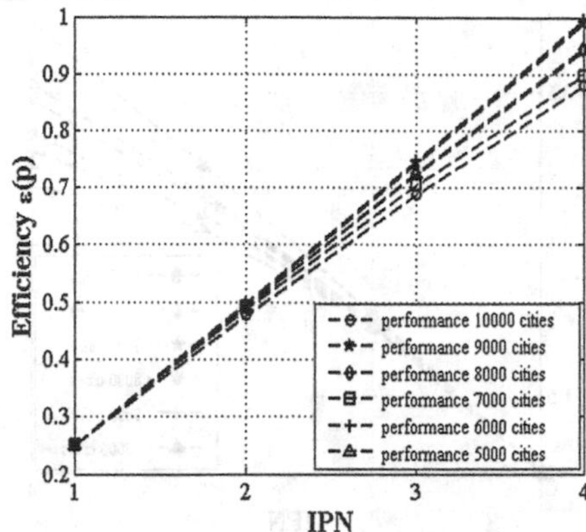

Fig. 4. Performance for the different numbers of cities (5000 to 10000).

cities that the seller has to go, and the amount of nodes or stops of the tour that the parallel elastic net needs. The results in Fig. 4 show the evaluation of the performance for the different numeric calculations in function of the amount of processors or information processing nodes (IPN), amount of cities and nodes.

The parallel elastic net with a net of 12500 nodes for 5000 cities with 4 IPN resulted in a 0.9899 efficiency $\varepsilon(p)$. Having 15000 nodes for 6000 cities with 4 IPN resulted in $\varepsilon(p)$ 0.9897. 17500 nodes for 7000 cities with 4 IPN resulted in 0.9946 $\varepsilon(p)$. 20000 nodes for 8000 cities with 4 IPN resulted in 0.99195 $\varepsilon(p)$. 22500 nodes for 9000 cities with 4 IPN resulted in 0.9966 $\varepsilon(p)$; and finally, having 25000 nodes for 10000 cities resulted in 0.942275 $\varepsilon(p)$. In Fig. 5, the effects of Amdhal law are shown. Those curves allow evaluating the amount of data having a certain amount of IPN. In Fig. 5 the Speedup is shown.

A $S(p)$ whith an amount of 5000 cities for 4 IPN resulted in 3.9596, while an $S(p)$ with an amount of 5000 cities for a single IPN resulted 1.0, that mean that the $S(p)$ with 4 IPN was higher. We can also observe that a $S(p)$ with an amount of 6000 cities for 4 IPN resulted in 3.9588, while the $S(p)$ with an amount of 6000 cities for a single IPN was 1.0. It is known that known that the $S(p)$ for an amount of 7000 cities for 4 IPN resulted in 3.9784. $S(p)$ for an amount of 8000 cities for 4 IPN resulted in 3.9678. $S(p)$ for an amount of 9000 cities for 4 IPN resulted in 3.9861. Finally, a $S(p)$ for an amount of 10000 cities for 4 IPN resulted in 3.7691. We can see that at higher amounts of cities, the $S(p)$ increases, which indicates a speed in the calculation the optimization of the energy functional in Eq. 5, in the search for the seller optimal route, in a cluster Beowulf environment.

In Fig. 6 the time that it takes each computer in the cluster Beowulf for 5000; 6000; 7000; 8000; 9000 and 10000 citiesis observed. We can clearly appreciate that with 4 IPN the calculation in the convergence of the route of the seller is highly different than with less computers or IPN.

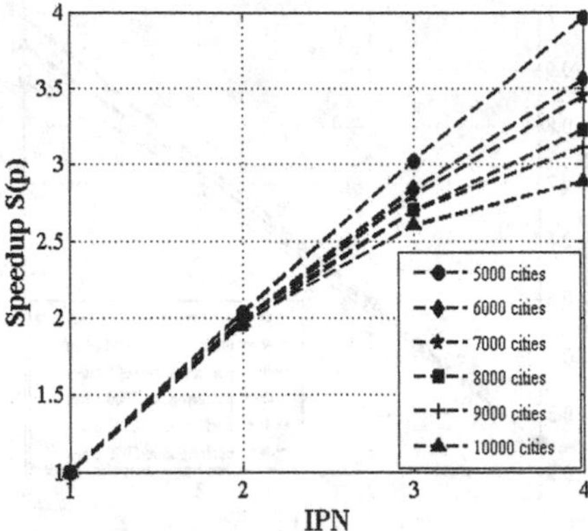

Fig. 5. Evaluation through Amdhal law for the different amounts of cities (5000 to 10000).

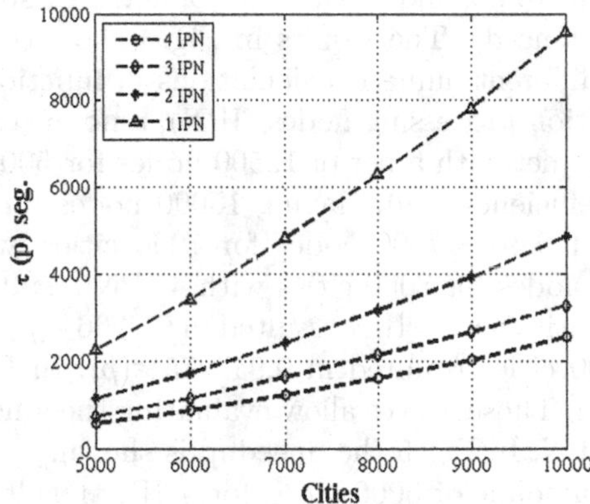

Fig. 6. Evaluation of the time of processing for 1, 2, 3, 4 IPN (5000 to 10000 cities).

In the following tables we can observe in detail the importance of parallelism with different numbers of IPN. Table 1 shows the calculation with 4 IPN. $\tau(p)$ indicates the calculation time with 4 IPN, but it does not show the amount of nodes or stops of the tour. $\varepsilon(p)$ indicates the efficiency of 4 IPN. $S(p)$ indicates the speedup with 4 IPN for different amount of cities. Finally α_c indicates the interaction among cities and nodes and α_t indicates the interaction among the nodes. In Tables 2, 3 and 4 we can observe the importance of a cluster Beowulf in the calculation processing and the impact of the algorithm in the parallel elastic net.

Table 1. Detail of differences in calculations with 4 IPN.

Cities	IPN	$\tau(p)seg.$	nodes	$\varepsilon(p)$	$S(p)$	α_t	α_c
5000	4	570	12500	0.9899	3.9596	0.01	0.5
6000	4	850	15000	0.9897	3.9588	0.01	0.5
7000	4	1161	17500	0.9946	3.9784	0.01	0.5
8000	4	1526	20000	0.9919	3.9678	0.01	0.5
9000	4	1956	22500	0.9966	3.9861	0.01	0.5
10000	4	2543	25500	0.9422	3.7691	0.01	0.5

Table 2. Detail for the differences in calculations with 3 IPN.

Cities	IPN	$\tau(p)seg.$	nodes	$\varepsilon(p)$	$S(p)$	α_t	α_c
5000	3	725	12500	0.7782	3.113	0.01	0.5
6000	3	1132	15000	0.7431	2.972	0.01	0.5
7000	3	1566	17500	0.7373	2.949	0.01	0.5
8000	3	2110	20000	0.7174	2.869	0.01	0.5
9000	3	2613	22500	0.7470	2.983	0.01	0.5
10000	3	23258	25500	0.7354	2.941	0.01	0.5

Table 3. Detail for the differences in calculations with 2 IPN.

Cities	IPN	$\tau(p)seg.$	nodes	$\varepsilon(p)$	$S(p)$	α_t	α_c
5000	2	1138	12500	0.4958	1.983	0.01	0.5
6000	2	1693	15000	0.4968	1.987	0.01	0.5
7000	2	2322	17500	0.4973	1.989	0.01	0.5
8000	2	3029	20000	0.4997	1.999	0.01	0.5
9000	2	3903	22500	0.4994	1.997	0.01	0.5
10000	2	4855	25500	0.4948	1.974	0.01	0.5

Table 4. Detail for the differences in calculations with 1 IPN.

Cities	IPN	$\tau(p)seg.$	nodes	$\varepsilon(p)$	$S(p)$	α_t	α_c
5000	1	2257	12500	0.25	1.0	0.01	0.5
6000	1	3365	15000	0.25	1.0	0.01	0.5
7000	1	4619	17500	0.25	1.0	0.01	0.5
8000	1	6055	20000	0.25	1.0	0.01	0.5
9000	1	7797	22500	0.25	1.0	0.01	0.5
10000	1	9585	25500	0.25	1.0	0.01	0.5

6 Discussion of Results and Conclusions

The free energy functional that describes the system of interaction among the stops of the tour and the cities; and the interactions among the stops of the tour (nodes) manages to indicate the minimization of the internal energy of the system and the maximization of the entropy, in order to find the optimal route for the seller. It should be noted that even when maintaining the convergence of free energy to a minimum, it is not guaranteed that it is really the global optimum. Even though, the efficiency is tremendously higher than with just one IPN.

This work achieves a significant difference in working with computers in a cluster Beowulf environment. Therefore, for 5,000 cities the seller can built a route base on 12500 nodes (no), with an efficiency $\varepsilon(p)$ of 0.9899 and a $S(p)$ of 3.9596 with control parameters of $\alpha_t = 0.01$ y $\alpha_c = 0.5$.

Whereas 1 IPN for 5000 cities, the efficiency is $\varepsilon(p) = 0.25$, and $S(p) = 1.0$, and the time it took was $\tau(p) = 2257\,\mathrm{s}$

Moreover, for 10000 cities, the seller can build a route based on 25000 nodes no, with an efficiency of $\varepsilon(p) = 0.9422$ and an $S(p) = 3.7691$ with a control parameter of $\alpha_t = 0.01$ and $\alpha_c = 0.5$. While for a IPN for 10000 cities, $\varepsilon(p) = 0.25$ and $S(p) = 1.0$. in the same way we can corroborate significant gains in the calculation of the TSP, as shown in Tables 1, 2, 3 and 4.

This work does not aim to make comparisons with other methods because it only works with the elastic net parallelized. Besides, we made experiments with more than 10000 cities for the TSP, because we pretend to demonstrate that the method is effective in solving the TSP in a cluster Beowulf environment

It is possible to guarantee the convergence of the energy functional by looking for equilibrium towards an optimization where we can diminish the internal energy and maximize the entropy.

Although it is true that with the technological advances that exist nowaday, like high performance computing, more numerical calculations can be made, so better results for the TSP can be obtained, this paper only works with elastic net used to find solutions for heuristic and combinatorial problems like in the cluster system for computers, parallel computing, grid computing and cloud computing [14–18].

References

1. Aho, A., Hopcroft, J., Ullman, J.: Data Structures and Algorithms. Addison-Wesley, Reading (1983)
2. Lawler, E.L., Lenstra, J.K., Rinnoy Kan, A.G., Shmoys, D.B. (eds.): The Traveling Salesman Problem: A Guided Tour of Combinatorial Optimization. Wiley, New York (1990)
3. Johnson, D., Papadimitriu, C.: Computational complexity. The traveling salesman problem, pp. 37–85 (1985)
4. Gorbunov, S., Kisel, I.: Elastic net for stand-alone RICH ring finding. Nucl. Instr. Meth. Phys. Res. Sect. A **559**(1), 139–142 (2006)

5. Osorio, R., Parada, V.: Un estudio comparativo de métodos basados en redes neuronales para resolver el problema del vendedor viajero. Memoria. USACH, Chile (1998)
6. Durbin, R., Szeliski, R.: Yuille: an analysis of the elastic net approach to the traveling salesman problem. Neural Comput. **1**, 348–358 (1989)
7. Durbin, R., Willshaw, D.: An analogue approach to the traveling salesman problem using an elastic net method. Nature **326**, 689–691 (1987)
8. Basse, S.: Algoritmos computacionales, computación y diseño, 3rd edn. Addison-Wesley, Mexico (2002)
9. Ball, K., Erman, B., Dill, K.: The elastic net algorithm and protein structure prediction. J. Comput. Chem. **23**(1), 77–83 (2002)
10. Lévano, M., Nowak, H.: New aspects of the elastic net algorithm for cluster analysis. In: Palmer-Brown, D., Draganova, C., Pimenidis, E., Mouratidis, H. (eds.) EANN 2009. CCIS, vol. 43, pp. 281–290. Springer, Heidelberg (2009)
11. Yuille, A.: Neural Comput. **2**(1) (1990)
12. Yuille, A.L., Kosowsky, J.J.: Neural Comput. **6**, 341 (1994)
13. Stolorz P.: In: Moody, J.E., Hanson, S.J., Lippmann, R.P. (eds.) Advances in Neural Information Processing Systems, vol. 4 pp. 1026–1032. Morgan Kaufmann, San Mateo (1992)
14. Bai, H., OuYang, D., Li, X., He, L., Yu, H.: MAX-MIN, ant system on GPU with CUDA. In: Proceedings of the Fourth International Conference on Innovative Computing, Information and Control, pp. 801–804 (2009)
15. Cecilia, J.M., García, J.M., Nisbet, A., Amos, M., Ujaldón, M.: Enhancing data parallelism for ant colony optimization on GPUs. J. Parallel Distrib. Comput. **73**(1), 42–51 (2013)
16. Fujimoto, N., Tsutsui, S.: A highly-parallel TSP solver for a GPU computing platform. In: Proceedings of the Seventh International Conference on Numerical Methods and Applications, pp. 264–271 (2010)
17. O'Neil, M.A.: Rethinking the parallelization of random - restart hill climbing a case study in optimizing a 2-opt TSP solver for GPU execution. In: GPGPU 2015. San Francisco, CA, USA (2015)
18. O'Neil, M.A.: A parallel GPU version of the traveling salesman problem. San Francisco, CA, USA (2015)

Comparison of Methods for Automated Feature Selection Using a Self-organising Map

Aliyu Usman Ahmad(✉) and Andrew Starkey

School of Engineering, University of Aberdeen, Aberdeen, UK
r01aual4@abdn.ac.uk

Abstract. The effective modelling of high-dimensional data with hundreds to thousands of features remains a challenging task in the field of machine learning. One of the key challenges is the implementation of effective methods for selecting a set of relevant features, which are buried in high-dimensional data along with irrelevant noisy features by choosing a subset of the complete set of input features that predicts the output with higher accuracy comparable to the performance of the complete input set. Kohonen's Self Organising Neural Network MAP has been utilized in various ways for this task. In this work, a review of the appropriate application of multiple methods for this task is carried out. The feature selection approach based on analysis of the Self Organising network result after training is presented with comparison of performance of two methods.

Keywords: Clustering · Self-organising neural network MAP · Feature selection · Engineering optimisation

1 Introduction

Clustering is one of the most widely used data analysis methods for numerous practical applications in emerging areas [1]. Clustering entails the process of organising objects into natural groups by finding the class of objects such that the objects in a class are similar to one another and dissimilar from the objects in another class. Data clustering methods have been applied on numerous applications in engineering. In automotive engineering for example, clustering has been used in various ways [2–6] for utilization of product manufacturing processes to enhance development performance. Other applications of clustering in engineering include engineering design, quality assurance, process control and manufacturing system design [2, 4, 7–10]. A clustering algorithm usually considers all input parameters in an attempt to learn as much as possible about the given objects.

The Self-Organising Neural Network MAP (SOM) by Kohonen [11] has been widely used as one of the most successful clustering methods with strong data exploration and visualization capabilities [12]. The most extensive application of the SOM in engineering is in the monitoring and identification of machine conditions and complex processes [6, 10, 13]. The SOM's mapping preserves a topological relation by maintaining neighborhood relations such that patterns that are close in the input space are mapped to neurons that are close in the output space, and vice-versa.

C. Jayne and L. Iliadis (Eds.): EANN 2016, CCIS 629, pp. 134–146, 2016.
DOI: 10.1007/978-3-319-44188-7_10

One of the biggest drawbacks of the SOM algorithm is its in-ability to automatically identify the features that are relevant for analysis and discard the irrelevant inputs that negatively distort the analysis result [14]. This feature however would prove extremely useful in many applications in Engineering and in other problem domains. In an attempt to resolve this, researchers [15–17] have worked on the improvement of the algorithm by a feature weighting method during training with the application of the steepest descent optimization method for the identification of important inputs for clustering. The core of the weighted method lies in attempting to describe the contribution of each feature in the clustering algorithm in order to improve the clustering result.

This paper investigates the application of the weighted method and also a standard SOM approach in order to identify the key features in a number of artificially produced datasets.

2 Neural Network Clustering Methods

2.1 Self-organising Neural Network MAP

The Self Organising Neural Network Map (SOM) is an unsupervised artificial neural network learning method trained to produce a low-dimensional representation of a high dimensional input samples [12].

A typical SOM consists of the computational (Map) layer and the input layers as shown in Fig. 1 below;

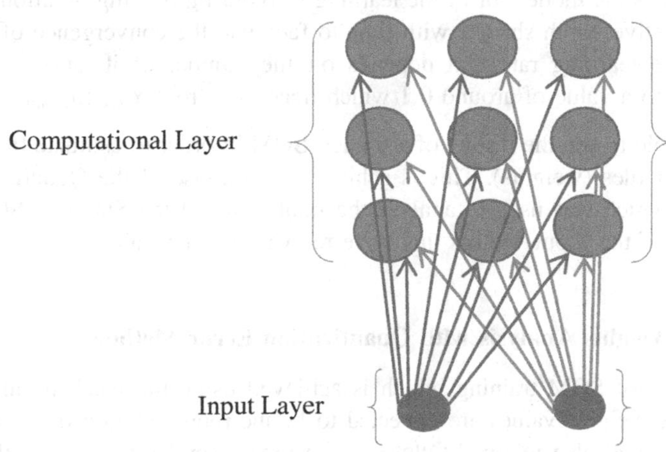

Fig. 1. A 2-dimensional self organising map architecture

The input layer comprises of the source nodes representing the sample's features/attributes. There are as many weights for each node as there are number of features (dimensions) in the input layer, represented in the form of an input vector, i.e. $x = [x_i^1, x_i^2, \ldots, x_i^d]$ for an input sample where d denotes sample dimensions and i the sample number and n denoting the total number of samples.

The computational layer (MAP) consists of neurons placed in nodes of a 2 dimensional grid (lattice), each neuron is identified by its index position, i.e. j, on the map and associated with a weight vector, i.e. $W_j = \{w_{ji} : j = 1, \ldots, n; i = 1, \ldots, d\}$, the size of which is equal to the dimension of the input vector. The set of weights W parameters are determined by iteratively minimizing the cost function below;

$$R(C, W) = \sum_{i=1}^{N} \sum_{j=1}^{|W|} \kappa_{j,c(x_i)} \quad ||x_i - w_j||^2 \tag{1}$$

At every n^{th} training step, The Gaussian neighbourhood function was calculated for the map; this is expressed as;

$$K_{j,c\,(x_i)\,(n)} = \alpha(n) \cdot \mathrm{e}\,(-\frac{\delta^2_{j,c\,(x_i)}}{2\sigma(n)^2}) \tag{2}$$

Where

- $K_{j,c(x_i)(n)}$ is the neighborhood function between each unit (j) on the map and the winning unit c (x_i) at the nth training step
- $\delta_{j,c(x_i)}$ Is the distance (Euclidean) from the position of unit (j) to the winning unit c (x_i) on the map.
- $\sigma(n)$ is the effective width of the topological neighborhood at the nth training step, this serves as the moderator of the learning step during training iterations. The size of the effective width shrinks with time to facilitate the convergence of the map.
- $\alpha(n)$ is the learning rate that depends on the number of iterations (n), this is initialised to a value of around 0.1 which decreases from α_{max} to α_{min}

It is possible to use the results of a trained SOM in order to estimate the relevance of feature variables (weights). This is achieved by the use of the Quantization Error method [18] which was used to analyse the final result of the Standard SOM for the identification of the Input vectors that were relevant for the training.

2.2 SOM Weights Analysis with Quantization Error Method

On completion of SOM training which is achieved using the batch training method [11], the node weights values are expected to be the representation of their matching input samples, and also relatively close to the input samples mapped to their neighboring nodes and relatively far from the input samples mapped to distant nodes.

Let M_j be set of the training samples x_i mapped to node j, and the quantization error for node j is calculated after SOM training as;

$$E_j = \sum\nolimits_{M_j} \left\| x_i - w_j \right\|^2 \tag{3}$$

The weight features with lowest quantization error are expected to be the features whose corresponding input sample features are most relevant when comparing the samples against their winning nodes. A further analysis was carried out on the quantization error values for all the node weights in order to automatically separate the group of the relevant inputs from the irrelevant inputs, a parametric statistical test with median split was carried out on the quantization error values to differentiate the group of high values (as irrelevant features) from the group of low values (as relevant features). Since there is no reliance on a hard coded threshold value to determine irrelevant and relevant features this means that this step could be used for any amount of data features and results in a fully automated step for this aspect of the process.

2.3 Weighted Self Organising Neural Network MAP

The weighted SOM (WSOM) function proposed by [17] is another method designed to compute the relevance of feature variables (weights) automatically during the training process. This approach entails the use of additional random weights that are multiplied against the input weights as a metric for measuring the relevance of the observations during training, and since the comparison is done one sample at a time, the updating method for the WSOM is incremental rather than batch as in the standard SOM.

Let $\Re^d$ be the Euclidean data space and $E = \{x_i; i = 1, \ldots, N\}$ a set of observations, where each observation $x_i = (x_i^1, x_i^2, \ldots, x_i^d)$ is a vector in $\Re^d$.

Each node j has prototype weights $w_j = (w_i^1, w_i^2, \ldots, w_i^d)$, and a single random weight is assigned to for each input attribute such that; $\pi_d = (\pi_1, \pi_2, \ldots \pi_d)$.

This method attempts to find the relevance of all the weights of a single vector which are applied against the whole set of input weights, but is not able to determine the relevance of an individual weight of each node j in a trained SOM.

The set of Weights W and π parameters are determined by iteratively minimizing the cost function below;

$$R_{gvw}(C, W, \pi) = \sum_{i=1}^{|E|} \sum_{j=1}^{|W|} \kappa_{j,c(x_i)} \quad \left\| \pi_d \otimes x_i - w_j \right\|^2 \tag{4}$$

The cost function $R_{gvw}(W, \pi)$ as described in Eq. 4. The algorithm is optimised by finding the $\underset{W,\pi}{min}\; R_{gvw}(W, \pi)$. The process begins by initially starting with some random values for W, π then these values are modified in order to reduce $R_{gvw}(W, \pi)$, until the minimum of the cost function is reached.

The method uses the Steepest Descent algorithm in order to optimise its cost function;

$$\{$$

$$R_j := R_j - \alpha \frac{\partial}{\partial R_j} R_{gvw}(W, \pi) \quad (\text{For}\, j = W\, \text{and}\, j = \pi) \tag{5}$$

$$\}$$

The gradient descent minimization of the function can be implemented as;

$$w_j(n+1) := w_j(n) - (n) - \alpha(n)\kappa_{j,c\,(x_i)}(n)\ \kappa_{j,c\,(x_i)}\left(w_j - \pi_g \otimes x_i\right) \tag{6}$$

$$\pi_g(n+1) := \pi_g(n) - (n) - \alpha(n)\kappa_{j,c\,(x_i)}(n)\ \kappa_{j,c\,(x_i)}\ x_i\left(\pi_g \otimes x_i - w_j\right) \tag{7}$$

The Steepest Descent algorithm which is utilised by the WSOM method searches for the minimum of a function by computing the gradient of the function, starting at a random point P_0, and moving from P_i to P_{i+1} in the direction of the local downhill gradient $-\nabla f(P_i)$ for each iteration of line minimization.

The Steepest descent method is guaranteed to find a solution for quadratic functions, which are convex shaped functions with a single minimum that is equal to the global minimum [19] (as illustrated in Fig. 2). For problems beyond quadratic functions with multiple local minimums (such as Schwefel Function, Fig. 3), the gradient descent finds the solution of a function based on the first identified local minimum and therefore ignores other local minimums, and does not necessarily and cannot be guaranteed to find the global minimum of the given function. It is therefore important to confirm that the cost function for the WSOM method results in a single global minimum that can be found by the steepest descent approach.

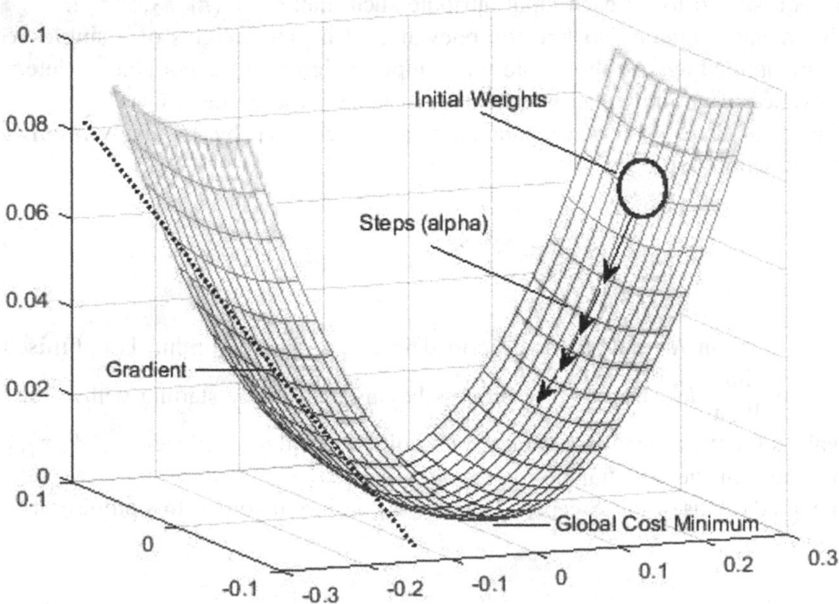

Fig. 2. Steepest decent method for a quadratic functions

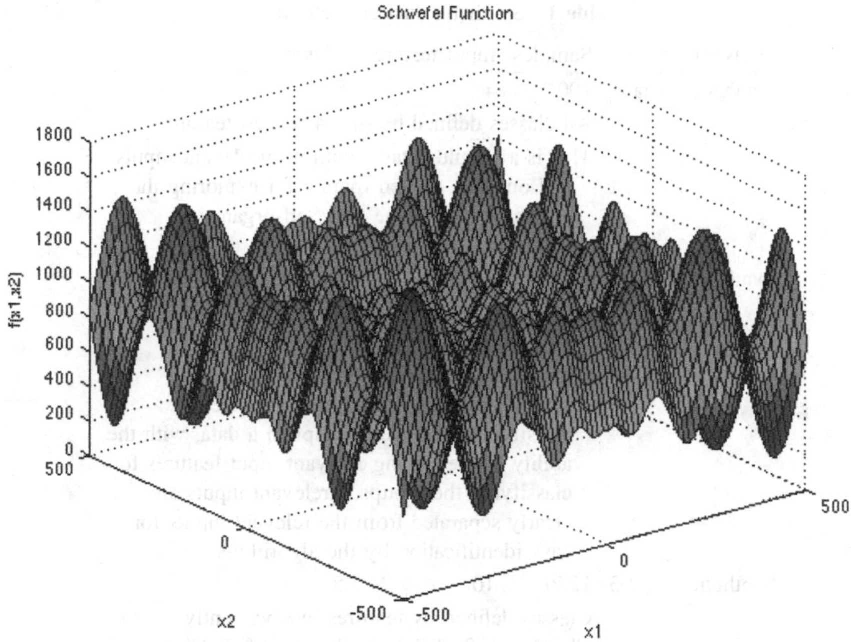

Fig. 3. Problems beyond quadratic functions: Schwefel function [20]

For a full description of the WSOM process, the reader is directed to [16, 17] In the WSOM approach, the relevance of an input vector is indicated by the global weights with irrelevant vectors having global weights close to 0 and relevant vectors having global weights different from 0. The relevance of an input vector can be measured by this method only if the given data sample is normalized to the same scale.

3 Experiment

3.1 Synthetic Datasets

In order to assess the efficacy of the WSOM method against a standard SOM implementation, a number of synthetic datasets were developed which had different features, starting with a simple dataset with a small number of attributes, and moving to datasets with a larger number of inputs and additional noise. All data sets had equal class distribution (i.e. same number of samples for each class), and was normalised. These datasets are discussed more fully in the tables below. The use of synthetic data rather than real data sets of this type is very important as it allows a full assessment of how well the techniques work and what type of problems they can solve, and where they may encounter difficulties, if any (Table 1).

Table 1. Synthetic datasets definition

Dataset name	Samples	Input features	Classes
Synthetic_Data01	100	4	5
	All classes defined by first 4 related features		
	This is a simple dataset with no irrelevant inputs and outliers, created mainly for exploring the cost functions of the two Self-organising algorithms		
Synthetic_Data02	1220	7	5
	All classes defined by first 4 related features Irrelevant features: 5,6,7		
	This dataset was created to evaluate the self-organising system's performance in classifying concealed groups in a data, with the ability of identifying relevant input features for classifying the groups. Irrelevant inputs are clearly separated from the relevant inputs for easy identification by the algorithms		
Synthetic_Data03	1220	10	5
	Classes defined by features independently Class1 = 1, 2, & 3, Class2 = 4, 5, & 6, Class3 = 2, 3, 4 & 5, Class4 = 6, 7, & 8, Class5 = 1, 4, & 8. Noise features; Features 9 & 10		
	In addition to Synthetic_Data02, definition of classes were distributed among variables, to identify the self-organising method's ability of identifying the degree of relevance of the input features for classification		

3.2 Experiment Design

As both methods rely on a random process, the performance of the algorithms were measured based on results of 10 runs for each of the methods on the Synthetic Datasets.

To check whether the weighted SOM cost function (Eq. 4) for Synthetic_Data01 is a quadratic function, to be suitable for the steepest descent optimization approach, the cost function was optimised with the simulated annealing algorithm on Synthetic_Data01; a stochastic search method that aims to expose all possible minimums of the function by random search in space, with initial temperature $(T_o) = 10.0$, cooling rate $(\alpha) = 0.99$ and maximum iteration $(Maxtime) = 1000$.

If a cost function has a single global minimum, the best combination of weight values for different runs of the algorithm will be expected to be within the same region, and to have a positive linear correlation when compared against each other. Otherwise, if the cost function has multiple local minimums then the best combination of weight values would be in different regions for different runs of the algorithm and will not be correlated.

The normalised correlation matrix of the final weights from the simulated annealing algorithm was computed to show the similarity of the weights produced from 6 runs of the algorithm against each other, the same experiment was carried out on the standard SOM cost function for comparison.

When undertaking the correlation analysis for the WSOM method, the raw SOM weights cannot be used directly but must be divided by the corresponding global weights π. This step is required since the global weights and WSOM node weights are linked as described in the cost function, and it is possible due to the random nature of the process that different values could be arrived at for these variables whilst still mapping against similar input samples and the node weights could therefore be different from one run to the next.

In addition, the problem with direct comparison of weights at the same index for the 6 different simulated annealing runs is that nodes are not necessarily localized to a specific index. In a single run, a node might appear in the first index, while in a separate run, the same node might appear in a different index. As such, direct comparison of the weights infers the comparison of random un-related nodes, which are most likely to be not-correlated at all times.

To overcome this problem, the indexes of all the nodes weights was re-arranged to correspond to the best matching positions for all the nodes from the 6 different simulated annealing runs before carrying out the correlation test on the weights.

Let E be set of weight values for a given SOM run ($w_n^i; n = 1, \ldots N$), were N is the total number of weights and W is set of weight values for other SOM runs to total number of SOM runs R. The index position I of a node in a given SOM when compared against the SOM with weight values E is computed as Eq. 8. For completeness, this was executed for all the SOM runs.

$$I = \sum_{n=1}^{|E|} \sum_{m=1}^{|W...R|} min \left\| w_n^i - w_m^j \right\|^2 \quad (8)$$

A null hypothesis test $Ho : \rho = 0$ was conducted with 0.5 significance level to investigate the relationship between the final node weights (i.e. to see if they are correlated or not). Nodes are correlated if their correlation coefficient is different from zero, and therefore means there is linear relationship between the nodes. There is no correlation for nodes with correlation coefficients close to zero.

4 Results

The analysis of the correlation results can be seen in Figs. 4 and 5 the bars in the plot represent the correlation coefficients values ρ for a given run of the self-organising process compared against another. The red line on the plots at 0.5 and −0.5 indicate the respective boundaries for positive and negative correlations, with no correlation shown at or around 0.

The bars above the red line indicate the pairs of node weights with correlations significantly different from 0, which implies that there is a significant linear relationship

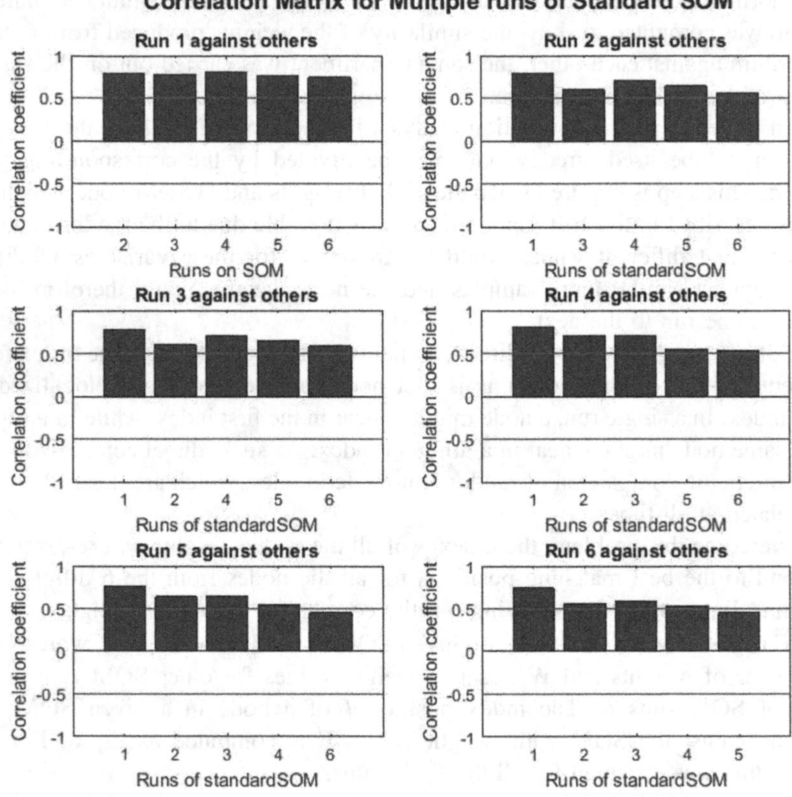

Fig. 4. Correlation matrix for multiple runs of standard SOM

between the weights of the final runs. (i.e. the SOM weights are broadly equivalent and the values can be said to be the same). On the other hand, the bars below the line indicate the pairs of node weights with a correlation coefficient that is not significantly different from 0, which implies that there is no significant linear relationship between the weights. (i.e. the SOM weights are not the same and contain different values).

The correlation matrix for the standard SOM (Fig. 4) weights shows that almost all pairs of weights (4 out of 6) have correlations significantly different from zero which proves positive correlation among weights. On the other hand, the correlation matrix for the weighted SOM (Fig. 5) shows that only 2 out of the 6 weights are correlated, which indicates that this method has resulted in different solutions being found.

In summary it can be concluded that the simulated annealing algorithm with the standard SOM cost function finds similar solutions recurrently in all the different runs. On the other hand, the algorithm with the weighted SOM cost function finds different solutions for most of the runs, which is most likely to be as a result of multiple local minimums in the cost function.

Further analysis was undertaken on the performance of the two methods by investigating their performance on the three synthetic data sets, and in particular

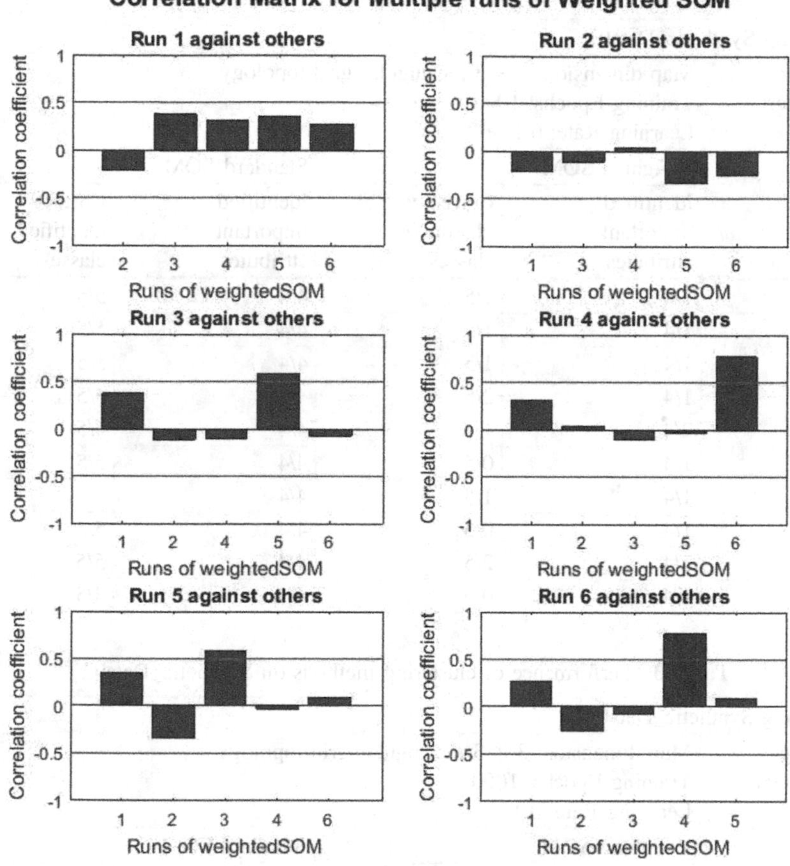

Fig. 5. Correlation matrix for multiple runs of weighted SOM

whether they correctly identified the important attributes, and whether the classes were also separated by the final SOM with all samples mapped to a single node, this is required as the SOM training was initialized with number of nodes within the range of the true number of classes. These results can be seen in Tables 2, 3 and 4.

5 Discussion and Recommendation

As seen in Table 2, the standard SOM was able to correctly identify all the classes in most of the runs for simple data with no irrelevant inputs, and was also able to identify all inputs as important due to low quantization error between weights to their mapped input samples. Unlike the standard SOM, the weighted SOM failed to identify the classes for the same simple data set with no irrelevant inputs. The weighted SOM method also performed poorly by failing to correctly identify clusters and differentiate the relevant input vectors from the irrelevant input vectors on the Synthetic_Data02.

Table 2. Performance of clustering methods on Synthetic_Data01

Clustering Synthetic_Data01				
Training parameters	Map dimension: 3 × 3 rectangular grid topology Training Epochs: 1000 Learning Rate: 0.1			
Runs	Weighted SOM		Standard SOM	
	Identified important attributes	Correctly identified classes	Identified important attributes	Correctly identified classes
Run 1	1/4	1/5	4/4	5/5
Run 2	1/4	1/5	4/4	5/5
Run 3	1/4	1/5	4/4	5/5
Run 4	1/4	2/5	4/4	5/5
Run 5	2/4	1/5	4/4	4/5
Run 6	1/4	0/5	4/4	5/5
Run 7	1/4	1/5	4/4	5/5
Run 8	1/4	0/5	4/4	5/5
Run 9	1/4	2/5	4/4	5/5
Run 10	1/4	0/5	4/4	4/5

Table 3. Performance of clustering methods on Synthetic_Data02

Clustering Synthetic_Data02				
Training parameters	Map dimension: 3 × 3 rectangular grid topology Training Epochs: 1000 Learning Rate: 0.1			
Runs	Weighted SOM		Standard SOM	
	Identified important attributes	Correctly identified classes	Identified important attributes	Correctly identified classes
Run 1	1/4	0/5	4/4	2/5
Run 2	0/4	0/5	4/4	1/5
Run 3	2/4	1/5	4/4	2/5
Run 4	1/4	0/5	3/4	1/5
Run 5	2/4	0/5	4/4	1/5
Run 6	3/4	1/5	4/4	2/5
Run 7	1/4	1/5	4/4	3/5
Run 8	2/4	0/5	4/4	2/5
Run 9	0/4	0/5	4/4	3/5
Run 10	1/4	0/5	4/4	2/5

However, the Standard SOM with Quantization error method after training clearly identified the relevant vectors for the training on this dataset. Both methods performed poorly in correctly identifying the clusters in the data as the result of the influence of the irrelevant inputs during the training.

Table 4. Performance of clustering methods on Synthetic_Data03

Clustering Synthetic_Data03				
Training parameters	Map dimension: 3 × 3 rectangular grid topology Training Epochs: 1000 Learning Rate: 0.1			
Runs	Weighted SOM		Standard SOM	
	Identified important attributes	Correctly identified classes	Identified important attributes	Correctly identified classes
Run 1	0/8	0/5	2/8	3/5
Run 2	0/8	0/5	4/8	1/5
Run 3	1/8	1/5	2/8	1/5
Run 4	0/8	0/5	3/8	0/5
Run 5	1/8	0/5	1/8	1/5
Run 6	2/8	0/5	1/8	0/5
Run 7	1/8	1/5	5/8	1/5
Run 8	1/8	0/5	1/8	2/5
Run 9	0/8	0/5	2/8	1/5
Run 10	1/8	0/5	2/8	1/5

For the more complicated dataset with overlapping class definition (Synthetic_Data03), the analysis of the Standard SOM's training result with the Quantization Error also failed to identify what was important for the training, as presented in Table 4.

As discussed in Sect. 2.3. The steepest descent algorithm is guaranteed to find the local minimum for quadratic functions with a single global minimum, whereas for functions with multiple local minimums, the gradient descent finds the solution of the function based on the first identified local minimum ignoring other local minimums, and therefore is not suitable for the proposed WSOM cost function as our results clearly demonstrate that multiple minimums exist in the solution space defined by the WSOM cost function.

The quantization error between the weights values and their matched classified input samples shows more potential for identifying important features however these results show that this approach will only work in certain types of data. It is interesting to note that for the second synthetic dataset, that the quantization method correctly identifies the important features despite not being able to correctly classify the groupings in the data. This is clearly not desirable and further work is required to develop methods that are better capable of identifying important features and to correctly undertake the classification of the groupings within the data when they are hidden within noisy data.

References

1. Tajunisha, S., Saravanan, V.: Performance analysis of k-means with different initialization methods for high dimensional data. Int. J. Artif. Intell. Appl. (IJAIA) **1**, 44–52 (2010)
2. Yin, S., Huang, Z.: Performance monitoring for vehicle suspension system via fuzzy positivistic C-means clustering based on accelerometer measurements (2015)
3. Gebauer, H.: Identifying service strategies in product manufacturing companies by exploring environment–strategy configurations. Ind. Mark. Manag. **37**, 278–291 (2008)
4. Ivanov, V.: A review of fuzzy methods in automotive engineering applications. Eur. Transp. Res. Rev. **7**, 1–10 (2015)
5. Krishnan, V., Ulrich, K.T.: Product development decisions: a review of the literature. Manag. Sci. **47**, 1–21 (2001)
6. Liu, H., Li, Y., Li, N., Liu, C.: Robust visual monitoring of machine condition with sparse coding and self-organizing map. In: Liu, H., Ding, H., Xiong, Z., Zhu, X. (eds.) ICIRA 2010, Part I. LNCS, vol. 6424, pp. 642–653. Springer, Heidelberg (2010)
7. Pham, D., Afify, A.: Clustering techniques and their applications in engineering. Proc. Inst. Mech. Eng. Part C **221**, 1445–1459 (2007)
8. Khoshnevisan, B., Bolandnazar, E., Barak, S., Shamshirband, S., Maghsoudlou, H., Altameem, T.A., Gani, A.: A clustering model based on an evolutionary algorithm for better energy use in crop production. Stoch. Environ. Res. Risk Assess. **29**, 1921–1935 (2015)
9. Nguyen, S.D., Nguyen, Q.H., Choi, S.: A hybrid clustering based fuzzy structure for vibration control–Part 2: an application to semi-active vehicle seat-suspension system. Mech. Syst. Sig. Process. **56**, 288–301 (2015)
10. Maren, A.J., Harston, C.T., Pap, R.M.: Handbook of Neural Computing Applications. Academic Press, Cambridge (2014)
11. Kohonen, T.: The self-organizing map. Proc. IEEE **78**, 1464–1480 (1990)
12. Yin, H.: The self-organizing maps: background, theories, extensions and applications. In: Fulcher, H., Jain, L.C. (eds.) Computational Intelligence: A Compendium. SCI, vol. 115, pp. 642–762. Springer, Heidelberg (2008)
13. Simula, O., Alhoniemi, E., Hollmen, J., Vesanto, J.: Monitoring and modeling of complex processes using hierarchical self-organizing maps (1996)
14. Shafreen Banu, A., Ganesh, S.H.: A study of feature selection approaches for classification, pp. 1–4 (2015)
15. De Carvalho, F.D.A., Bertrand, P., Simões, E.C.: Batch SOM algorithms for interval-valued data with automatic weighting of the variables. Neurocomputing **182**, 66–81 (2015)
16. Mesghouni, N., Temanni, M.: Unsupervised double local weighting for feature selection, vol. 1, pp. 413–417 (2011)
17. Grozavu, N., Bennani, Y., Lebbah, M.: From variable weighting to cluster characterization in topographic unsupervised learning, pp. 1005–1010 (2009)
18. De Bodt, E., Cottrell, M., Verleysen, M.: Statistical tools to assess the reliability of self-organizing maps. Neural Netw. **15**, 967–978 (2002)
19. Gonzaga, C.C., Schneider, R.M.: On the steepest descent algorithm for quadratic functions. 1–20 (2015)
20. Schwefel, H.: Numerische optimierung von computer-modellen mittels der evolutions strategie. Birkhäuser, Basel Switzerland (1977)

EEG-Based Condition Clustering using Self-Organising Neural Network Map

Hassan Hamdoun[1(✉)] and Aliyu Ahmad Usman[2]

[1] School of Natural and Computing Sciences, University of Aberdeen, Kings' College, Aberdeen, AB24 4UE, UK
hassan.hamdoun@abdn.ac.uk

[2] School of Engineering, University of Aberdeen, Kings' College, Aberdeen, AB24 4UE, UK
r01aua14@abdn.ac.uk

Abstract. Electroencephalography (EEG) has recently emerged as a useful neurophysiological biomarker for characterizing different physiological and pathological conditions of healthy and un-healthy brain activity measurements. However, the complexity and high temporal resolution of the EEG signal data has brought about the need for efficient and accurate automated methods for distinguishing mental tasks activities and the recording conditions. Distinguishing mental tasks with high accuracy is pertinent for early detection and clinical diagnostic of several neurodegenerative diseases. Expert clinicians are needed in order to distinguish between mental tasks and EEG recording conditions, which is a manual process that is prone to inefficiencies and errors especially when the EEG data is miss-annotated at the recording stage. This paper proposes the application of a Self-organizing neural network Map (SOM) with Learning Vector Quantization (LVQ) for EEG Eyes Open (EO) and Eyes Closed (EC) condition classification. This was achieved with classification accuracy of 88.5 %. The proposed approach shows good performance and hence the method can be readily applied to other classification/clustering problems on brain measurements in the Brain Computer Interface (BCI) arena.

Keywords: SOM · LVQ · EEG · Classification · Dimensionality reduction

1 Introduction

Clinical diagnosis and investigations are increasingly dependent on the ability to record and analyse physiological signals. Electroencephalography (EEG) is one of those key signal recordings of patient's brain activity that clinicians and medical professionals rely on [1]. It has been suggested by several studies that EEG signals can be used to detect the severity of several diseases such as CJD, Alzheimer's, dementia, schizophrenia and epilepsy. For example, several research efforts have focused on detecting epileptic seizure and early signs of Alzheimer [2, 3]. EEG signals contain the complex brain activity for various frequency bands and the signal representing the underlying mental tasks the healthy control or subject (patient) is undertaking. EEG signals recorded from different scalp locations, although, looking similar, but contain different information

C. Jayne and L. Iliadis (Eds.): EANN 2016, CCIS 629, pp. 147–157, 2016.
DOI: 10.1007/978-3-319-44188-7_11

spanning different frequency bands. Expert clinicians, physicians and neuroscientists rely on their years of experience to distinguish between various EEG mental conditions e.g. muscle signal vs eye blinks, eyes open vs eyes closed, sleep, coma etc.

Typically an EEG recording lasts 30–45 min and includes annotations that describe the start and end of the condition/mental task the subject is performing. The EEG technician adds these annotations during the recording [4]. Once EEG is performed, there is a need for a timely clinical diagnosis on the EEG by the expert clinician in order to ensure swift medical attention and intervention to the subject. Lack or mislabeling of EEG recording annotations of the subject mental tasks is highly problematic and will impede the clinician ability to perform fast diagnosis. The speedy diagnostics is critical in epilepsy and other neuro-degenerative diseases whereby the neurologist is interested in certain recording length corresponding to a certain condition (e.g. Eyes Close (EC), Eyes Open (EO), etc.). This process is usually carried-out manually and requires expert clinical and neurophysiological knowledge and years of experience in analyzing such high dimensional complex EEG data. This manual intensive method is not efficient and hence the research community is still looking for an efficient automated method to detect the differences between EEG mental tasks and conditions with high accuracy.

Several studies on EEG signals relies on the successful detection of the EO and EC conditions e.g. frequency analysis- for which Hilbert-Haung Transform (HHT) [5]. The HHT transform obtained accuracy of about 84 % in differentiation the transition between EO and EC conditions. Other time domain approaches relies on detecting the amplitude increase associated with eye lid movements during very small periods of the signal. [6] provided 87 % accuracy based on the continuous amplitude increase in the signal artifacts. Other power spectrum approaches are based on detecting the change in Alpha-wave power during the EC condition [7]. While those techniques provide good accuracy, they become computationally expense fo date with high dimensions. We use SOM, on the other hand, because it has powerful visualization capabilities for high dimensional data.

This paper proposes the application of a Self-organizing neural network Map (SOM), which is a powerful visualization tool for clustering high dimensional data, for EEG Eyes Open (EO) and Eyes Closed (EC) condition classification. It is well known that the alpha band power increases during the EC condition and hence in this paper we use frequency domain power features to discriminate between the EEG recordings [7]. The occipital areas of the skull get usually activated in Alpha and Beta bands. The Alpha-band has significant more power in the frontal, parietal, and occipital areas of the skull when eyes are closed. Hence in this paper we propose the join use of the discriminatory power spectrum features of the EEG signal with the capability of the SOM when dealing with high dimensional data. The implementation of the SOM is carried out in Matlab on a 208 healthy subjects EEG datasets recorded using EO and EC condition for a 1 min duration. The Learning Vector Quantization (LVQ) technique is combined with the SOM to obtain enhanced performance. The EEG datasets were recorded and contributed by Physionet using the BCI2000 instrumentation system described in [8]. This paper focuses on EC and EO classification using the power and flexibility of the SOM method. However, it is anticipated that there exist scenarios where there are discriminatory features present in high dimensional brain measurement data, where our proposed approach can be applied. In such scenarios, high accuracy rate classification of EEG

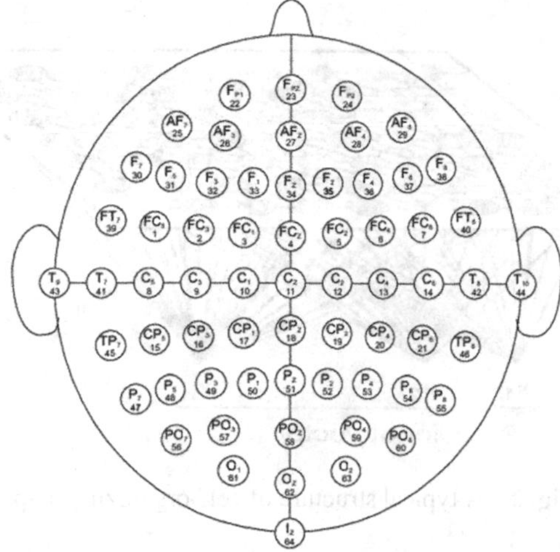

Fig. 1. Location and nomenclature of 64 electrodes as per the international 10–10 system

conditions will transform the applications of EEG for medical diagnostic and can also integrate with the growing Brain Computer Interface (BCI) e.g. controlling a wheelchair and prosthesis [9].

This paper is organized as follows. Section 2 reviews EEG signal acquisition and analysis. Section 3 presents the clustering technique based on SOM map and Learning Vector Quantization (LVQ) technique. Section 4 describes the experiment design and data collection. Section 5 describes the results and discussion. Conclusions are drawn in Sect. 6.

2 EEG Signal Acquisition

EEG signals record the differences of the voltage from two locations on the scalp over time. The EEG signals has an amplitude in the range of 1–100 μV with frequency in the range of 0.5 to 10 Hz [10]. EEG can be recorded with different sampling rates (typical range is between 200 Hz–2048 Hz) offering high temporal resolution of the brain activity. Such high temporal resolution allows the decomposition of EEG signal into different frequency waveforms that help describe different mental states. In order to compare EEG signal results over time, a standardized locations on the scalp are used e.g. 10–10 or 10–20 systems [11]. The nomenclature and locations of the electrodes for the EEG datasets analysed in this paper is shown in Fig. 1.

2.1 EEG Signal Analysis and Mental States

EEG signal useful features need to be extracted to help in clinical diagnostic. An EEG signal is preprocessed and presented in terms of its rhythmic activity [12]. The common extracted rhythms are: (i) Alpha (α) wave: appears in healthy adults while awake, relaxed and their eyes closed. It occurs in the frequency range of 8–13 Hz with a voltage range

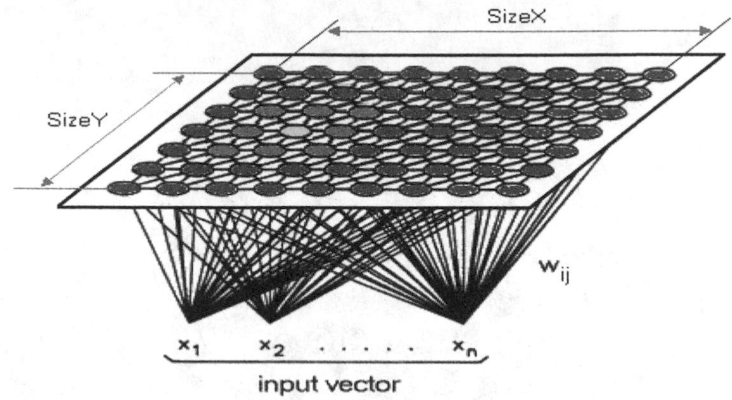

Fig. 2. A typical structure of self-organizing map

of 20–200 μV. When opening the eye, alpha rhythm diminishes as attention and stimulus desynchronize the frequency, referred to as Alpha blockage. Alpha wave is usually observed in the posterior region of the head. (ii) Beta (β) wave: with a frequency range from 13–30 Hz, and lower amplitude than Alpha (ranging from 5–10 μV. β wave appears with extra excitation and vigilance. Beta waves are observed in the parietal and frontal region of the scalp. (iii) Theta (θ): with frequency range from 4–7 Hz, θ waveform is prominent during sleep, arousal and idling. It is recorded across the temporal and parietal region of the scalp with amplitude range of 5–10 μV. One type of θ wave activity is associated with decreased alertness, cognitive impairment and dementia.

(iv) Delta (δ) wave: this is the lowest frequency wave and is less than 3.5 Hz with amplitude range from 20–200 μV. δ wave occurs during deep sleep and with serious organic brain diseases. δ wave can be recorded in the frontal region of the scalp in adults and in the posterior region in children. (v) Gamma (γ) wave: with frequency range from 30–100 Hz, γ wave is recorded in the somatosensory cortex during short-term memory to recognize objects, sounds and in pathological cases due to cognitive decline.

3 Self-Organizing Map

The Self-organizing neural network Map (SOM) is an unsupervised clustering method that has been widely used as tool for visualization of high dimensional data to a lower dimension, usually 2-dimensional space by producing a representation of the input sample in a grid of nodes as introduced in 1982 by Teuvo Kohonen [13].

The SOM can be seen as a nonlinear mapping of a high dimensional input samples manifolds to a lower dimensional array, similar to classical vector quantization. Having a set of input samples $x_i = [x_{i1}, x_{i2}, \ldots x_{in}]^T \in \Re^n$ each element of the input dimension is associated with a weight vector for each node i on the map as $W_i = [W_{i1}, W_{i2} \ldots W_{in}]^T \in \Re^n$. This is illustrated in Fig. 2. A distance measure is used between x and W_i denoted $d(x, W_i)$ to match a sample to a node index winner c from Eq. 1.

$$c = \arg_i^{min}\{d(x, W_i)\} \tag{1}$$

The SOM has surpassed numerous clustering algorithms because of its powerful lattice preserving properties, that arranges the nodes that are similar to one another closely and dissimilar nodes far from one another on the grid using the neighborhood function during the training and has been applied in various disciplines ranging from engineering to health, social sciences and business [14–19]. A further description of the SOM algorithm and training can be found in [13].

3.1 Learning Vector Quantization (LVQ)

As discussed in Sect. 3. The basic aim of the SOM is the representation of a high dimensional data into a lower dimensional grid, this is similar to the vector quantization theory, the aim of which is dimensionality reduction. LVQ is a supervised self-organizing process that uses class information for repositioning of the Voronoi vectors (node weights) to improve the SOM's classification. This is achieved using a two layered training stages as shown in Fig. 3.

After EEG data collection, the raw EEG signals are transformed using feature extraction. The transformed features are then calculated for the selected EEG electrodes locations on the scalp. The set of those features then form the transformed input samples for the SOM training. The SOM training aims to fit the input samples to weights and produces a set of voronoi vectors as best matching units that determines the class membership of the input samples. The voronoi vectors and the true class labels of the training set are then used by the LVQ training algorithm to produce an adjusted class labels resulting in higher classification accuracy. In the LVQ training, a number of reference vectors w_i in the input space are assigned to each class represented by a given node i. Group of input samples x are decided to share the same class of the nearest reference vector [20]. Let t represents a time domain during training; the reference vectors are updated by Eq. 2 below;

$$w_k(\mathrm{t}+1) = w_k(\mathrm{t}) - \alpha(t)\big(x - w_k(t)\big) \tag{2}$$

$$w_l(\mathrm{t}+1) = w_l(\mathrm{t}) + \alpha(t)\big(x - w_l(t)\big) \tag{3}$$

Such that $0 < \alpha(t) < 1,$ *and* $\alpha(t)$ determines the learning rate that decreases monotonically with time. The reference vectors w_i and w_j are the best matching vectors (i.e. nearest to x) with x *and* w_j belonging to the same class and x *and* w_i belonging to different classes, x is required to fall in the mid-plane of w_i and w_j which are updated to satisfy condition of Eq. 4;

$$\min\left(\frac{d_i}{d_j}, \frac{d_j}{d_i}\right) > s \tag{4}$$

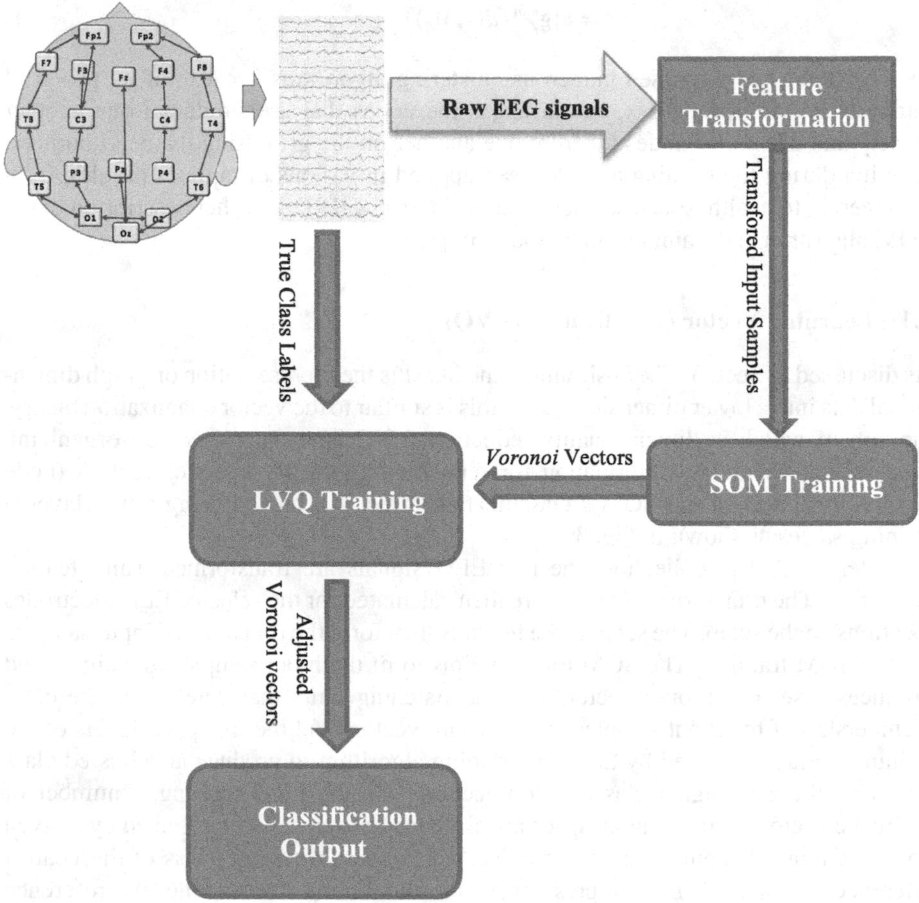

Fig. 3. Learning vector quantization applied on the self-organizing map

such that $d_i = |x - w_i|, d_j = |x - w_j|$. Eq. 3 aims to shift the decision boundaries towards Bayes limits with repulsive forces from x. Equation 4 ensures that the reference vectors keeps on with the approximation of the class distribution.

4 Experiment Design

The EEG data is transformed by extracting Alpha band features for 15 electrodes out of those shown in Fig. 1, for each of the 218 healthy subjects; resulting in data of 218×45 dimension. The selected electrodes are (Fp2, Fp1, F8, F7, F4, F3, C4, C3, P4, P3, O2, O1, Fz, Cz, Pz). These electrodes has been selected because they cover the frontal (Fp1,Fp2), occipital (O1,O2) and parietal regions (P3 and P4) which, as noted in Sect. 1 above, have more increase in the Alpha-band power during the EC condition. Our data comprises of both an EO and EC condition for all the 109 healthy subjects divided into two classes of 109 EO and 109 EC. Three Alpha band features are extracted from the raw EEG data for each selected electrode by applying Fourier Transform [21]; converting from time domain to frequency domain characteristics. These features include the absolute power, relative power and the peak frequency. Relative power is the area under the curve of the power spectrum and hence directly reflects the presence of alpha rhythm, absolute power reflects the peak power observed, whereas the peak frequency reflects the variation of the Alpha band frequency for each condition. EEG data is collected from physionet [22] for which 109 datasets have an EEG recordings of 1 min EO and 1 min EC duration.

The dataset is split into training and test sets of 110 and 108 subjects respectively using the K-fold cross validation method with k equal to 2. The SOM is then applied on the training dataset while the accuracy was computed using the test set. SOM implementation is performed in Matlab and for the classification problem in hand, there are only two classes. Hence a SOM map size of 2 neurons with hexagonal grid topology and using a random weight/bias rule training method over 1000 epocs is used to separate the EO and EC conditions. Following the flow in Fig. 3, LVQ is then applied on the trained SOM in order to adjust the voronoi weights for higher accuracy. The confusion matrix is then computed using the true EO and EC labels of EEG data. Mean Square Error (MSE for classification performance during the training phase of the SOM.

5 Results

We obtained the confusion matrix for our EEG data classifier, an error matrix that is used to describe the performance of binary classification problems for which the true values are known. The columns of which represents the instances of the predicted (target) EO or EC class and the rows represent the instances in the actual obtained class. The

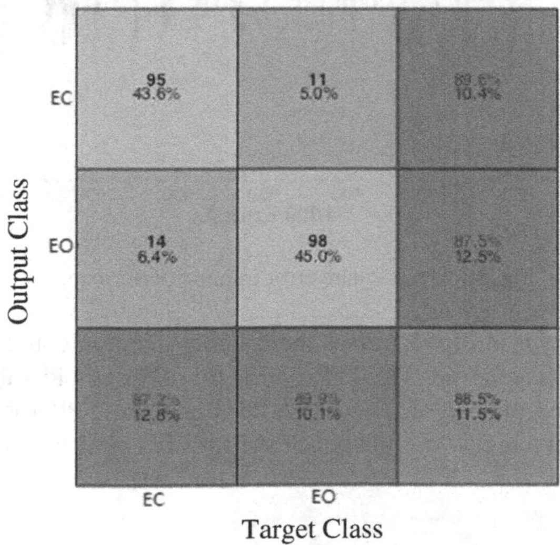

Fig. 4. Confusion matrix output of the EEG classification

confusion matrix is important as several measures of classification accuracy can be obtained from it. The EO/EC classification confusion matrix is shown in Fig. 4. Table 1, shows the classification accuracy measures derived from the confusion matrix. The classification accuracy from the confusion matrix is found to be 88.5 % which is high accuracy that would help greatly in automating clinical diagnostic. The misclassification rate (error rate) is 11.5 %. The True Positive rate (TPR)–out of the actual EO class, how often does the classifier predict EO- is found to be 89.9 %, a good performance for the intensive Alpha rhythm EO class. The False Positive Rate (FPR) – out the of the actual EC class, how often the classifier does predicts EO- is found to be 10.1 %, a reasonable value for the application needs. The specificity –out of the actual EC, how often the classifier predicts it is EC- is found to be 87.2 %, again good performance for the EC condition. Finally the precision of the EO condition is found to be 89.9 %.

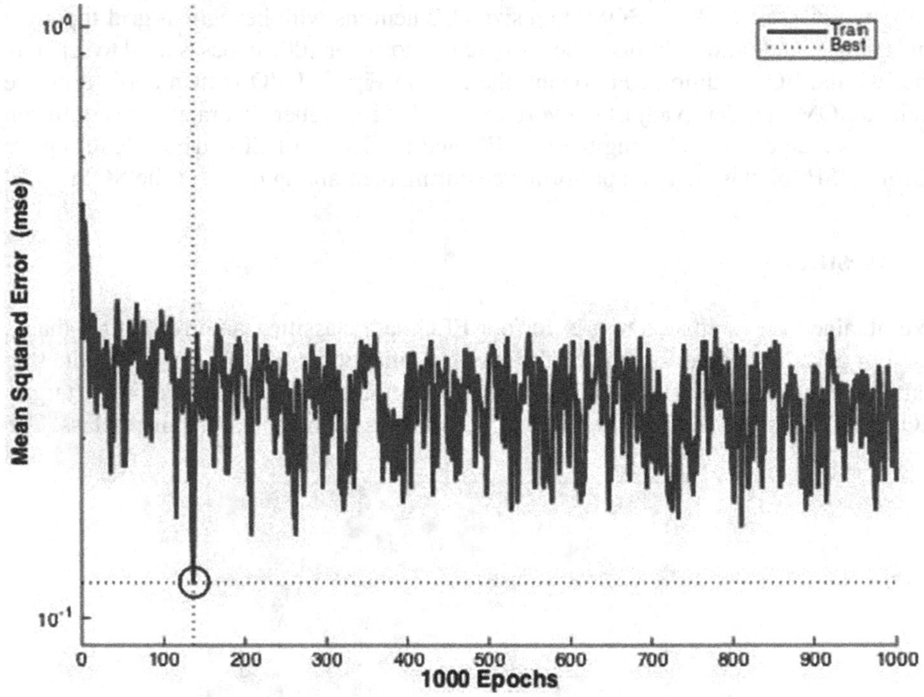

Fig. 5. Mean square error training performance

The MSE is shown in Fig. 5, versus the training iteration count. As we can see a minimum of 0.114 is achieved. The TPR v FPR for all threshold values is depicted in Fig. 6. Our proposed method achieves high TPR and low FPR for both EO and EC conditions making it a highly useful application for EEG clinical diagnostic.

Table 1. Classification accuracy measurements

Measure	Definition	Obtained value (%)
Accuracy	(TP + TN)/total subjects	88.5
Misclassification rate (error rate)	(FP + FN)/total subjects	11.5
True Positive Rate (Sensitivity or recall)	TP/actual EO	89.9
False positive rate	FP/actual EC	10.1
Specificity	TN/actual EC	87.2
Precision	TP/predicted EC	89.9

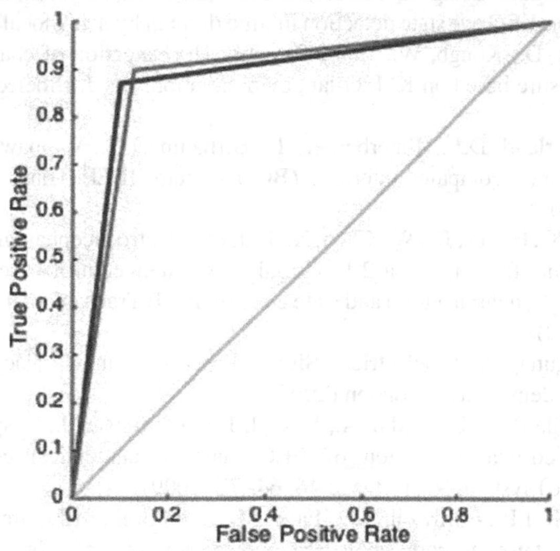

Fig. 6. Receiver operating conditions (roc) curve of classifier

6 Conclusion

This paper implemented a Self-organizing neural network Map (SOM) application for the condition classification of EEG data. We collected data from the physionet repository, calculated Alpha waveform features and then classified according to the eyes open and eyes closed conditions. The LVQ method was applied in conjunction with the SOM. Results showed classification accuracy of 88.5 % for eyes open and eyes closed conditions for a group of 218 healthy subjects. The application of joint SOM and LVQ demonstrated performance improvements and flexibility in the selection of the learning parameters for the training.

References

1. Mbuya, S.: The role of neuro-electrophysiological diagnostic tests in clinical medicine. East Afr. Med. J. **83**, 52–60 (2006)
2. Al-Qazzaz, N.K., Ali, S.H., Ahmad, S.A., Chellappan, K., Islam, M.S., Escudero, J.: Role of EEG as biomarker in the early detection and classification of dementia. Sci. World J. **2014**, 1–16 (2014). Article ID 906038
3. Liu, Y., Sourina, O., Nguyen, M.K.: Real-time EEG-based emotion recognition and its applications. In: Gavrilova, M.L., Tan, C., Sourin, A., Sourina, O. (eds.) Transactions on Computational Science XII. LNCS, vol. 6670, pp. 256–277. Springer, Heidelberg (2011)
4. Niedermeyer, E., da Silva, F.L.: Electroencephalography: Basic Principles, Clinical Applications, and Related Fields. Lippincott Williams & Wilkins, Philadelphia (2005)
5. Thuraisingham, R.A., Tran, Y., Craig, A., Nguyen, H.: Frequency analysis of eyes open and eyes closed EEG signals using the Hilbert-Huang Transform, pp. 2865–2868 (2012)
6. AKBEN SB Online EEG eye state detection in time domain by using local amplitude increase
7. Sakaia, M., Weia, D., Kongb, W., Daib, G., Hub, H.: Detection of change in alpha wave following eye closure based on KM2O-langevin equation. Int. J. Bioelectromag. **12**, 89–93 (2010)
8. Schalk, G., McFarland, D.J., Hinterberger, T., Birbaumer, N., Wolpaw, J.R.: BCI2000: a general-purpose brain-computer interface (BCI) system. IEEE Trans. Biomed. Eng. **51**, 1034–1043 (2004)
9. Huang, D., Qian, K., Fei, D., Jia, W., Chen, X., Bai, O.: Electroencephalography (EEG)-based brain–computer interface (BCI): a 2-D virtual wheelchair control based on event-related desynchronization/synchronization and state control. IEEE Trans. Neural Syst. Rehabil. Eng. **20**, 379–388 (2012)
10. Sörnmo, L., Laguna, P.: Bioelectrical Signal Processing in Cardiac and Neurological Applications. Academic Press, London (2005)
11. Koessler, L., Maillard, L., Benhadid, A., Vignal, J.P., Felblinger, J., Vespignani, H., Braun, M.: Automated cortical projection of EEG sensors: anatomical correlation via the international 10–10 system. Neuroimage **46**, 64–72 (2009)
12. Snyder, S.M., Hall, J.R., Cornwell, S.L., Falk, J.D.: Addition of EEG improves accuracy of a logistic model that uses neuropsychological and cardiovascular factors to identify dementia and MCI. Psychiatry Res. **186**, 97–102 (2011)
13. Kohonen, T.: The self-organizing map. Proc. IEEE **78**, 1464–1480 (1990)
14. Chakraborty, B., Menezes, A., Dandapath, S., Fernandes, W.A., Karisiddaiah, S., Haris, K., Gokul, G.: Application of hybrid techniques (self-organizing map and fuzzy algorithm) using backscatter data for segmentation and fine-scale roughness characterization of seepage-related seafloor along the western continental margin of India. IEEE J. Oceanic Eng. **40**, 3–14 (2015)
15. Yu, H., Khan, F., Garaniya, V.: Risk-based fault detection using self-organizing map. Reliab. Eng. Syst. Saf. **139**, 82–96 (2015)
16. Rigamonti, M., Baraldi, P., Zio, E., Alessi, A., Astigarraga, D., Galarza, A.: A self-organizing map-based monitoring system for insulated gate bipolar transistors operating in fully electric vehicle, vol. 6 (2015)
17. Merkevičius, E., Garšva, G., Simutis, R.: Forecasting of credit classes with the self-organizing maps **33** (2015)
18. Merényi, E., Mendenhall, M.J., O'Driscoll, P.: Advances in Self-Organizing Maps and Learning Vector Quantization. Advances in Intelligent Systems and Computing. Springer, Switzerland (2016)

19. Mans, R., Schonenberg, M., Song, M., van der Aalst, W., Bakker, P.: Process Mining in Healthcare (2015)
20. Hammer, B., Villmann, T.: Generalized relevance learning vector quantization. Neural Netw. **15**, 1059–1068 (2002)
21. Murugappan, M., Murugappan, S.: Human emotion recognition through short time Electroencephalogram (EEG) signals using Fast Fourier Transform (FFT), pp. 289–294 (2013)
22. Goldberger, A.L., Amaral, L.A., Glass, L., Hausdorff, J.M., Ivanov, P.C., Mark, R.G., Mietus, J.E., Moody, G.B., Peng, C.K., Stanley, H.E.: PhysioBank, PhysioToolkit, and PhysioNet: components of a new research resource for complex physiologic signals. Circulation **101**, E215–E220 (2000)

Cyber-Physical Systems and Cloud Applications

Intelligent Measurement in Unmanned Aerial Cyber Physical Systems for Traffic Surveillance

Andrei Petrovski[1(✉)], Prapa Rattadilok[1], and Sergey Petrovskii[2]

[1] School of Computing Sciences and Digital Media, The Robert Gordon University, Aberdeen, UK
{a.petrovski,p.rattadilok}@rgu.ac.uk

[2] School of Electric Stations, Samara State Technical University, Samara, Russian Federation
petrovski@rambler.ru

Abstract. An adaptive framework for building intelligent measurement systems has been proposed in the paper and tested on simulated traffic surveillance data. The use of the framework enables making intelligent decisions related to the presence of anomalies in the surveillance data with the help of statistical analysis, computational intelligent and machine learning. Computational intelligence can also be effectively utilised for identifying the main contributing features in detecting anomalous data points within the surveillance data. The experimental results have demonstrated that a reasonable performance is achieved in terms of inferential accuracy and data processing speed.

Keywords: Intelligent measurement · Traffic surveillance · Data anomalies · Computational intelligence · Artificial neural networks · Cyber physical system

1 Introduction

One of the main purposes of intelligent measurement systems (IMS) is to model the relationship between information that is required ('primary characteristics'), and the information which may be readily derived from (processed) sensor outputs such as target tracks ('secondary variable'). An IMS is capable of providing frequent 'on-line' estimates of primary characteristics on the basis of their correlation with the data, obtained from available sensors, measured in real time. As such, an IMS can help to reduce the need for measuring devices, improve system reliability, and develop tight control policies.

There are several advantages of IMS in comparison with traditional instrumentation [3]:

- Such measurement systems give more insight into the process under observation through capturing the information hidden in data.
- They are an emergent technology that allows users to improve productivity, become more energy and cost efficient.

C. Jayne and L. Iliadis (Eds.): EANN 2016, CCIS 629, pp. 161–175, 2016.
DOI: 10.1007/978-3-319-44188-7_12

- They can be easily implemented on existing hardware; moreover, various model-building algorithms can be used to adapt the IMS when an operating environment changes.
- They involve little or no capital cost such as the cost of installation, management of the required infrastructure, and commissioning.

The range of tasks fulfilled by IMS is quite broad – not only can IMS be used as a substitute or complement to physical sensors, but they can also perform monitoring and control of the process under observation, and can provide off-line operational assistance (e.g. design, diagnosis, knowledge refinement) [2].

The key challenge in building an IMS is to find a suitable structure for the inference model(s), using which a good estimator of the primary characteristics could be found. A basic rule in estimation is not to estimate what is already known or can be inferred from the data available. In other words, it is important to be able to utilise prior knowledge and physical insights about the process under observation/analysis when selecting the model structure. It is customary to distinguish between three levels of prior knowledge [4]:

- *White-box* models: the structure and parameters of the model are known or can be obtained from physical insights or basic principles;
- *Grey-box* models: some physical insights are available, but several model parameters remain to be determined from observed data;
- *Black-box* models: no physical insight is available, but the chosen model structure belongs to generic classes (e.g. artificial neural networks) that are known to have good flexibility and have been successfully applied in various problem domains.

Most of the existing IMS utilise black-box models operate on sensor data and produce estimates of essential (or primary) characteristics of the system under observation – for example, an unmanned aerial system (UAS) as shown in Fig. 1. (N.B. The

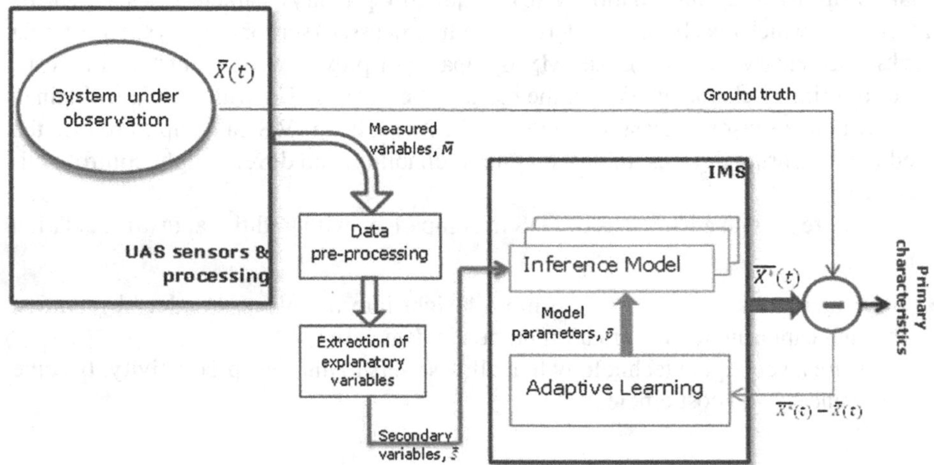

Fig. 1. IMS framework for the unmanned aerial vehicle

red arrow on the diagram representing ground truth information is desirable for more effective leaning, but not mandatory for the operation of an IMS.)

Having determined the relationship between the primary characteristics and the secondary variables, it becomes possible to obtain reasonable estimates of the former much faster and at a lower cost.

Also, the ability to infer primary characteristics raises the level of “information intelligence” coming from the UAS, enabling thereby to shift the workload of ground operators from “target detection to target analysis” and to optimise the throughput of data communication channels. Taking a road traffic example where it is desired to identify dangerous drivers represented by the state vector $\overline{X}(t)$, the ‘dangerous driver’ categorisation would be the primary characteristic, while secondary variables could include such quantities as driving speed or lane discipline.

A UAS in this context can be considered an autonomous cyber physical system that is used to acquire large amount of data about complex and changing environments, to perform interpretation and fusion of the data, and to present the information gathered or inferred in a synthetic and compact form highlighting the features of interest in the environment explored. The situation awareness of a UAS is determined by its operating conditions, various inputs obtained from essential sensors, as well as control adjustments received from a ground station. The situation awareness in terms of determining abnormal traffic conditions is an example of a primary characteristic that is difficult to measure directly. However, the large amount of data coming from on-board sensors or received from a ground station can be referred to as secondary variables. Due to the nature of UAS operation, the states of many secondary variables reflect the states of primary characteristics. For instance, surveillance data obtained from various sensors can indicate, and even identify, unusual or dangerous behaviour of drivers on the road [6].

Heterogeneous data acquiring sensors on-board of an unmanned aerial vehicle, which is part of the UAS, also add complexity in the form of analytical challenges, especially when there exist time and cost differences in processing data from different sources. Selecting suitable data acquisition sources, e.g. data that can be processed approximately in order to obtain representative samples, can help in time critical situations. Additional data acquisition sources that involve longer data processing but are more accurate or detailed, can be applied later to provide adaptive measurement features.

With the vast amounts of data, traditional data acquisition and data processing methods have become inefficient or sometimes inappropriate, especially in a real time environment. Computational Intelligence (CI) techniques have been successfully applied to problems in various application domains [1, 5]. These techniques however require accurately labelled training data to provide reliable and accurate specification of the context in which a UAS operates. For example, drivers may behave differently in the different road conditions (e.g. icy, wet, and foggy). The term “driver(s)” used throughout this paper refers to drivers of vehicles on the road (i.e. in the simulated model) under the surveillance of a UAS. The context enables the system to highlight potential anomalies in the data so that intelligent and autonomous control of the underlying process can be carried out.

Anomalies are defined as incidences or occurrences, under a given circumstances or a set of assumptions, that are different from the expectance. By their nature, these incidences are rare and often not known in advance. This makes it difficult for the computational intelligence techniques to form an appropriate training dataset. Moreover, UAVs often operate in different or dynamic environments [11]. This can further aggravate the lack of training data by the increased likelihood of intermittent anomalies. Computational intelligence techniques that are used to tackle dynamic problems should therefore be able to adapt to environmental/contextual changes [6].

The research work described in the presented paper is aimed at using machine learning algorithms for addressing 'Situational Assessment'. The immediate application area is the development and evaluation of such algorithms for a UAS application carrying out wide area surveillance of a tract of ground.

The detection of unusual profiles or anomalous behavioural characteristics from sensor data is especially complicated in security applications where the threat indicators may not be known in advance [8]. Data-driven modelling in such cases can yield insights on usual and baseline profiles, which in turn can be used to isolate unusual profiles when new data are observed in real time.

In general terms, therefore, the problem being tackled can be defined as finding the most effective ways of using measured data obtained from multiple sensors on board an aerial vehicle, in order to address the inherent difficulty in precisely defining and quantifying what constitutes anomalies. The presence of several sources of variability in anomalous patterns (for example, traffic density, vehicle types, features of terrain, etc.) and the limited availability, or even absence, of training datasets aggravate the difficulty of the problem being addressed [10].

The desired outcome of the work would be to devise a solution framework for intelligent processing of data obtained from multiple UAS sensors. This framework, described in Sect. 2 of the paper, is built with the premise that all the data sources considered together are capable of capturing the important features that could lead to a reliable anomaly detection, to efficient extraction and to intelligent interpretation of these features, which could in turn significantly reduce the number of false alarms generated as a result of the UAS operation.

To handle the challenges presented by the problem being addressed an incremental approach was adopted as a three-stage development of intelligent measurement systems. The first stage (anomaly detection) processes the available data by extracting the most representative features (referred to as 'secondary' variables) that characterise potential anomalies – this process is described in Sect. 3. For anomaly detection a mixture of statistical analysis and computational intelligence (CI) techniques has been adopted. The choice of detection techniques depends on the amount of historical data and the availability of insights on 'normal' system profiles – at the start of the detection process preferences are given to statistical techniques utilising probabilistic measure of data anomalies. As more data is being obtained, anomalous patterns/profiles start appearing, which can be detected more effectively with the help of CI techniques.

The features selected are then used to build inferential models, demonstrated in Sect. 4, that are utilised in the second stage (anomaly modelling) to interpret the new incoming data for real-time decision. In the second stage of building data-driven inference models two types of classifiers have been used – conventional classifiers utilising clustering algorithms, which do not require training data sets, and computational intelligence methods that carry out supervised learning of anomalous data patterns (in particular, artificial neural network (ANN)).

Finally, when the operating conditions of the system/process under observation change, both the secondary variables and the inference models are adapted in the third stage (anomaly modification) to provide the means of adjusting the IMS within dynamic operating environments. This final stage of adaptive measurement by the IMS is implemented using an automated machine learning algorithm, described in our previous work [7], that continuously tunes the inference models built for processing measured data and the representative features of data anomalies.

The proposed approach to intelligent measurement is evaluated on simulated and benchmark datasets – the main conclusions and proposed areas of further research are summarized in Sect. 5.

2 Inferential Measurement Systems

The impediments caused by unavailability or ineffectiveness of conventional measurements can negatively affect "situational assessment", but the problem can be alleviated, at least partially, by developing an intelligent measurement system (IMS) that performs intelligent sensing through the use of "soft" sensor technology. Intelligent sensing is a relatively new capability of measurement systems that supports such features as long mission duration, reliability and availability, real-time operation in hazardous and changing environments, as well as flexibility of use. These requirements lead to measurement systems with increasingly autonomous functionalities based on decentralised and distributed system architecture, effectively utilising available instrumentation data. Figure 2 illustrates a generic framework for building an IMS, proposed in [7].

Modelling using Computational Intelligence (CI) has become a versatile tool for enhancing the capabilities and efficiency of inferential measurement systems [5]. This type of modelling utilises the computational capabilities of modern computing devices (smart sensors, DSP-based microcontroller, and microprocessors) to effectively process the acquired input and infer the desired information. The AI-based techniques are applicable at various layers of IMS – from the data acquisition (sensor) layer, through to the layer of instrument calibration and customisation, then to the layer of process modelling, control and optimisation, and finally to the knowledge acquisition layer [6]. The wide spectrum of possible applications is due to the capabilities of an IMS to gain insight into the behaviour of complex dynamic systems by means of data-driven modelling, a systematic approach to which is described in this section.

The underlying principle of "soft" sensing is in estimating unmeasured variables, properties or parameters by using a model of a process under investigation, or of a part thereof, that correlates the measurements of interest (primary characteristics in Fig. 1) with more immediate (secondary) variables. As the name suggests, the model used by "soft" sensors is usually implemented in software; the secondary variables for such

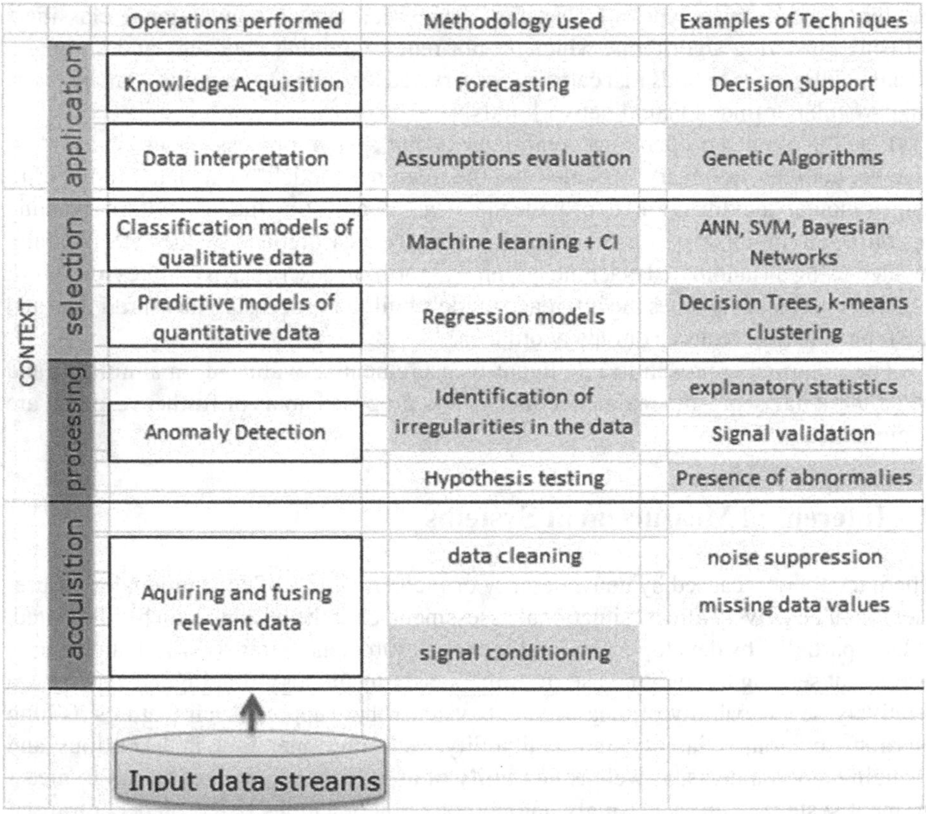

CONTEXT	Operations performed	Methodology used	Examples of Techniques
application	Knowledge Acquisition	Forecasting	Decision Support
	Data interpretation	Assumptions evaluation	Genetic Algorithms
selection	Classification models of qualitative data	Machine learning + CI	ANN, SVM, Bayesian Networks
	Predictive models of quantitative data	Regression models	Decision Trees, k-means clustering
processing	Anomaly Detection	Identification of irregularities in the data	explanatory statistics
			Signal validation
		Hypothesis testing	Presence of abnormalities
acquisition	Aquiring and fusing relevant data	data cleaning	noise suppression
			missing data values
		signal conditioning	
	Input data streams		

Fig. 2. Generic framework for building Intelligent Measurement Systems (IMS)

sensors are the controlled inputs, disturbances, and other intermediate variables affecting the process/application of interest [9].

In the course of the project the following tasks have been addressed:

- *Data pre-processing*: this step is performed only for building the inference models based on supervised machine learning techniques. Essentially, the step involves annotating the input data streams with the "ground truth" values needed for training certain Computational Intelligence (CI) algorithms (discussed later in the paper). Once the necessary training has been carried out, the data pre-processing activity becomes unnecessary, but can still be used if the reduction of noise in the data streams or filling in missing values are desired.
- *Selection of secondary* variables: it is important to choose the appropriate secondary variables (also referred to as data filters) to be used in building the inference model(s) – the number of these variables affects the time and complexity of inference, as well as the size of the data set needed for model development. The main objective of this step is to make use of the least number of secondary variables to develop a model of sufficient accuracy.

- *Building inference models*: Once a set of potential secondary variables is selected and their values are determined (this might involve passing the original data streams coming from sensors through several data filters), inferential models can be obtained using various data-driven modelling paradigms. At this stage it is important to strike the right balance between the accuracy and generalizability (i.e. minimising the effect of overtraining), and simplicity of the inference models. This is often achieved by varying the number of secondary variables (e.g. number of input nodes of ANNs) used in building the models through running screening and regression experiments (explained in more detail later in the paper).
- *Evaluating and tuning the inference models*: the inference models built have been validated on previously unseen data using a cross-validation approach. After the validation the inference model parameters (e.g. the window size of a data filter) can be dynamically adjusted if the operating environment changes (e.g. significant increase in traffic density) or the objectives of inferential modelling are modified (e.g. switching from the identification to classification mode of operation). The process of dynamic parameter adjustment is shown by the block at the bottom in Fig. 1, and is performed by a meta-learning layer of the developed IMS using a genetic algorithm (one of the Computational Intelligence techniques adopted within the proposed framework) [7].

2.1 Context Acquisition Level

In the presented research work, it is assumed that raw input data are pre-processed by having been already passed through the stages at the Data Acquisition level in Fig. 2 (e.g. data cleaning, fusing) and therefore this level is not considered. The only exception is the data discretisation activity, which can also be attributed to context processing level.

2.2 Context Processing Level

The Context Processing level in Fig. 2 utilises statistical and mathematical techniques of characterising raw input data. Depending on the complexity of the application domain, statistical methods can be used with the raw input data in order to identify anomalies within the input data stream; alternatively, statistical analysis may be used to prepare the raw input data for processing by computational intelligence techniques in identifying the pattern(s) of interest (or anomalies).

At this level, measurable variables are used to create secondary variables by applying different data filters and window sizes. For example, a secondary variable of *speed* may be defined as the change in distance travelled over *a* period of time, where *change* represents a data filter, *period of time* represents a window size ($W_1 = 10$ *sec*) applied to the measurable variable *distance*. Secondary variables can also be obtained by nesting data filters (with corresponding window sizes) one within another. For example, a composite secondary variable, based on the one exemplified above, could be defined as an average over the observed length of the road of the changes in travelled distance in a specified period of time. The applications of data filters and window sizes onto measurable variables are carried out by the Context Processing (see Fig. 2). Context agnostic data filters can also be created that characterise interactions between objects within the system under observation (e.g. relative distances or speeds) or the operating environment (e.g. object density).

Context Processing might also involve data annotation, which provides ground truth for the training of supervised learning techniques and for evaluating the accuracy of both supervised and unsupervised learning. Ground truth labels can be obtained by using some form of statistical thresholds (e.g. 3σ interval for normally distributed data), by manual annotation, or by obtaining the labels directly from a simulation model.

2.3 Context Selection Level

Once the data anomalies have been identified, they are then passed onto the Context Selection level. Classification of anomalies and the predictions of their effects are achieved by applying machine learning in order to build inference models. Additional raw or processed input data may be required at this level.

The Inference Model builder operates in the following way:

- The structure of the model specifies which learning technique $\mathcal{L}$ is going to be used with the chosen secondary variables $S_j, j \subset \overline{1,m}$.
- The specified learning technique checks the need for data conditioning and training datasets.
- The selected secondary variables determine the measurable variables M_i, $i \subset \overline{1,n}$ and the data filters with corresponding window sizes $f_{w_k}^k, k \subset \overline{1,l}$.
- This process minimises the amount of data collected and processed while the inference models (represented as tuples $(\mathcal{L}, \overline{\text{S}}, \overline{\text{p}})$, where $\overline{\text{S}}$ is the vector of secondary variables and $\overline{\text{p}}$ is a vector of parameters for the learning technique $\mathcal{L}$ (for example as an error acceptance rate for ANN) being built and evaluated.

The number of selected secondary variables m directly influences the structure, complexity and usability of the inference model, and thus needs to be optimized in accordance with the size of data samples.

2.4 Context Application Level

The Context Application level supports autonomous operation of the IMS by reducing the importance of human involvement in adjusting the model to changing operating conditions. As was mentioned previously, this task is achieved with the help of genetic algorithms, which autonomously select the optimal parameters on the inference through the effective use of evolutionary processes adopted from nature.

Based on the way the intelligence is obtained, intelligent measurement systems can be categorised either by the function they perform (calibration, error compensation, data validation, anomaly detection, adaptation, decision making, etc.) or by the technique(s) used (statistical, symbolic, ANN-based, fuzzy logic, and the like) [2]. Having chosen the secondary measurands to be used, the processed data together with the inference models build are then passed on to an autonomously chosen supervised or unsupervised learning algorithm. These learning algorithms are used to identify and classify the patterns of interest in the analysed data streams, which reflect dynamic operating environment.

3 Data Filtering for Intelligent Instrumentation

The analysis of surveillance information in general, especially related to situation awareness, is a complex process that, given the amount and heterogeneous nature of data, is prone to data overload. This results in an inability to support real-time processing and analysis of surveillance data. This is especially true when using mobile platforms where datalink and bandwidth issues are significant [12, 13].

3.1 Problem Specification

In order to design and build an intelligent measurement system a testing dataset derived from a MATLAB vehicle simulation model (developed and evaluated by our industrial collaborator) was used in this research. This model is capable of mimicking the behaviour of various types of drivers; typical examples are the normal and "cowboy" drivers. Normal drivers are those that observe road discipline, which regulates that no undertaking is acceptable, and that the vehicles shall move to the left lane whenever possible. The "cowboy" drivers are those that might violate these constraints.

The simulation model provides ground truth 'normal' and 'cowboy' labels; the characteristics of particular drivers within a type are subject to distributions rather than being entirely deterministic – frequencies and instances of exhibited behaviours are context dependent (e.g. traffic density, behaviours of other close vehicles). Therefore, a "cowboy" driver may or may not exhibit the salient features of his behaviour during the observation period.

In total, five driver types are considered – three of these are additional 'abnormal' types (viz. slow, cautious and boy racer). The slow and cautious drivers are similar to the normal driver in that they both follow the lane discipline. Cautious drivers, however, tend to leave a larger gap in front of them, whereas the slow drivers move more slowly, as well as react, brake and accelerate more gently. The "boy racers" are similar to the "cowboy" drivers in that both types do not always follow the lane discipline; what distinguishes them is that the "boy racers" drive faster, braking and accelerating harder, than the "cowboy" drivers.

Such a vehicle simulation model creates a data source rich enough to be used for making intelligent measurement of the driver type. In particular, the presence of several types of anomalous drivers makes it sensible to conduct the inference process in a number of phases: identification, classification and prediction. The identification phase minimises the volume of data and the data processing cost by analysing only a small set of measured data using anomaly identification techniques, such, for instance, as outlier detection. Identified potential anomalies are then passed onto the classification phase, where they are separated out into different types.

As a means of understanding the potential of the IMS techniques developed in the general context, the aims of such evaluation are to use the datasets generated by this vehicle simulation model in order to:

1. Identify anomalous drivers (i.e. all driver types different from the "normal") – the identification phase of IMS operation.
2. Appropriately classify these anomalous drivers into the corresponding types.

3.2 Choosing Secondary Measurands

There are a number of simulation parameters that can be adjusted within the MATLAB traffic simulation model. Some of the simulation parameters directly affect the behaviour of simulated drivers (i.e. *speed ranges*, driver *reaction time*). The other parameters determine the environment – in our case the characteristics of the road (i.e. *lane width* and *number of lanes*), which indirectly influence how each driver behaves.

The task of choosing the right set of variable to measure (i.e. measurands), which provide reliable inference capabilities, is not trivial. Therefore, selection of an appropriate set of secondary (i.e. based on applying filtering to directly measurable data inputs) measurands is a vital step in building an inference measurement system, affecting its accuracy, complexity and generalizability of the inference operation(s).

A conventional methodology of choosing a set of input variables is based on conducting a 'screening' experiment aimed at establishing the significance of each input in terms of inference capabilities of an IMS. This experiment is done by setting the high and low levels for six main variables within the vehicle simulation model: *lane*, *average speed*, *traffic rate*, *road length*, *road width* and *reaction time*. The proportion of normal vs. anomalous drivers was fixed as 80:20. There are sixteen trials in total, i.e. half-factorial screening experiment has been carried out.

Given the difficulty of empirically selecting secondary measurands for building an inference model(s), a more systematic approach has been proposed in the course of this work that is capable of not only choosing the most appropriate input data streams and associated data filters, but also of automatically determining the most effective learning algorithms for adapting the IMS to operate in changing environments. The results of the screening experiments are summarised in Table 1.

The results in Table 1 are obtained using four different statistical data filters (i.e. AVERAGE, VARIANCE, MIN, MAX) on three measurable variables (*distance travelled (along road)*, *lateral movement* or frequency of changing lanes, and *total number* of vehicles). The results shown are obtained using balanced training datasets, which use the equal number of training examples for each driver type (unbalanced training datasets use unequal number of training examples).

The F-test and t-test have been applied to analyse the statistical significance of different features, represented by low p-values (which represent the probability of obtaining the observed differences in accuracy purely by inherent randomness of experiments). Low p-values $p \leq 0.05$ indicate that the differences in model performance are

Table 1. Significance level of secondary measurands

Input data	p-values							
	Neural Networks		SVM		Bayesian networks		K-Means	
	Balan.	Unbal	Balan.	Unbal	Balan.	Unbal	Balan.	Unbal
lane	0.012	0.216	0.006	0.088	0.001	0.132	0.036	0.105
average speed	0.053	0.388	0.076	0.193	0.013	0.219	0.214	0.476
traffic rate	0.321	0.916	0.346	0.168	0.194	0.341	0.912	0.171
road length	0.132	0.439	0.012	0.196	0.002	0.228	0.172	0.313
road width	0.021	0.437	0.035	0.251	0.002	0.322	0.256	0.195
reaction time	0.588	0.847	0.121	0.777	0.588	0.346	0.269	0.176
lane*avg speed	0.097	0.421	0.032	0.354	0.002	0.644	0.205	0.569
lane*traffic rate	0.020	0.415	0.013	0.152	0.002	0.126	0.107	0.486
lane*road length	0.083	0.338	0.100	0.200	0.004	0.241	0.171	0.234
lane*road width	0.022	0.616	0.017	0.796	0.008	0.869	0.150	0.694
lane*reaction time	0.104	0.759	0.283	0.518	0.025	0.829	0.913	0.236
avg speed*road length	0.468	0.488	0.186	0.842	0.003	0.864	0.079	0.927
avg speed*reac.time	0.062	0.573	0.028	0.998	0.003	0.646	0.845	0.821

attributed to systematic factors (significant parameters are highlighted in yellow, low p-values are shown in red):

- *Lane* is the only variable in this experiment that is shown to significantly affect the accuracy measure of all learning techniques.
- Another variable that has a significant effect on the accuracy rate of supervised learning techniques is the *road width*;
- The interactions between *lane* with *road width* and *traffic rate* significantly affect the effectiveness of supervised learning – see the p-values highlighted in red in the table below.

These significance values exhibit a degree of correlation with the design of the vehicle simulation model, where lane discipline is a major characteristic that distinguishes different types of drivers.

4 Building Inferential Capabilities Within IMS

The analysis of surveillance information in general, especially related to situation awareness, is a complex process that, given the amount and heterogeneous nature of data, is prone to data overload. This results in an inability to support real-time processing and analysis of surveillance data. This is especially true when using mobile platforms where datalink and bandwidth issues are significant [12, 13].

In this study, the data to be acquired and processed by an intelligent measurement system comes from various sensors on-board a UAS, such as radar, electro-optical/infrared, GPS and Inertial Navigation Systems (INS). Apart from on-board input data streams, additional contextual input data can also be taken into account. The choice of which contextual input to apply can be automatically tailored using the computational intelligence techniques.

Four learning techniques are currently available within the IMS and are used for building the models – three of which are CI-based: artificial neural network (ANN), support vector machine (SVM), Bayesian network (BN), and K-means classifier. These techniques are implemented in JAVA and the Encog machine learning library [12]. Built in statistical analyses include Difference, Average, Variance, Standard deviation, Summation, Min and Max.

The simulated data set includes: X and Y locations of each vehicle on the road over the surveillance distance of a 6 kilometre road with three lanes, as well as the ground truth labels of driver types.

An inference model can be represented as a tuple $(\mathcal{L}, \overline{\mathrm{S}}, \overline{\mathrm{p}})$, where $\overline{S}$ is the vector of secondary measurands, $\overline{p}$ is a parameter vector of the learning technique $\mathcal{L}$, specifying such values as, for example, an error acceptance rate for artificial neural networks. The process of building an inference model is, in fact, an application of the learning technique $\mathcal{L}$ with its set of parameters $\overline{p}$ to the vector of chosen secondary variables $\overline{S}$ that provides both training and testing data inputs.

Having built the inference models corresponding to all the learning techniques used, this case study explores the influence of salient features of the modelled system on the performance of the IMS. As an example, one salient feature of the traffic simulation model is the ratio of abnormal and normal drivers, which in our experiments varies from 5 % to 25 %. The dependence of inference accuracy on this ratio for each learning technique implemented by the IMS are shown in Figs. 3 and 4.

Therefore, a multi-tiered IMS that uses computational intelligence techniques should be able to enhance situation awareness of a UAV, especially in a real-time environment. Once anomalies are identified from direct measurements, additional data from both easily accessible and detail-rich data sets can be added to improve the system classification and prediction performance.

For balanced training (Fig. 3), the numbers of training samples representing normal drivers is limited by the number of samples representing abnormal drivers of a particular type, which are relatively small in the case-study.

For unbalanced training (Fig. 4), the size of the training dataset representing normal drivers can exceed that of the abnormal ones. All other experimental parameters,

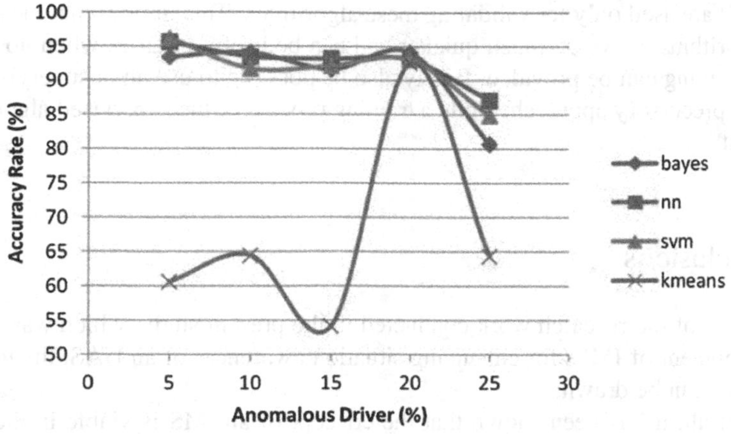

Fig. 3. Multiple data sources fused by an IMS

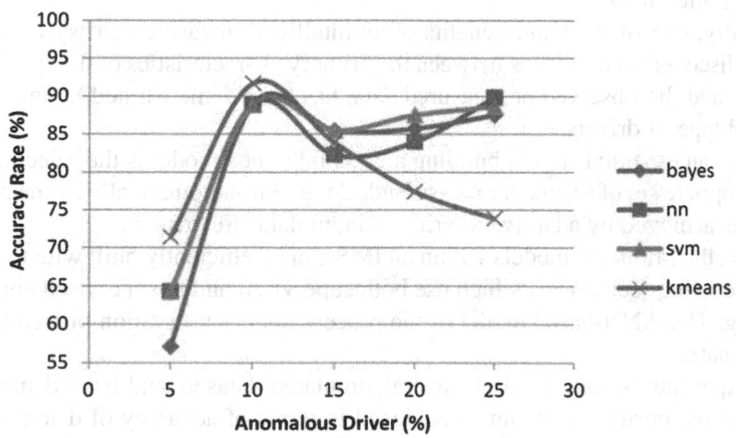

Fig. 4. The effect of different surveillance distances on system accuracy

including the size of testing datasets and the ratio of normal vs. abnormal drivers, are the same.)

As can be observed from the two figures above, the performance of supervised learning techniques is by and large similar, especially for smaller ratios of the numbers of abnormal and normal drivers, denoted as $\lambda_{ab/n}$.

SVM outperforms other supervised learning techniques when $\lambda_{ab/n}$ is small (<10 %), whereas for large values of $\lambda_{ab/n}$ (>20 %), ANNs become the best choice of supervised learning used for building inference models.

Unsupervised learning generally shows worse performance, but can also reach quite high inference accuracy. Despite their inconsistency in inference accuracy, unsupervised algorithms (unlike their supervised counterparts) do not require training. The ground truth labels obtained from MATLAB simulation (i.e. driver types – "normal" and "cowboy") are used only for validating these algorithms. This implies that the unsupervised algorithms converge much quicker and can be useful in cases when no (or very limited) training can be provided. It may also be possible to use an unsupervised algorithm as a precursory approach, while a training process of the supervised algorithms is carried out.

5 Conclusions

On the basis of the research work conducted in the present study, which was aimed at the development of IMSs for enhancing situation awareness of an UAS, the following conclusions can be drawn:

First of all, it has been shown that the concept of an IMS is viable in the chosen context – it has been demonstrated that the implementation of a framework for building such measurement systems is a feasible task, even with limited amounts of data available for making inferences.

Secondly, one of the main benefits of an intelligent measurement system, i.e. the ability to discover relationships between the primary characteristics of the system being monitored and the observed or measured data, has been demonstrated by inferring the behavioral type of drivers.

Thirdly, an essential step in building a good inference model is the selection of the most appropriate set of secondary measurands done semi-automatically by the proposed IMS that is achieved by adaptive filtering of input data streams.

Finally, the inference models within an IMS can be efficiently built with the help of machine learning techniques, which use both supervised and unsupervised approaches to learning. The ANN-based model of the process under observation proved to be the most adequate.

The experiments conducted on several simulated datasets and have demonstrated that reasonable performance can be achieved in terms of accuracy of data processing and its speed. For comprehensive evaluation of the developed IMS aimed at enhancing situational awareness of a UAS, however, it would be desirable to deploy the system on a mobile computing platform and to feed it with real-time sensor data, related to traffic surveillance. Experimenting with such a setup will inevitably bring some programming and engineering issues to the forefront, addressing which would reinforce system usability.

Acknowledgment. The authors would like to acknowledge the contribution of their industrial partner – Selex ES, a subsidiary of Finmeccanica Company – for providing the traffic simulation model, funding and general support for the work on this research project.

References

1. Warne, K., Prasad, G., Rezvani, S., Maguire, L.: Statistical and computational intelligence techniques for inferential model development: a comparative evaluation and a novel proposition for fusion. Elsevier: Eng. Appl. Artif. Intell. **17**, 871–885 (2004). doi:10.1016/j.engappai.2004.08.020
2. Buyan, M.: Intelligent Instrumentation: Principles and Applications. CRC Press, Boca Raton (2010). ISBN 9781420089530
3. Khatibisepehr, S., Huang, B., Khare, S.: Design of inferential sensors in the process industry: a review of Bayesian methods. J. Process Control **23**, 1575–1596 (2013)
4. Sjoberg, J., et al.: Nonlinear Black-Box Modelling in System Identification: A Unified Overview. Automatica. Elsevier, Amsterdam (1995)
5. Khan, Z., Ali, A.B.M.S., Riaz, Z. (eds.): Computational Intelligence for Decision Support in Cyber-Physical Systems. Springer, Heidelberg (2014)
6. Rattadilok, P., Petrovski, A., Petrovski, S.: Anomaly monitoring framework based on intelligent data analysis. In: Yin, H., Tang, K., Gao, Y., Klawonn, F., Lee, M., Weise, T., Li, B., Yao, X. (eds.) IDEAL 2013. LNCS, vol. 8206, pp. 134–141. Springer, Heidelberg (2013)
7. Petrovski, S., Bouchet, F., Petrovski, A.: Data-driven modelling of electromagnetic interferences in motor vehicles using intelligent system approaches. In: Proceedings of IEEE Symposium on Innovations in Intelligent Systems and Applications, INISTA 2013, Albena, Bulgaria, pp. 1–7 (2013)
8. Lun, Y., Cheng, L.: The research on the model of the context-aware for reliable sensing and explanation in cyber-physical system. Procedia Eng. **15**, 1753–1757 (2011)
9. Park, K.J., Zheng, R., Liu, X.: Cyber-physical systems: Milestones and research challenges. Comput. Commun. **36**(1), 1–7 (2012)
10. Kromanis, R., Kripakaran, P.: Support vector regression for anomaly detection from measurement histories. Adv. Eng. Inform. **27**(2013), 486–495 (2013)
11. Valavanis, K.P.: Advances in Unmanned Aerial Vehicles: State of the Art and the Road to Autonomy. Springer, Heidelberg (2007)
12. Encog Library. http://www.heatonresearch.com/encog

Predictive Model for Detecting MQ2 Gases Using Fuzzy Logic on IoT Devices

Catalina Hernández(✉), Sergio Villagrán, and Paulo Gaona

Universidad Distrital Francisco José de Caldas, Bogotá, Colombia
{cmhernandezr,savillagranm}@correo.udistrital.edu.co,
pagaonag@udistrital.edu.co

Abstract. This paper shows the design, implementation and analysis of a fuzzy system for monitoring and alert generation for gas detection in enclosed spaces, which can be very useful either at home or industrial environments. Furthermore, this could be a useful application in the fields of Home Automation which may be developed by integrating devices and technologies of The Internet of Things. Such application consists of the provision of sensors, which constantly receive signals on gases in the environment. Subsequently, the information is analyzed by a fuzzy system that determines when to generate alert notifications, identifying the times when levels are high, either by incendiary or high pollution situations. The prototype consists of connecting an MQ-2 sensor with a Raspberry Pi, which receives the information provided and analyses it by fuzzy logic, thus determining in which cases it is necessary to alarm at sensitive events, generating alert emails and historical data.

Keywords: Internet of Things (IOT) · Gas sensor · Fuzzy logic · MQ-2 · Raspberry Pi

1 Introduction

Through Internet of Things (IoT), multiple applications in intelligent environments can be performed [7]. Disciplines and approaches can range from monitoring human health by ultrasonic sensors and sphere [4] to wireless multi-sensor networks in order to provide surveillance to a means [3]. Besides this, it is possible through radar sensors to either determine the speed of a moving object [5] or establish resource management and efficiency control of an engine [2].

By using various techniques such as fuzzy logic or neural networks, it is possible to perform monitoring, controlling and understanding of system-based sensors for smart environments and decision-making [6]. Thus, in diverse areas of action, the fuzzy logic is a tool that provides the formulation of rules of inference to facilitate processes of modelling and understanding of systems in which there is a high degree of uncertainty and imprecision [10]. Based on control system principles, expert systems or logical control, the fuzzy logic takes the subjectivity or language in terms of a mathematical model [11] and identifies a number of aspects that facilitate the parameterization of systems for their proper functioning. This can be achieved by defining membership

C. Jayne and L. Iliadis (Eds.): EANN 2016, CCIS 629, pp. 176–185, 2016.
DOI: 10.1007/978-3-319-44188-7_13

functions to represent fuzzy sets containing elements found partially or to a limited extent [12]. Accordingly, this is how the fuzzy logic becomes one of the tools that facilitates the identification of control variables which may notoriously improve the modelling of real-time applications.

Therefore, the purpose of this article is to perform a predictive model for gas detection that allows to determine times of risk or danger by generating alerts. The motivation for conducting this study is to exploit the advantages of fuzzy logic in order to identify a model which may facilitate the recognition of such situations, by analyzing and improving existing solutions that may cause failures and which are associated with precision and reliability.

This study addresses a gas flow environment for which it seeks to implement a system of gas detection sensors MQ-2 which, through fuzzy logic, is capable of sending early warnings to nearby users. Based on a similar experience, necessary fundamentals and knowledge are taken in pursuit of increasing the sensitivity of fire detection devices and reducing false alarms [1].

The following article is organized as follows: Sect. 2 provides a contextualization of related studies and background information. Section 3 presents the model proposed or the methodology. Section 4 addresses the predictive model which has been designed and proposed, whereas Sect. 5 offers the case study. Section 6 shows some preliminary results as well as the conclusion, being future work and references analyzed in Sects. 7 and References respectively.

2 Background

Fire is the phenomenon that manifests itself through the combustion of light, flame and heat. The proportion of each of these elements determines its nature [1]. This discovery was a momentous occasion for humankind as it has been an indispensable resource for the man in their many activities over the centuries. However, it is important to recognize that it has not only been essential and valuable but also unstable and dangerous to that which surrounds it according to the application environment.

Sowah worked on a case study in which some types of sensors for fire detection were integrated [1]. Such group comprised a room temperature sensor, another one for density of gases and a third one that measured the intensity of flames. The data captured was then followed by an overall analysis through parameters that allowed observations of the information from the sensors simultaneously. Subsequently, these parameters were defined by fuzzy sets for each input in the system.

As for the intensity of flames, the established sets were as follows: close, not-so-far and far. With regard to temperature, the following sets were distinguished: cold, normal and hot. Concerning gas, the ones associated were: low, medium and high density. Output regions were: fire, potential fire and non-existent fire. The total number of rules provided for decision-making in the fuzzy system was 27.

The results found by Sowah state that it is not necessary for the three sensors to display alarming measures of an incendiary situation due to the established rules [1]. This can be simply given by elevated levels in two of the three sensors, taking into account a margin of error for the temperature sensor of 5°C. For this study, a 90 % of effectiveness in detecting fires was obtained. It is also essential to consider that gas sensors pose problems with nonlinearity, since they are scarcely selective and sometimes insensitive to determine and differentiate various types of gases in situations where there are blends and combinations. However, according to Parthasarathy, their effectiveness may increase if they are integrated and put to operate together, linking and analyzing the information provided as a multisensor system [8].

Fuzzy logic, neural networks and genetic algorithms have shown great capacity and versatility in solving problems of identification and control of large amounts of data in complex and nonlinear systems [15]. Methods based on rules, which are very common in human thinking, allow modelling of variables with approximations and imprecisions. The information obtained is soon after analyzed and processed by fuzzy sets. On the other hand, fuzzy logic allows subjective concepts to be modelled with values that are expressions of colloquial language, which means that anyone can perceive and understand the information by means of using the classical set theory with the help of membership functions as a part of the of fuzzy set theory. [9].

Zadeh argues that fuzzy logic is not fuzzy as such but accurate in its imprecision, leading to approximate reasoning [10]. It is an instrument in which two human capacities converge: on the one hand, making rational decisions in an environment of imprecision and uncertainty, and on the other hand, performing physical and mental tasks without any measurements or calculations. Fuzzy logic is a very useful tool to understand a world that is real and largely diffused.

From this overview, the following section describes the approach used by the working model, which expresses the methodology that was carried out for the design and implementation of the predictive model for gases through the fuzzy logic.

3 Proposed Working Model

To develop the predictive model for gas, it was required to establish the process of interaction between different elements and technologies that were able to define the inputs and outputs in the system, classify them in a context and identify each of the essential stages in the implementation. By examining the overall process, it was necessary to establish a methodology to conduct the case study -which is structured in two stages- as shown in Fig. 1.

The first phase comprises the data extraction, which is executed by the reading of values recorded by the sensor, received by the Raspberry Pi and entered in a txt file. Phase two is based on the analysis of the information captured. Subsequently, a fuzzy system is designed, the input and output sets are defined as well as the different rules which will be applied. Besides this, it is important to establish what output parameters may trigger events called alarms.

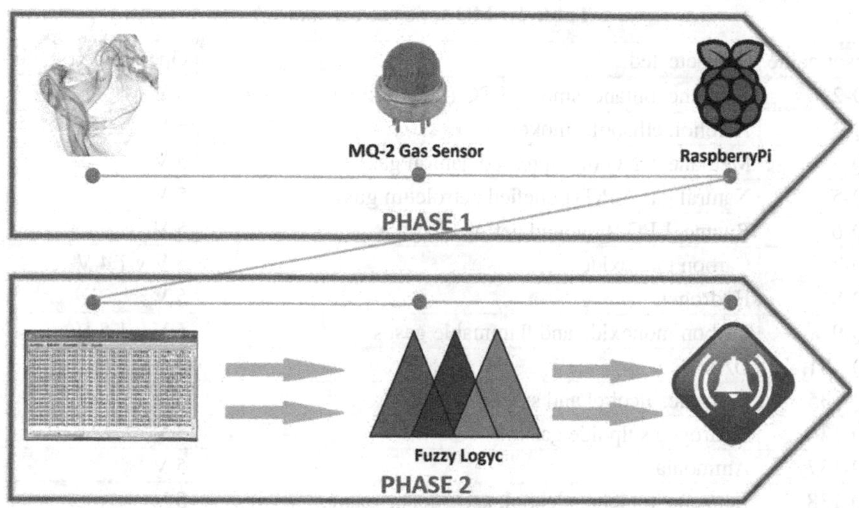

Fig. 1. Methodology.

3.1 Element Selection Criteria

In order to choose the gas sensors which are used, a search is conducted to evaluate efficiency, precision and cost. Sinha examines the MQ series of sensors, which specializes in gas detection and offers various alternatives, each one dedicated to one or more gases under various conditions [14]. This information is presented in Table 1.

From this classification, and according to studies undertaken by Jiru who states that MQ-2 sensors can be used for monitoring the presence of combustible and flammable gases, it is therefore understood that this is due to the basic principle of their operation [13]. The study involves the use of the adsorption property that tin has, which is a semiconductor element that comprises the sensor and attracts oxygen anions when gases are present in the air, thus varying its conductivity and resistance. Taking these characteristics into consideration, the sensor chosen for the case study is the MQ-2.

On the other hand, the advantages offered by the MQ-2 sensor lies in the recognition of several gases that can be found in homes or industrial environments, being effective in situations where leakage is present in such areas and in particular for the development of applications including home automation topics. It is a highly sensitive gas detector, which is able to sense gases such as: butane, propane, methane, among others. In addition to this, it discovers and detects the presence of smoke and combustible gases in concentrations of 300 to 10,000 ppm, which makes it an inexpensive sensor with a fast response time. It should be worth mentioning as well that the MQ-2 includes in the output analogue signals which are required to be transformed digital. This process can be achieved through an ADC0832 converter that returns its equivalent numerical values.

The Raspberry Pi is an excellent alternative for connecting sensors if considering the fact that it is a computer arranged on a single circuit. That said, it is the most economical device among the various existing machines and allows setting and encoding functionality through its input or output GPIO pins.

Table 1. MQ series of sensors.

Sensor name	Gas detected	Operating voltage
MQ-2	Methane, butane, smoke, LPG (liquefied petroleum gas)	5 V
MQ-3	Alcohol, ethanol, smoke	5 V
MQ-4	Methane, CNG (compressed natural gas)	5 V
MQ-5	Natural gas, LPG (liquefied petroleum gas)	5 V
MQ-6	Butane, LPG (liquefied petroleum gas)	5 V
MQ-7	Carbon monoxide	5 V y 1.4 V
MQ-8	Hydrogen	5 V
MQ-9	Carbon monoxide and flammable gases	5 V y 1.5 V
MQ-131	Ozone	6 V
MQ-135	Benzene, alcohol and smoke	5 V
MQ-136	Hydrogen sulphide gas	5 V
MQ-137	Ammonia	5 V
MQ-138	Benzene, toluene, alcohol, acetone, propane, formaldehyde, hydrogen	5 V
MQ-214	Methane, natural gas	6 V
MQ-216	Natural gas, coal gas	6 V
MQ-303A	Alcohol, ethanol, smoke	0.9 V
MQ-306A	Butane, LPG (liquefied petroleum gas)	0.9 V
MQ-307A	Carbon monoxide	0.2 V y 0.9 V
MQ-309A	Carbon monoxide and flammable gases	0.2 V y 0.9 V

If the reasons for choosing the various physical devices are taken into consideration, it is therefore necessary to continue with the approach of the predictive model designed by fuzzy logic, which is described in the next section.

4 Proposed Predictive Model

Fuzzy logic uses the properties of fuzzy sets so as to establish the relationship of an element using the membership functions that define it. This provides greater possibilities than considering specific and limited values. Fuzzy logic allows intermediate values to be defined in a range, thus generating other possible responses which are situated at the extreme points. Given these advantages offered by the technique, it is considered appropriate for modelling a gas system that receives and generates entries associated with many values, so a fuzzy logic model could classify, define and group them into sets as well as defining rules for their analysis and results.

The type of model chosen is the Mamdani, one of the best-known prototypes and the first to be subjected to a test. Its main feature is the handling of fuzzy IF-THEN rules, in which the input and output variables are related. As a result, this determines the interaction of the entries in the established fuzzy sets and allows analyzing their interaction and equivalence on an output set; finding output relations in the form of: If X1 is C1 and C2 X2, then Y is C3; being X1, X2 system inputs and output Y1 and C1, C2,

C3 fuzzy sets belonging to the model designed. The operations performed are the ones associated with set theory, such as union, intersection, Cartesian product, etc.

For parameterization of input variables, two fuzzy sets corresponding to two gases are designed. These define the two extremes of the interval, as shown in Fig. 2.

Fig. 2. Input fuzzy sets.

As it can be seen, membership functions are of type S and Z respectively. Each of them represents a gas which is specified in the section of the case study. Figure 3 presents fuzzy sets in which possible outputs of the modelling system are parameterized.

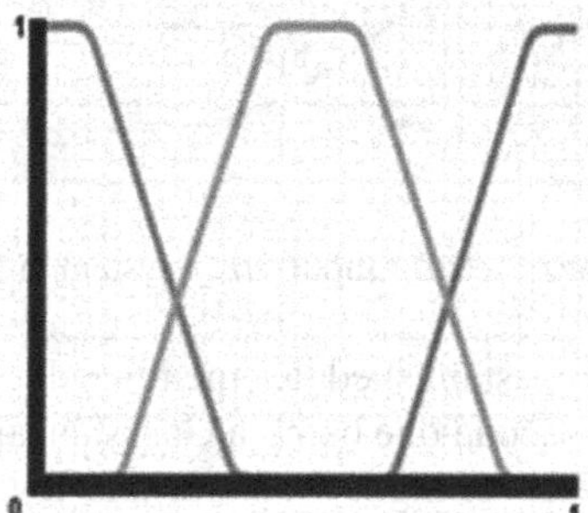

Fig. 3. Output fuzzy sets. (Color figure online)

Output sets are summarized in (blue), medium (orange) and high (purple) respectively, indicating the possibility of fire or the presence of noxious gas. The design of rules is subjected to inlet gas in the case of application, so their combination and interaction give rise to the possible outputs of the fuzzy system.

Once the designed model is known, the application of the case study takes place. In this, the inputs and rules are specified, as well as the context in which the practice is performed.

5 Case Study

Three inputs were defined for the case study. They correspond to three connected MQ-2 sensors, which performed the parallel reading of surrounding gases dispersed in the environment. A script in python was responsible for obtaining and processing the signals given by each sensor and converted by the integrated circuit, compiling in a txt the list of data obtained from the work accomplished. As previously presented, two fuzzy sets were set for each input, associating them to hydrogen and butane respectively. The graphical part of the experience with regard to the fuzzy logic was performed using the Matlab software.

Figure 4 shows the graph corresponding to the fuzzy sets whose domain is between 0 and 254 (these being the possible values shown by the MQ2 sensor, given by the analogue-to-digital converter). In addition to this, there are two-phase functions, each representing the possible gas the sensor may capture: hydrogen or butane. If the sensor produces an increase in the value, it is more likely for this gas captured from the air to be butane and, at the same time, it is therefore less likely for it to be hydrogen.

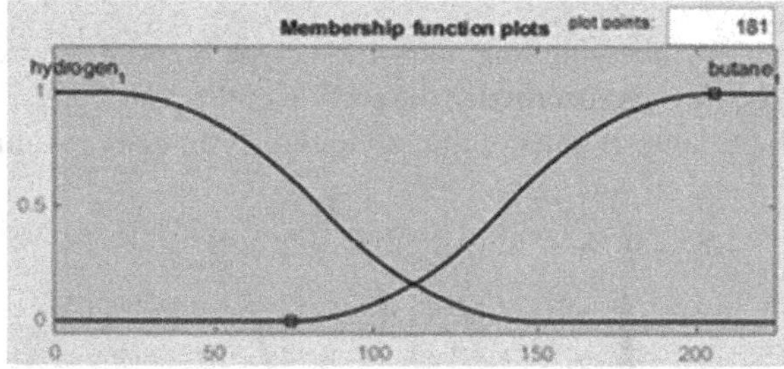

Fig. 4. Fuzzy sets for input fuzzy system in Matlab.

Subsequently, outputs are established as mentioned previously, being of three different sets: one type S, the second one being a Gaussian, and the last Z type, defining three possibilities of gas presence: high, medium, and low. Figure 5 shows the sets arranged in Matlab.

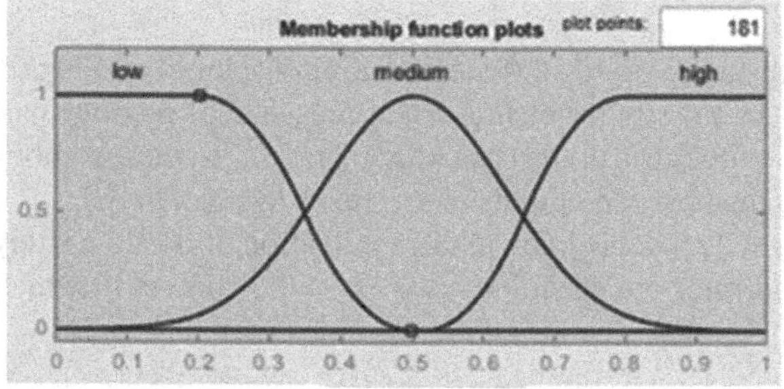

Fig. 5. Outputs fuzzy system.

Figure 6 shows the rules formulated in the designed fuzzy system, in which the IF-THEN relationships between the input and output sets can be seen. A total of eight rules were generated, corresponding to the combination of three input variables, each with two possible values.

1. If (Sensor_1 is hydrogen_1) and (Sensor_2 is hydrogen_2) and (Sensor_3 is hiydrogen_3) then (Level_of_Smoke is low) (1)
2. If (Sensor_1 is butane_1) and (Sensor_2 is hydrogen_2) and (Sensor_3 is hiydrogen_3) then (Level_of_Smoke is medium) (1)
3. If (Sensor_1 is butane_1) and (Sensor_2 is butane_2) and (Sensor_3 is hiydrogen_3) then (Level_of_Smoke is high) (1)
4. If (Sensor_1 is butane_1) and (Sensor_2 is butane_2) and (Sensor_3 is butane_3) then (Level_of_Smoke is high) (1)
5. If (Sensor_1 is hydrogen_1) and (Sensor_2 is hydrogen_2) and (Sensor_3 is butane_3) then (Level_of_Smoke is medium) (1)
6. If (Sensor_1 is hydrogen_1) and (Sensor_2 is butane_2) and (Sensor_3 is hiydrogen_3) then (Level_of_Smoke is medium) (1)
7. If (Sensor_1 is butane_1) and (Sensor_2 is hydrogen_2) and (Sensor_3 is butane_3) then (Level_of_Smoke is high) (1)
8. If (Sensor_1 is hydrogen_1) and (Sensor_2 is butane_2) and (Sensor_3 is butane_3) then (Level_of_Smoke is high) (1)

Fig. 6. Fuzzy rule-based system.

Based on the case study and the various tests, the analysis of results takes place, which will be explained in the next section.

6 Preliminary Results

The fuzzy logic system shows a stability-oriented result, being the output a combination of the eight rules that had been previously established. Such rules are proportional to the input values, which means that the higher the data recorded by the sensors, the more likely the output is to trigger a potential fire and/or high level of gas alert. Figure 7 shows the output given by the fuzzy controller in the case study, which can take values from 0 to 1, indicating lower or higher amount of gas in the environment.

Fig. 7. Preliminary results of application case study.

The figure above shows a signal that rapidly increases in magnitude, demonstrating high levels of gas for a period of time that remains constant and gradually descend to be close to zero. The output given by the controller shows that the fuzzy system performs a proper analysis; measurements made mostly expressed high concentrations of gas, so the fuzzy system responded promptly to the high values entered.

7 Conclusion

As stated in the preliminary results, it may be a good idea to experience with two membership functions in order to discriminate the inputs. However, this may be improved through the definition of larger sets, as establishing new settings may lead to performing a more specific and therefore accurate analysis. In addition to this, the number of rules may increase and as a result, the fuzzy system could improve its analysis as its knowledge base and fuzzy reasoning improve. As future work, it is necessary to further testing of the reliability of the sensors to determine the effectiveness of gas detection levels by comparing the actual and detected amounts.

Furthermore, it is essential to take into consideration that a determining factor for these applications is the spatial arrangement of the sensors as they may represent inconsistencies and erroneous analysis of the place. If sensors are placed too close, they may take similar measures of a gas concentration which may not represent a risk as such. Conversely, if they are widely separated, they could not take sufficient quantities. Moreover, it is also possible to state that the effectiveness of these warning systems also depends on their physical configuration, so that fidelity and efficiency of measures may be improved by taking advantage of the features or benefits provided by the space. In order to do this, different types of methods such as algorithms, ant colony, particle swarm optimization, among others, can be used. They could help to analyze specific features of the environment and identify each sensor as a node in an interconnected network, or as a graph in its simplest essence.

Fuzzy logic systems are an excellent alternative for modelling applications in which unique and exact values are not adequate to establish analysis in a large framework of possibilities. That said, and being an expert system, it is therefore recommended to examine this kind of information. The situation described is one of many examples in which different technologies converge to provide a possible solution. The Internet of Things offers tools for multiple needs in different environments, so it has become a network in which various disciplines and sciences interact so that they can design, integrate and propose even better and more effective solutions every day.

References

1. Sowah, R., Ofoli, A., Krakani, S., Fiawoo, S.: Hardware Module Design of a Real-Time Multi-sensor Fire Detection and Notification System Using Fuzzy Logic (2014)
2. Mathew, T., Sam, C.: Closed Loop Control of BLDC Motor Using a Fuzzy Logic Controller and Single Current Sensor (2013)
3. Manjunatha, P., Verma, A.K., Srividya, A.: Multi-sensor Data Fusion in Cluster Based Wireless Sensor Networks Using Fuzzy Logic Method (2008)
4. Hata, Y., Kobashi, S., Taniguchi, K., Nakajima, H.: Human Health Monitoring System of Systems with Fuzzy Logic by Sensor Network (2009)

5. Guo, L., Galarza, L., Fan, J., Choi, C.: High Accuracy Three-Dimensional Radar Sensor Design Based on Fuzzy Logic Control Approach (2013)
6. Liu, Y., Chen, M., Wang, M.L., Dokmeci, M.: Sensing characteristics of RNA oligomer coated SWNT gas sensors (2011)
7. Russell, L., Goubran, R., Kwamena, F.: Personalization Using Sensors for Preliminary Human Detection in an IoT Environment (2015)
8. Parthasarathy, R., Kalaichelvi, V., Sundaram, S.: A Novel Fuzzy Logic Model for Multiple Gas Sensor Array (2015)
9. Jonda, S., Fleischer, M., Meixner, H.: Temperature control of semiconductor metal-oxide gas sensors by means of fuzzy logic (1995)
10. Zadeh, L.A.: Is there a need for fuzzy logic. J. Inf. Sci. **178**, 2751–2779 (2008)
11. Ramot, D., Friedman, M., Langholz, G., Kandel, A.: Complex Fuzzy Logic **11**, 450–461 (2003)
12. Yu, Y.: Rule based fuzzy logic inferencing (1994)
13. Jiru, P.: Design of intelligent monitoring system (2013)
14. Sinha, N., Pujitha, K., Alex, J.: Xively Based Sensing and Monitoring System for IoT (2015)
15. Rajati, M.R., Khaloozadeh, H., Pedrycz, W.: Fuzzy logic and self-referential reasoning: a comparative study with some new concepts **41**, 331–357 (2014)

A Multi-commodity Network Flow Model for Cloud Service Environments

Ioannis M. Stephanakis[1](✉), Syed Noor-Ul-Hassan Shirazi[2], Antonios Gouglidis[2], and David Hutchison[2]

[1] Hellenic Telecommunication Organization S.A. (OTE), 99 Kifissias Avenue, 151 24 Athens, Greece
stephan@ote.gr

[2] InfoLab21, School of Computing and Communications, Lancaster University, Bailrigg LA1 4WA, UK
{n.shirazi,a.gouglidis,d.hutchison}@lancaster.ac.uk

Abstract. Next-generation systems, such as the big data cloud, have to cope with several challenges, e.g., move of excessive amount of data at a dictated speed, and thus, require the investigation of concepts additional to security in order to ensure their orderly function. Resilience is such a concept, which when ensured by systems or networks they are able to provide and maintain an acceptable level of service in the face of various faults and challenges. In this paper, we investigate the multi-commodity flows problem, as a task within our $D^2R^2 + DR$ resilience strategy, and in the context of big data cloud systems. Specifically, proximal gradient optimization is proposed for determining optimal computation flows since such algorithms are highly attractive for solving big data problems. Many such problems can be formulated as the global consensus optimization ones, and can be solved in a distributed manner by the alternating direction method of multipliers (ADMM) algorithm. Numerical evaluation of the proposed model is carried out in the context of specific deployments of a situation-aware information infrastructure.

Keywords: Resilience · Big data cloud · Multi-commodity flow networks · Distributed algorithms · Consensus optimization · Alternating direction method of multipliers (ADMM)

1 Introduction

Cloud computing delivers computing services from large, highly virtualized network environments to many independent users, using shared applications and pooled resources. One may distinguish amongst Software-as-a-Service (SaaS) where software is offered on-demand through the internet by the provider and it is parametrized remotely (e.g., on-line word processors, spreadsheets, Google Docs and others); Platform-as-a-Service (PaaS) where customers are allowed to create new applications that are remotely managed and parametrized, and offer tools for development and computer interface restructuring (e.g., Force,

C. Jayne and L. Iliadis (Eds.): EANN 2016, CCIS 629, pp. 186–197, 2016.
DOI: 10.1007/978-3-319-44188-7_14

Google App Engine and Microsoft Azure), and Infrastructure-as-a-Service (IaaS) where virtual machines, computers and operating systems may be controlled and parametrized remotely (e.g., Amazon EC2 and S3, Terremark Enterprise Cloud, Windows Live Skydrive, Rackspace Cloud, GoGrid, Joyent, AppNexus, etc.). The aforementioned service models of the cloud can be offered in three difference deployment models, i.e., public, private and hybrid. In public cloud systems, everyone may register and use the services. Private ones are accessible through a private network. Lastly, hybrid clouds refer to a combination of the previous two and usually used in the case where sensitive data is required to be kept in the private network and non-core applications are deployed in the public. The key functionality of the private deployment model is the ability to use and release resources from public clouds as and when required. This is used to handle sudden demand surges ('flash crowds') and is known as 'cloud-bursting'.

Cloud computing is an on-demand service whose size depends upon users needs and should feature scale flexibility. It is built upon such network elements as switches supporting novel communication protocols, specific servers based on Virtual Machine (VM) technology and dynamic resource management as well as Network-Attached-Storage (NAS). Specific software platforms may be used for service orchestration in cloud environments (e.g., OpenStack). Several next-generation implementations require widespread connectivity, security and a successful combination with machine-to-machine M2M applications[1] and cloud computing. Integration platforms are important facilitating the convergence of IoT[2], cloud computing, analytic, and big data. They support links among cloud applications and they tie together the distributed devices at one end of a network pipe with enterprise applications and analytic at the other end. Integration platforms shorten the development cycle for connecting devices to the cloud or enterprise systems. Other applications of cloud computing may be seen in the area of critical infrastructures. An example of that is the use of cloud computing services to perform analysis of the data conveyed between the various components of a Supervisory Control and Data Acquisition (SCADA) network [14].

Cloud systems and services can be applicable in a wide range of applications, as described above. Therefore, this further motivates us towards the investigation of concepts additional to security in order to ensure their orderly function. Resilience is such a concept that can ensure that a network or system can provide and maintain an acceptable level of service in the face of various faults and

[1] A typical M2M architecture includes an application domain, a network domain, an M2M device domain and one or more direct connections or gateways from the M2M area network to the network domain. M2M device area networks can use a variety of communication technologies (RFID, ZigBee, M-BUS, IEEE 802.15, 6LoWPAN), thus a gateway layer becomes important. The solutions for communication between the gateway and M2M applications include LTE, WiMAX, xDSL, and WLAN. In the application domain, clients will often include dashboards for data virtualization, status monitoring, reconfiguration and other functions.

[2] Among the consortia working on standards for IoT are AllSeen Alliance, HyperCat Consortium, and Industrial Internet Consortium. There are also initiatives such as the Eclipse M2M Industry Working Group and ITU-T Focus Group M2M initiative.

challenges to normal operations [20]. In order to accomplish the previous requirement we have proposed a resilience strategy entitled $D^2R^2 + DR$, i.e., Defend, Detect, Remediate, Recover, and Diagnose and Refine [20]. The first four consist processes of an internal loop process and the latter two of an off-line outer loop. In more detail, it is: Defend against challenges and threats to normal operation; Detect when an adverse event or condition has occurred; Remediate the effects of the adverse event or condition; Recover to original and normal operations; Diagnose the fault that was the root cause; and Refine behaviour for the future based on past $D^2R^2 + DR$ cycles. Based on our resiliency strategy, we further developed an architectural framework for resilience, which is able to operate in the context of cloud systems and used for diagnosing anomalies [18].

In this paper, we further investigate the multi-commodity network flow in the context of our resilience strategy and towards ensuring network resilience in cloud services. Multi-commodity network flow models provide the tools for optimal network design and dimensioning in telecommunications given a list of traffic nodes (sources or sinks) [17]. The basic mathematical models used to formulate and solve optimal network design problems make use of graph-theoretic and/or linear programming-based models (see e.g., [2,4,13]). The set of all possible topologies for the network to be constructed will typically be described by means of a given (undirected) graph $G = [V, E]$ where:

- the node set V represents the various traffic sources/sinks to be interconnected;
- the edge set E corresponds to the various pairs of nodes, which may be physically connected by installing transmission links.

A single-commodity flow between a source and a sink is a **M** vector, $\varphi = (\varphi_1, \varphi_2, \ldots, \varphi_M)$ such that $|\varphi_u|$ represents the amount of transmission resource used on edge $e = (i, j)$. The aforementioned commodity model is adopted to accommodate cloud computing and in-network processing [9,11]. A walk-based as well as an edge based formulation is adopted. Applying such models in the context of distributed and parallel computing in the cloud is challenging. The sheer volume of data in next generation implementations requires advanced analytic capable of exploiting the big data and the computing power of the cloud. Scaling up to 50 and 200-billion connected devices requires innovative security solutions. Hybrid architectures focus on security at endpoints and when data is in transit: device security, cloud security, and network security. Virtualization must be done with resilient virtual machines (VM), resilient single-tenant and multi-tenant servers, and resilient software defined networks (SDN).

The structure of the remainder of this paper is: Sect. 2 elaborate on distributed proximal algorithms and on ADMM in cloud networks. A commodity network model for maximizing information processing in the cloud is presented in Sect. 3. An evaluation of the model is provide via simulations in Sect. 4. Conclusions are presented in Sect. 5.

2 Parallel Algorithms for Optimizing Big Data

2.1 Big Data Analytics and Distributed Proximal Algorithms

The information explosion propelled by the advent of online social media, Internet and global-scale communications has rendered big data analytics as well as data-driven statistical learning increasingly important [19]. Dealing with large-scale data sets poses formidable challenges. The sheer volume and dimensionality of data make it impossible to run analytics and traditional inference methods using standalone processors [3,16]. Decentralized learning with parallelized multi cores is preferred [5,12], while the data themselves are stored in the cloud or distributed file systems as in MapReduce/Hadoop [10]. Distributed signal processing can be used within the context of sensor networks as well (see for example [1]). Optimizing large scale data may be expressed as:

$$F^{*} \overset{def}{=} \min_{x}\{F(x) := f(x) + g(x) : x \in R^{p}\} \tag{1}$$

where f and g are convex functions. Efficient numerical methods to obtain x in the context of large scale problems arising in big data applications are, namely, first order methods, randomization as well as parallel and distributed computing [8].

- First-order methods: First-order methods obtain low- or medium-accuracy numerical solutions by using only first-order oracle information from the objective, such as gradient estimates. They handle important non smooth variants of Eq. 1 by making use of the proximal mapping principle. They feature nearly dimension-independent convergence rates, they are theoretically robust to the approximations of their oracles, and they typically rely on computational primitives that are ideal for distributed and parallel computation.
- Randomization: Randomization approaches stand out among many other approximation techniques since they enhance the scalability of first order methods. We can control their expected behaviour. Key ideas include random partial updates of optimization variables, replacing the deterministic gradient as well as proximal calculations with cheap statistical estimators, and speeding up basic linear algebra routines via randomization.
- Parallel and distributed computation: First-order methods naturally provide a flexible framework to distributive optimization tasks and perform computations in parallel. Surprisingly, one can further augment these methods with approximations to enormously scalable asynchronous algorithms with decentralized communications.

The three aforementioned classes of algorithms complement each other and offer surprising scalability benefits for big data optimization. For closed proper convex functions, one may define the proximal operator of the scaled function λf, where $\lambda > 0$, as:

$$prox_{\lambda f}(\mathbf{u}) \overset{def}{=} \text{argmin}\left(f(\mathbf{x}) - \frac{1}{2\lambda}\| \mathbf{x} - \mathbf{u} \|_2^2\right) \tag{2}$$

This is also called the proximal operator of f with respect to λ. The parameter λ controls the extent to which the proximal operator maps towards the minimum of f, with larger values of λ associated with mapped points near the minimum, and smaller values giving smaller movement towards the minimum. A proximal algorithm is an algorithm for solving a convex optimization problem that uses proximal operators of the objective terms. Proximal algorithms have been used for multi-commodity network flow optimization [15].

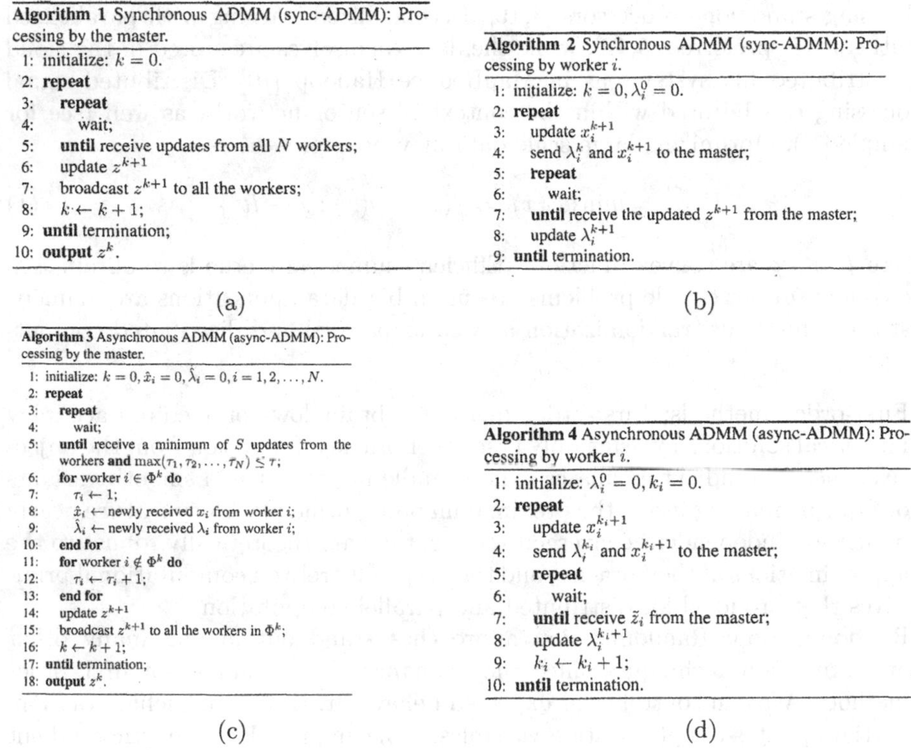

Algorithm 1 Synchronous ADMM (sync-ADMM): Processing by the master.

initialize: $k = 0$.
repeat
  repeat
    wait;
  until receive updates from all N workers;
  update z^{k+1}
  broadcast z^{k+1} to all the workers;
  $k \leftarrow k + 1$;
until termination;
output z^k.

(a)

Algorithm 2 Synchronous ADMM (sync-ADMM): Processing by worker i.

initialize: $k = 0, \lambda_i^0 = 0$.
repeat
  update x_i^{k+1}
  send λ_i^k and x_i^{k+1} to the master;
  repeat
    wait;
  until receive the updated z^{k+1} from the master;
  update λ_i^{k+1}
until termination.

(b)

Algorithm 3 Asynchronous ADMM (async-ADMM): Processing by the master.

initialize: $k = 0, \hat{x}_i = 0, \hat{\lambda}_i = 0, i = 1, 2, \ldots, N$.
repeat
  repeat
    wait;
  until receive a minimum of S updates from the workers **and** $\max(\tau_1, \tau_2, \ldots, \tau_N) \lesssim \tau$;
  for worker $i \in \Phi^k$ **do**
    $\tau_i \leftarrow 1$;
    $\hat{x}_i \leftarrow$ newly received x_i from worker i;
    $\hat{\lambda}_i \leftarrow$ newly received λ_i from worker i;
  end for
  for worker $i \notin \Phi^k$ **do**
    $\tau_i \leftarrow \tau_i + 1$;
  end for
  update z^{k+1}
  broadcast z^{k+1} to all the workers in Φ^k;
  $k \leftarrow k + 1$;
until termination;
output z^k.

(c)

Algorithm 4 Asynchronous ADMM (async-ADMM): Processing by worker i.

initialize: $\lambda_i^0 = 0, k_i = 0$.
repeat
  update $x_i^{k_i+1}$
  send $\lambda_i^{k_i}$ and $x_i^{k_i+1}$ to the master;
  repeat
    wait;
  until receive $\tilde{z}_i$ from the master;
  update $\lambda_i^{k_i+1}$
  $k_i \leftarrow k_i + 1$;
until termination.

(d)

Fig. 1. Synchronous and asynchronous processing from master and workers for ADMM

2.2 Synchronous and Asynchronous Consensus ADMM in Cloud Information Networks

Many machine learning problems can be formulated as the global consensus optimization problem, which can then be solved in a distributed manner by the alternating direction method of multipliers (ADMM) algorithm. The global variance consensus optimization problem [5,6] in the context of minimization $f(x)$ reads:

$$\min_{x_1, \ldots, x_N, z} \quad f(x) + g(z) = \sum_{i=1}^{N} f_i(x_i) : x_i = z, i = 1, 2, \ldots, N \tag{3}$$

where z is the so-called consensus variable, and x_i is node i local copy of the parameter to be learned. The aforementioned problem may be reformulated as augmented *Lagrangian* optimization:

$$L(\{x_i\}, z) = g(z) + \sum_{i=1}^{N} f_i(x_i) + \langle \lambda_i, x_i - z \rangle + \frac{\beta}{2} \| x_i - z \|^2 \quad (4)$$

where λ_i are the Lagrangian multipliers, $\beta > 0$ is the penalty parameter, and $\langle , \rangle$ denotes the inner product. At the k-th iteration, the values of x_i and z denoted (x_i^k and z^k) are updated by minimizing L with respect to x_i and z. Unlike the methods of multipliers, these are minimized in an alternating manner, which allows the problem to be more easily decomposed:

$$x_i^{k+1} = \underset{x}{\operatorname{argmin}}\, f_i(x) + \langle \lambda_i^k, x \rangle + \frac{\beta}{2} \| x - z^k \|^2 = prox_{f_i/\beta}(z_i^k - \lambda_i^k) \quad (5)$$

$$z^{k+1} = \underset{z}{\operatorname{argmin}}\, g(z) + \sum_{i=1}^{N} -\langle \lambda_i^k, z \rangle + \frac{\beta}{2} \| x_i^{k+1} - z^k \|^2 = prox_{g/\beta}(x^{k+1} + \lambda^k) \quad (6)$$

$$\lambda_i^{k+1} = \lambda_i^{k+1} + \beta\big(x_i^{k+1} - z^{k+1}\big) \quad (7)$$

The updates can be easily implemented in a distributed system with one master and N workers [21]. Each worker i is responsible for updating (x_i, λ_i) using the above equations. The updated x_i^{k+1} are then sent to the master, which is responsible for updating the consensus variable z as well as distributing its updated value back to the workers. Updating may be performed in a synchronous or an asynchronous manner (see algorithms in Fig. 1).

3 A Commodity Network Model for Maximizing Information Processing in Cloud Implementations

One may adopt the proximal-point method in order to optimize network flows in a cloud environment in an iterative fashion:

$$x^{k+1} = \underset{X \in S}{\operatorname{argmin}} \left(f(x) + \frac{1}{2\lambda} (x - x^k)^T (x - x^k) \right) = prox_{\lambda f}(x^k) \quad (8)$$

where S is a convex set, $f(x)$ is a convex function and $1/2\lambda$ is a constant. One may assume N clusters featuring a distributed processing power given by $\pi = (\pi_1, \pi_2, \ldots, \pi_N)^T$. Vector x consists of network flows $x = [x_{s1}, x_{s2}, x_{s3}, \ldots]$ where subscript s runs over all network edges $s \in \{a, b, c \ldots\}$. The constraint optimization problem for one proximal-point iteration reads:

$$x = \min_{X \in S} \left(\left[\frac{c_1}{\lambda_1}, \frac{c_2}{\lambda_2}, \ldots, \frac{c_n}{\lambda_N} \right] \mathbf{CF}_{clusters} x + \beta (x - z)^T (x - z) + {\theta_s}^T (\mathbf{CF}_s x - s_{in}) \right.$$
$$\left. + {\lambda_e}^T (x - \mathbf{BW}) + {\lambda_p}^T (\mathbf{CF}_c x - P) \right) \quad (9)$$

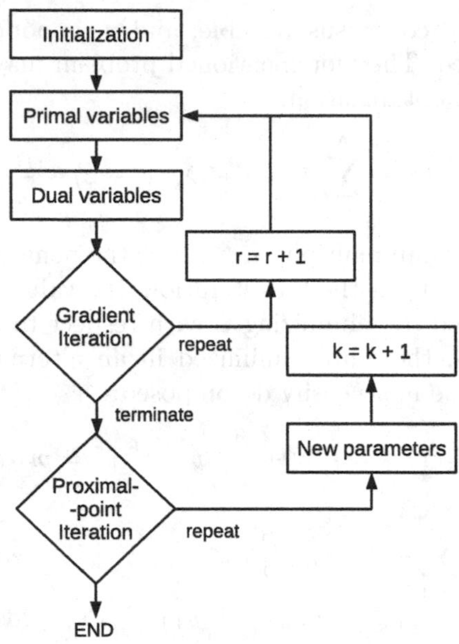

Fig. 2. Block diagram of the dual gradient algorithm.

where z is the approximation at the previous iteration step and $\mathbf{CF}_{clusters}$ is the incidence matrix for cloud clusters such that $\mathbf{CF}_{clusters}x$, gives the incident flows at all cloud processing nodes. Flow $(\mathbf{CF}_{clusters}x)_j$ is processed by processor π_j within time $T_j \sim (1/\pi_j)(\mathbf{CF}_{clusters}x)_j$. Cluster processing utilization is analogous to processing time. It is assumed that processing cost per site is proportional to **CP** utilization, i.e. the term $[c_1/\lambda_1, c_2/\lambda_2, \ldots, c_n/\lambda_N]\mathbf{CF}_{clusters}x$ is equal to the total processing cost to be minimized according to Eq. 9. Dual variable θ_s accounts for flow incidence conditions at sensor nodes, i.e., $\mathbf{CF}_{sensors}x = s_{in}$ whereas dual variables λ_e and λ_p account for upper edge bandwidth limits $x_{edge_l} \leq \mathbf{BW}_{edge_l}, l = 1, \ldots, L$ and cluster processing capabilities. According to Slaters conditions (see for example [7]) strong duality holds for the optimization problem (i.e. the optimal values of the dual and the primal problem are equal. We carry out successive optimization over primal and dual and variables according to the block diagram in Fig. 2. Primal variables as well as dual variables are estimated iteratively several times during the execution cycle of an iteration step. Current estimation of x^k is used as the proximal point z for the next estimation x^{k+1} (see Eq. 8). As an alternative approach, one may use distributed processing and synchronous ADMM and solve Eq. 9 using Eqs. 5, 6 and 7 according to algorithms (a) and (b) in Fig. 1. Dual variables are estimated a number of times during each iteration step after the estimation of the primal variables z and x_i. Global parameter z is updated by the master so that flow incidence conditions at sensor nodes are satisfied, $\mathbf{CF}_{sensors}z = s_{in}$ and, finally, $z^k = \sum_{i=1}^{N} x_i^k$.

4 Numerical Simulations

Computing time T_j is assumed to be normalized to CP utilization at cluster node j. Total computing cost is assumed to be analogous to total processing time, i.e., $c_1T_1 + c_2T_2 + c_3T_3 + c_4T_4 \propto$ total processing cost. Two distinct cases of assigning processor costs are investigated. Case 1 assumes that $c_1 = c_2 = c_3 = c_4 = 100\ units$ whereas Case 2 assumes that $c_1 = 80\ units, c_2 = 60\ units, c_3 = 100\ units$ and $c_4 = 90\ units$. Convergence behaviour of the proposed proximal algorithm for z equal to x_k is depicted in Fig. 4. Figure 4a depicts total cost, Fig. 4b depicts flow equilibrium conditions at sensor nodes during the execution of the algorithm for Case 1 and Fig. 4c depicts processor utilization per cluster for Case 1. Similar results are illustrated for Case 2 in Figs. 4d, e and f. Similar results of solving Eq. 9 for Case 1 and Case 2 using distributed processing and synchronized ADMM are depicted in Fig. 5. Figures 5d and h present the difference of $z^k - \sum_{i=1}^{4} x_i^k$ over 1,000 iterations. Both approaches give similar total cost values for Case 1 and Case 2.

Table 1. Terminal devices nominal source flows (**Mbps**)

s_1=1	s_2=1	s_3=2	s_4=3	s_5=0.5	s_6=0.5	s_7=4	s_8=3
s_9=2	$s_1$0=2	$s_1$1=2	$s_1$2=1	$s_1$3=1.5	$s_1$4=5	$s_1$5=2	$s_1$6=1
$s_1$7=2	$s_1$8=1	$s_1$9=1	$s_2$0=2	$s_2$1=4	$s_2$2=2.5	$s_2$3=3.5	$s_2$4=2
$s_2$5=2.5	$s_2$6=1	$s_2$7=1	$s_2$8=2	$s_2$9=3.5	$s_3$0=5	$s_3$1=1	$s_3$2=1

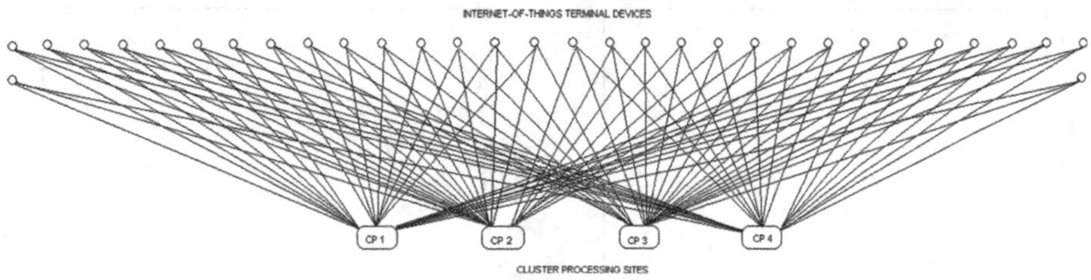

Fig. 3. Cloud interconnections

Numerical simulations are carried out for artificial data for four (4) processing sites (clusters) and a total of thirty-two (32) terminal devices (sensors) connected to cloud nodes. Each terminal device (sensor) is connected to a main processing cluster and two backup processing clusters. Link capacities vary from 0.5 Mbps to 5 Mbps. Each terminal device produces original source flows featuring values ranging from 0.5 Mbps to 5 Mbps (see Table 1). It is assumed that each flow may be diverted totally or partly from one processing site to the other. Each of the four (4) computing sites is capable of processing a total sum of flows, i.e. $\pi_1 = 20$ Mbps for **CP1**,$\pi_2 = 10$ Mbps for **CP2**,$\pi_3 = 30$ Mbps for **CP3** and $\pi_4 = 25$ Mbps for **CP4**). Cloud interconnections are illustrated in Fig. 3.

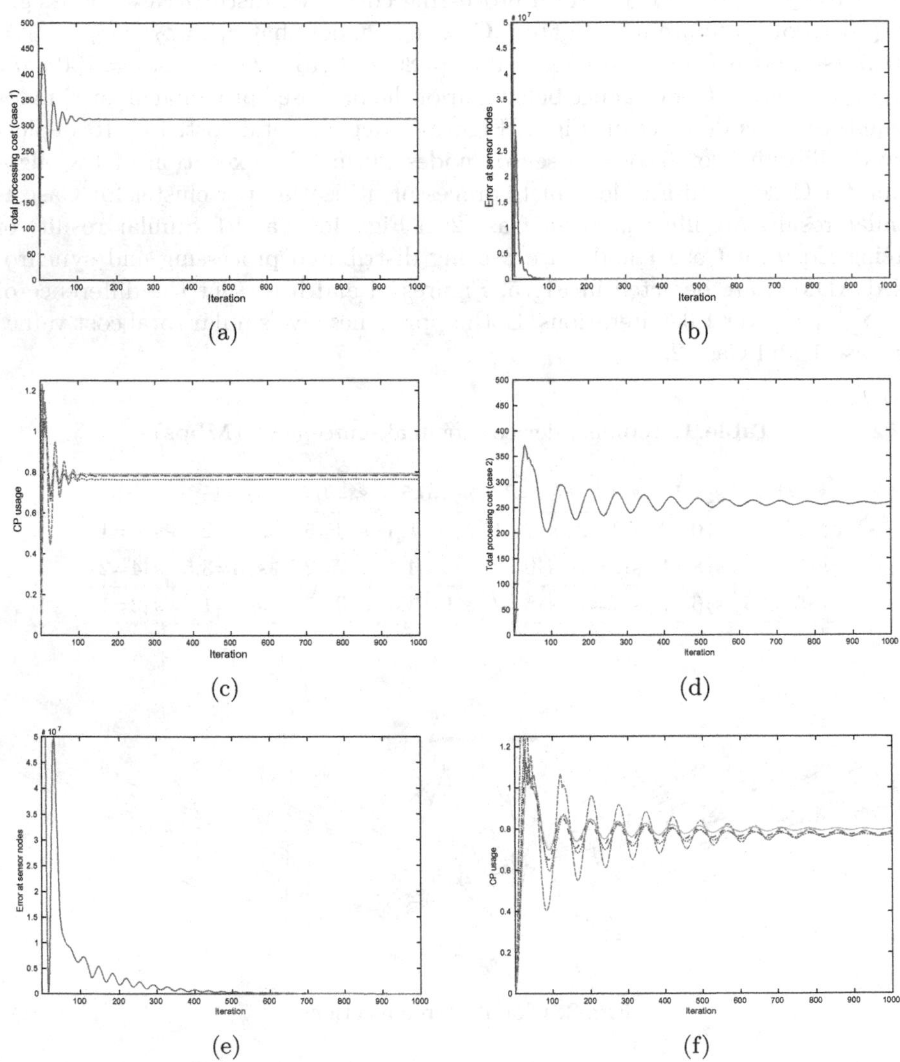

Fig. 4. Iterative solution of Eq. 9 in the proximity of the previous approximation, i.e.,$z = x_{k-1}$ (total processing cost for four processing clusters in (a) and (d), total error at sensor nodes in (b) and (e) and CP utilization for each of the four processing sites in (c) and (f))

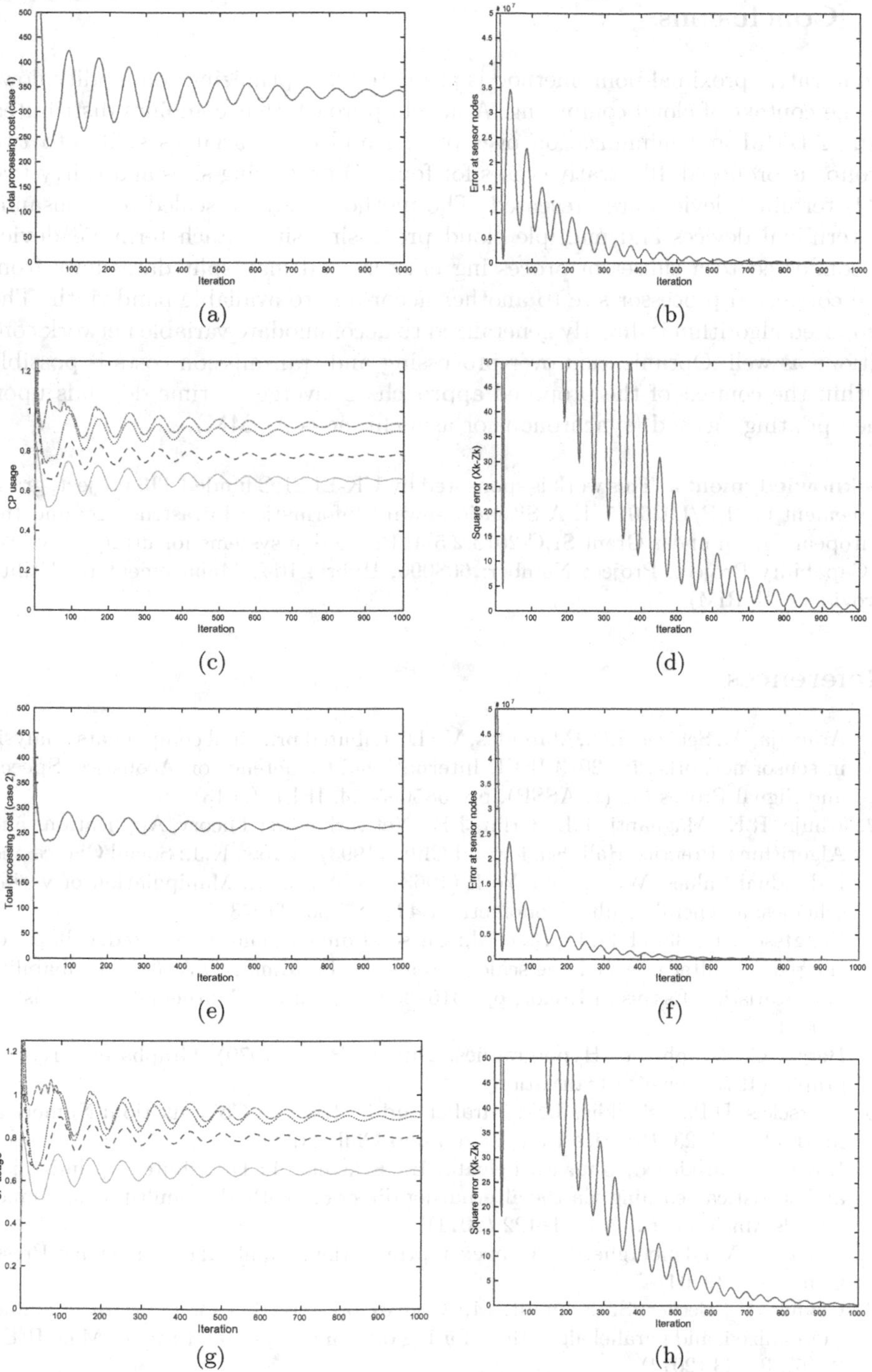

Fig. 5. Solution of Eq. 9 using distributed processing - sync ADMM (total processing cost for four processing clusters in (a) and (e), total error at sensor nodes in (b) and (f))

5 Conclusions

An iterative proximal-point method is presented for optimizing commodity flows in the context of cloud computing. A novel approach that combines distributed sync ADMM and minimization over primal and dual variables split into two groups is proposed. Illustrative cases for four (4) processing sites and thirty-two (32) terminal devices are presented. The methods may be scaled to thousands of terminal devices and multiple cloud processing sites. Each terminal device is connected to a subset of processing clusters and may split data flows from one connected processor site to another according to available bandwidth. The proposed algorithm is directly generalized to accommodate variable network conditions as well. Optimization over processing and transmission costs is possible within the context of the proposed approach. Convergence time depends upon the updating method (synchronous or asynchronous ADMM).

Acknowledgments. This work is sponsored by UK-EPSRC funded TI3 project, grant agreement no. EP/L026015/1: A Situation-aware Information Infrastructure; and the European Union under Grant SEC-2013.2.5-4: Protection systems for utility networks – Capability Project, Project Number: 608090, Hybrid Risk Management for Utility Providers (HyRiM).

References

1. Aduroja, A., Schizas, I.D., Maroulas, V.: Distributed principal components analysis in sensor networks. In: 2013 IEEE International Conference on Acoustics, Speech and Signal Processing (ICASSP), pp. 5850–5854. IEEE (2013)
2. Ahuja, R.K., Magnanti, T.L., Orlin, J.B.: Network flows: Theory, Applications and Algorithms. Prentice-Hall, Englewood Cliffs (1993). Arrow, K.J.: Social Choice and Individual Values, Wiley, New York (1963). Gibbard, A.: Manipulation of voting schemes: a general result. Econometrica **41**, 587–602 (1973)
3. Bengtsson, T., Bickel, P., Li, B., et al.: Curse-of-dimensionality revisited: collapse of the particle filter in very large scale systems. In: Freedman, D.A. (ed.) Probability and Statistics: Essays in Honor, pp. 316–334. Institute of Mathematical Statistics (2008)
4. Berge, C.: Graphes et Hypergraphes. Dunod, Paris (1970). Graphs and Hypergraphs (1973) (english translation)
5. Bertsekas, D.P., Tsitsiklis, J.N.: Parallel and Distributed Computation: Numerical Methods, vol. 23. Prentice Hall, Englewood Cliffs (1989)
6. Boyd, S., Parikh, N., Chu, E., Peleato, B., Eckstein, J.: Distributed optimization and statistical learning via the alternating direction method of multipliers. Found. Trends Mach. Learn. **3**(1), 1–122 (2011)
7. Boyd, S., Vandenberghe, L.: Convex Optimization. Cambridge University Press, Cambridge (2004)
8. Cevher, V., Becker, S., Schmidt, M.: Convex optimization for big data: scalable, randomized, and parallel algorithms for big data analytics. Sig. Process. Mag. IEEE **31**(5), 32–43 (2014)
9. Charikar, M., Naamad, Y., Rexford, J., Zou, K.: Multi-commodity flow with in-network processing. www.cs.princeton.edu/~jrex/papers/mopt14.pdf

10. Dean, J., Ghemawat, S.: Mapreduce: simplified data processing on large clusters. Commun. ACM **51**(1), 107–113 (2008)
11. Feizi, S., Zhang, A., Médard, M.: A network flow approach in cloud computing. In: 2013 47th Annual Conference on Information Sciences and Systems (CISS), pp. 1–6. IEEE (2013)
12. Forero, P.A., Cano, A., Giannakis, G.B.: Consensus-based distributed support vector machines. J. Mach. Learn. Res. **11**, 1663–1707 (2010)
13. Gondran, M., Minoux, M., Vajda, S.: Graphs and Algorithms. Wiley, New York (1984)
14. Gouglidis, A., Noor-ul-Hassan, S., Simpson, S., Smith, P., Hutchison, D.: A framework for resilience management in the cloud. In: 2016 23rd Annual Conference on International Conference on Telecommunications (ICT). IEEE (2016)
15. Hanzalek, J.T.Z.: Distributed multi-commodity network flow algorithm for energy optimal routing in wireless sensor networks. Radioengineering **19**(4), 579 (2010)
16. Jordan, M.I., et al.: On statistics, computation and scalability. Bernoulli **19**(4), 1378–1390 (2013)
17. Minoux, M.: Multicommodity network flow models and algorithms in telecommunications. In: Resende, M.G.C., Pardalos, P.M. (eds.) Handbook of Optimization in Telecommunications, pp. 163–184. Springer, New York (2006)
18. Noor-ul-Hassan, S., Simpson, S., Oechsner, S., Mauthe, A., Hutchison, D.: A framework for resilience management in the cloud. e & i Elektrotechnik und Informationstechnik **132**(2), 122–132 (2015). http://dx.doi.org/10.1007/s00502-015-0290-9
19. Slavakis, K., Giannakis, G., Mateos, G.: Modeling and optimization for big data analytics: (statistical) learning tools for our era of data deluge. IEEE Sig. Process. Mag. **31**(5), 18–31 (2014)
20. Sterbenz, J.P., Hutchison, D., Etinkaya, E.K., Jabbar, A., Rohrer, J.P., Schller, M., Smith, P.: Resilience and survivability in communication networks: strategies, principles, and survey of disciplines. Comput. Netw. **54**(8), 1245–1265 (2010). http://www.sciencedirect.com/science/article/pii/S1389128610000824
21. Zhang, R., Kwok, J.: Asynchronous distributed ADMM for consensus optimization. In: Proceedings of the 31st International Conference on Machine Learning (ICML 2014), pp. 1701–1709 (2014)

Designing a Context-Aware Cyber Physical System for Smart Conditional Monitoring of Platform Equipment

Farzan Majdani(✉), Andrei Petrovski, and Daniel Doolan

School of Computing Science and Digital Media,
Robert Gordon University, Aberdeen, UK
{f.majdani-shabestari,a.petrovski,d.c.doolan}@rgu.ac.uk

Abstract. An adaptive multi-tiered framework, which can be utilised for designing a context-aware cyber physical system is proposed and applied within the context of assuring offshore asset integrity. Adaptability is achieved through the combined use of machine learning and computational intelligence techniques. The proposed framework has the generality to be applied across a wide range of problem domains requiring processing, analysis and interpretation of data obtained from heterogeneous resources.

Keywords: Context awareness · Cyber physical system · Asset integrity

1 Introduction

There exists a growing demand for intelligent and autonomous control in engineering applications. This is especially true when some constraints are present that cannot be satisfied by human intervention with regard to decision making speed in life threatening situations (e.g. automatic collision systems, exploring hazardous environments, processing large volumes of data). Because machines are capable of processing large amounts of heterogeneous data much faster and are not subject to the same level of fatigue as humans, the use of computer-assisted control in many practical situations is preferable. Cyber physical systems are the integration of information processing, computation, sensing and networking that allows physical entities to operate various processes in dynamic environments [5]. Many of these intelligent cyber physical systems involve human intervention at some point, either during the development process by embedding expert knowledge into the systems, or during operation by requiring humans to monitor, evaluate, and confirm/reject the systems inferences. The latter type of intervention is often associated with another salient feature of cyber physical systems dealing with the big data phenomenon. Big data has become a common research focus in the last decade due to the increasing volume, velocity, variety and veracity of data enabled by technological advancements and by a reduction in data acquisition costs. The integration of multiple data sources into

C. Jayne and L. Iliadis (Eds.): EANN 2016, CCIS 629, pp. 198–210, 2016.
DOI: 10.1007/978-3-319-44188-7_15

a unified system leads to data heterogeneity, often resulting into difficulty, or even infeasibility, of human processing, especially in real-time environments. For example, in real-time automated process control, information about a possible failure is more useful before the failure takes place so that prevention and damage control can be carried out in order to either completely avoid the failure, or at least alleviate its consequences. Computational Intelligence (CI) techniques have been successfully applied to problems involving big data in various application domains [4]. These techniques however require training data to provide reliable and reasonably accurate specification of the context in which a cyber physical system operates. The context enables the system to highlight potential anomalies in the data so that intelligent and autonomous control of the underlying process can be carried out. Anomalies are defined as incidences or occurrences, under a given circumstances or a set of assumptions, that are different from the expectance (for instance when Generator rotor speed of the gas turbine goes below 3000 rpm). By their nature, these incidences are rare and often not known in advance. This makes it difficult for the Computational Intelligence techniques to form an appropriate training dataset. Moreover, dynamic problem environments can further aggravate the lack of training data by occurrence of intermittent anomalies. Computational Intelligence techniques that are used to tackle dynamic problems should therefore be able to adapt to environmental/contextual changes. A multi-tiered framework for cyber physical systems with heterogeneous input sources is proposed in the paper that can deal with unseen anomalies in a real-time dynamic problem environment. The goal is to develop a framework that is as generic, adaptive and autonomous as possible. In order to achieve this goal both machine learning and computational intelligence techniques are applied within the framework, together with the online learning capability that allows for adaptive problem solving.

2 Cyber Physical Systems (CPS)

Rapid advances in miniaturisation, speed, power and mobility have led to the pervasive use of networking and information technologies across all economic sectors. These technologies are increasingly combined with elements of the physical worlds (e.g. machines, devices) to create smart or intelligent systems that offer increased effectiveness, productivity, safety and speed [5]. Cyber physical systems (CPS) are a new type of system that integrates computation with physical processes. They are similar to embedded systems but focus more on controlling the physical entities rather than processes embedded computers monitor and control, usually with feedback loops, where physical processes affect computations and vice versa. Components of cyber physical system (e.g., controllers, sensors and actuators) transmit the information to cyber space through sensing a real world environment; also they reflect policy of cyber space back to the real world [7]. Rather than dealing with standalone devices, cyber physical systems are designed as a network of interacting elements with physical inputs and outputs, similar to the concepts found in robotics and sensor networks.

The main challenge in developing a CPS is to create an interactive interface between the physical and cyber worlds the role of this interface is to acquire the context information from the physical world and to implement context-aware computing in the cyber world [6]. Figure 1 illustrates a conceptual framework for building context-aware cyber physical systems [9]. Each layer is dedicated to a certain context processing task, ranging from low-level context acquisition up to high level context application using either existing or acquired knowledge. Cyber physical systems may consist of many interconnected parts that must instantaneously exchange, parse and act upon heterogeneous data in a coordinated way. This creates two major challenges when designing cyber physical systems: the amount of data available from various data sources that should be processed at any given time and the choice of process controls in response to the information obtained. An optimal balance needs to be attained between data availability and its quality in order to effectively control the underlying physical processes. Figure 2 illustrates a systematic approach to handling the challenges related to context processing, which has been successfully applied by the authors to various real world applications [8,9]. As can be seen from Fig. 2, the suggested approach segregates processing of the input stream into three distinct phases. The Processing minimises the volume of data and the data processing cost by analysing only inputs from easy to process data sources using context identification techniques for finding anomalies in the acquired data. If any anomalies are detected at this stage, Alert 1 gets activated. This phase of the process is used to analyse real-time data and is a safe guard process on scenarios where the frameworks prediction fail to predict occurrence of unexpected changes in the environment. In the Prediction Phase, future values of each of the gas turbine's sensors get predicted, using a linear regression model. Moreover a new column is added which gets populated with the "predicted status" value for each data instance. In this phase if any of the future predicted value of the sensors goes beyond the set threshold, Alert 2 gets activated. The final step of the process Anomaly Detection, classifies the overall predicted future values to identify anomalies being present in the underlying process on the operation of the cyber physical system. If any anomalies are detected at this stage Alert 3 is triggered. Such an approach allows for the acquisition of data and/or activation of the necessary physical entities on an ad-hoc basis, depending on the outcome at each phase. Moreover, the accuracy attained at the specified phases can be enhanced by incorporating additional data sensors or additional environmental factors. Computational intelligence techniques and expert systems have been successfully applied to tackling many anomaly detection problems, where anomalies are known a priori. More interesting, however, is to detect previously unseen anomalies. Statistical analysis and clustering are examples of techniques that are commonly used when the characteristics of anomalies are unknown [1]. Figure 3 illustrates a more detailed process for the systematic approach where machine learning and computational intelligence techniques are combined to tackle the unknown anomalies and learn from the experience when similar anomalies occur

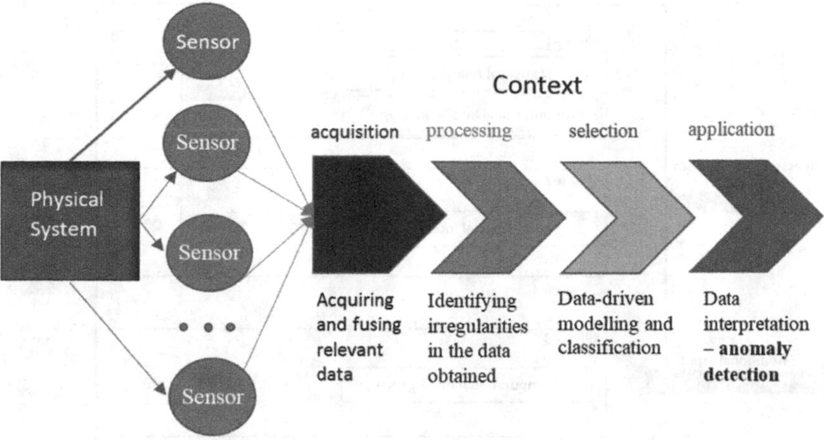

Fig. 1. Framework for designing context-aware CPS

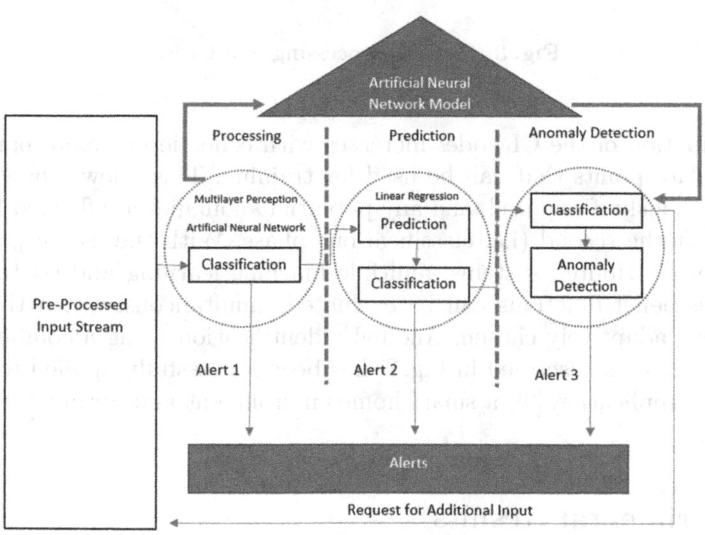

Fig. 2. Systematic approach to context processing

again. In Fig. 3, "b" represents a belief function of the output from both the statistical analysis and computational intelligence nodes, such that

$$f(t) = \sum_{i=1}^{n} w_i \mu_i(\overline{X}) + \sum_{j=1}^{m} w_j \eta_j(\overline{X}) \tag{1}$$

The weights (w_i and w_j) of this belief function are adaptively adjusted depending on how much knowledge related to the problem context has been obtained.

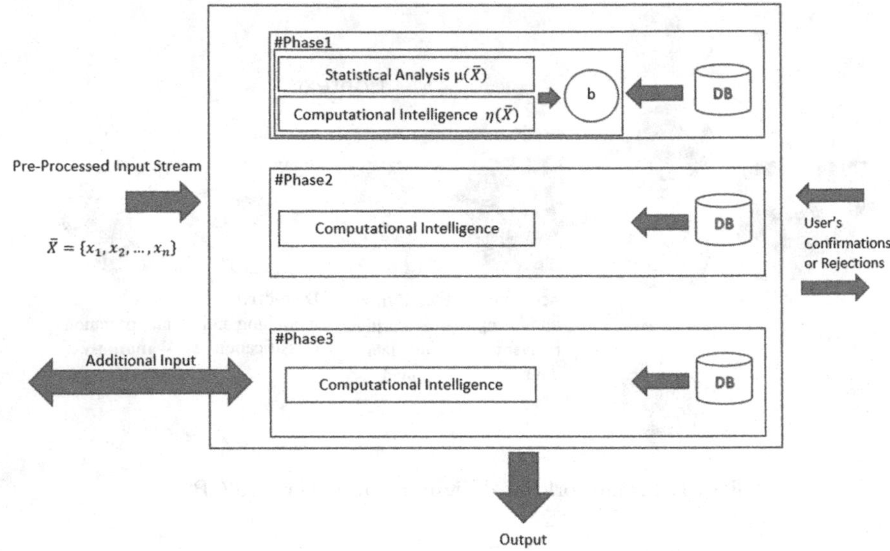

Fig. 3. Context processing in a CPS

The contribution of the CI nodes increases with collection of more normal and abnormal data points that can be used for training. This allows the system to run autonomously if required, and any potential anomalies are flagged for closer inspection at the second (i.e. classification) phase. With the use of parallelisation and/or distributed systems, multiple machine learning and CI techniques and various belief functions can be evaluated simultaneously with their parameters being adaptively chosen. Anomaly identification using a combination of such techniques, as described in Fig. 3, has been successfully applied to a traffic surveillance application [9], a smart home environment and automotive process control [8].

3 Experimental Results

3.1 Data Description

It is a common practice that most of the sensory data on a platform are stored in a historian system such as the PI system. PI is a form of historian system which act as a repository to store sensor information gathered from one or multiple installation. For this study we used historical sensor data of a gas turbine from an offshore installation in the North Sea. This data in real-time is transmitted offshore via satellite Internet. In this experiment about three months worth of data from a PI historian representing a total of 25 sensors from different parts of a gas turbine was used (see Fig. 4). Within this period system experienced 8 failures which are highlighted in Fig. 5 which are indicated by blue arrows.

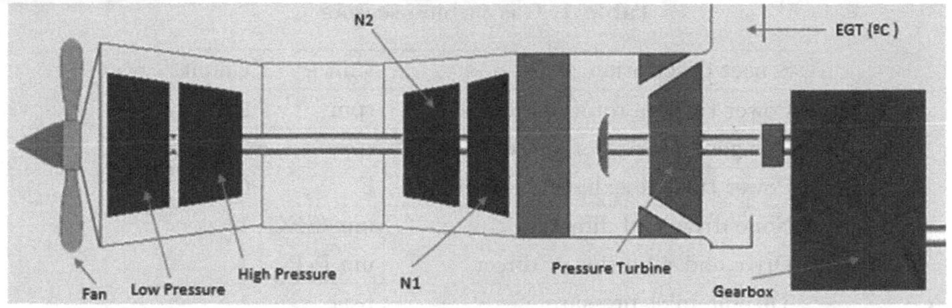

Fig. 4. Gas turbine process design

The sample data for the three months period includes around 217000 instances. Sensor used from the turbine are listed in Table 1. In addition to all the sensors we also had a turbine status which has each of the instances of the dataset labeled as either False, True or I/O timed out. False indicates the turbine failure state, True indicates the engine is running and I/O Timed out indicates when the engine is getting restarted or communication between the Pi historian and offshore is temporarily lost. The importance of having the I/O Timeout state is to prevent the system from sending an alarm when the system is actually in a state of reboot but not a failure.

3.2 Processing

The processing Phase of the proposed context-aware CPS implements computational intelligence to classify the input stream. To implement this Phase the Multilayer Perceptron is used, which is a feedforward Artificial Neural Network (ANN). Funahashi [2], Hoenike and Stunchcombe [3] and Qiu et al. [11] have all shown that only one hidden layer can effectively generate highly accurate results by also improving the processing time. Therefore an ANN Multilayer Perceptron with Backpropagation of error has been used to train the machine with 1, 2, 3 and 4 hidden layers 10 fold cross validation. The experiment had been continued up until 4 layer which eventually generated an excellent result. Table 2 lists the result gathered from the experiments with 1 to 4 hidden layers. Although by using only one hidden layer we have managed to classify 92.77 percent of the instances correctly, however by increasing the layers to 4 we have managed to classify 100 percent of the instances. Figure 6 illustrated the artificial neural network design. The input layer corresponds to the 25 input sensors of the gas turbine. The middle layers are used to form the relations between the neurons, their number being determined at runtime. The output neurons are the three classifications which indicates the status of the turbine.

Table 1. Gas turbine sensors

Sensor description	Unit	Count
Power turbine rotor speed	rpm	2
Gas generator rotor speed	rpm	2
Power turbine exhaust temperature	F	6
None drive end direction	mm/SEC	1
Drive end vibration X direct	um P-P	1
Turbine inlet pressure	psia	1
Compressor inlet total pressure	psia	1
Ambient temperature	F	1
Axial compressor inlet temperature	F	2
Mineral oil tank temperature	F	1
Synthetic oil tank temperature	F	1
OB bearing temperature	C	1
IB bearing temperature	C	1
IB thrust bearing temperature	C	1
OB thrust bearing temperature	C	1
Generator active power	Mwatt	1
Grid voltage	V	1

3.3 Prediction

The second phase of the proposed model is Prediction Phase, which is to predict the future values for the next 24 h of all 25 sensors. Times series was used to lag the data for 24 h followed with linear regression to predict the next 24. During this phase by looking at the historical data we have already set threshold for each of the sensors. Therefore if any of the predicted values for each of the sensors falls below or beyond the allowed threshold then Alarm 2 gets activated. Figure 7 illustrates the predicted results for all the 25 sensors.

3.4 Anomaly Detection

Since combination of all the sensors together reflect the status of the turbine, after predicting future value of all the sensors then all get merged into a single test dataset. The Artificial Neural Network model which has been trained as part of the Processing phase is used again, but this time to label the status of the turbine for each of the instances. After predicting the status of the turbine for all instances of the dataset, the developed framework iterate through all labels and if any of the instances are labeled as failed Alarm 3 gets fired. System then picks the time stamp of the predicted time and deduct it from the current

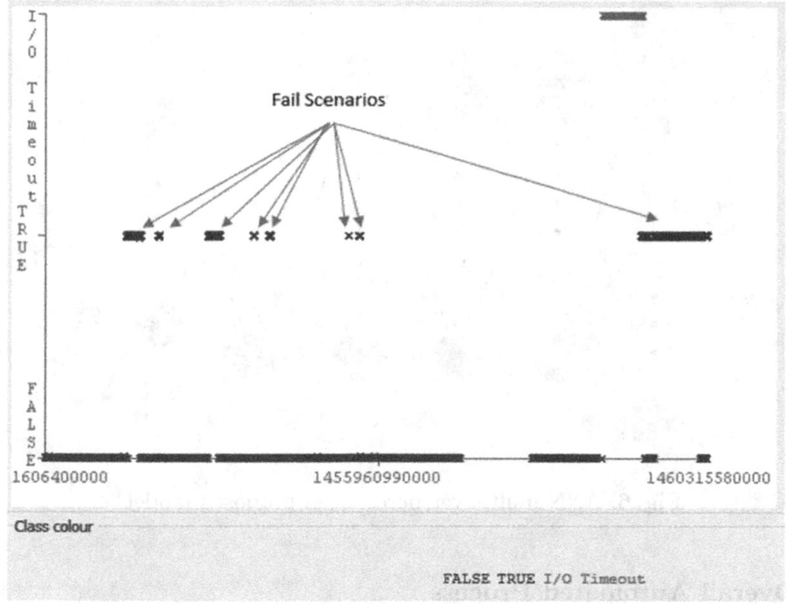

Fig. 5. Turbine's fail scenarios (Color figure online)

Table 2. ANN Multilayer Perceptron Optimisation

Layers count	One	Two	Three	Four
Correctly classified (%)	92.77	92.77	94.95	100
Incorrectly classified (%)	7.23	7.23	5.05	0
Kappa statistic	0.60	0.60	0.74	1
Mean absolute error	0.09	0.09	0.062	0
Root mean squared error	0.21	0.21	0.17	0
Relative absolute error (%)	57.32	57.79	39.10	0.34
Root relative squared error (%)	74.89	74.97	62.71	0.77
Coverage of cases (0.95 level) (%)	100	100	100	100
Mean rel. region size (0.95 level)	4.65	64.65	55.25	33.33

time to provide the estimated hours left until the system failure. In final step of the Anomaly Detection phase the total remainder hours gets included into an automatically generated email and sent out to the preset list of email addresses as well as playing an audio alarm on the PC.

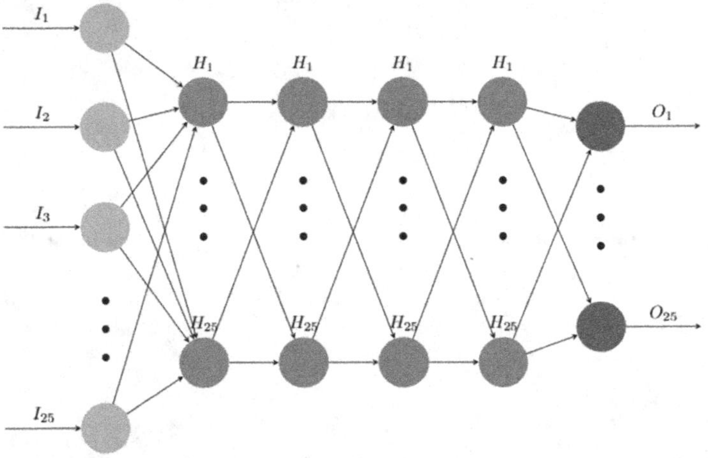

Fig. 6. ANN multilayer perceptron proposed model

3.5 Overall Automated Process

Initially Weka was used to run each of the phases separately. However in the final stage of the process we have actually formed the proposed framework using Knime. Knime is an open source data analytics, reporting and integration platform. Although there are other alternatives such as Weka's KnowledgeFlow and Microsoft Azure's Machine Learning. Knime was chosen since it has the capability of importing most of Weka's features through the addition of a plugin. Also being able to run java snippets and write the developed model into disk to free up space on memory it is a preferred option in comparison to Azure's Machine Learning. The dataset was divided into two sets of training and test data as illustrated in Fig. 8. Two months of data was used for training which included 8 cases of turbine failure with the remainder set aside for testing. The training dataset has been used to form an Artificial Neural Network Multilayer Perceptron (MLP) using Backpropagation of error. MLP is multilayer perceptron consists of multiple layers of artificial neurons which interact using weighted connections [10]. After training the model it was tested against the developed ANN MLP to classify the status of the engine. This implementation covered the Processing phase of the proposed System. This was followed by introducing times series lag and linear regression model to predict the next 24 h on the test dataset. By looking at the 8 failure situations thresholds were identified for each of the input sensors. Therefore if during the prediction stage any of the sensor's value go below or above the set threshold the second Alarm goes off. However this alarm is an amber rated alarm because that doesn't mean necessarily the turbine will fail. With all 24 h of predicted data for all the sensors gathered, in the final stage of the process all the predicted data is put together as a test dataset and is tested against the model developed in Processing Phase. If the status of the engine gets classified as False then the third and last alarm gets fired.

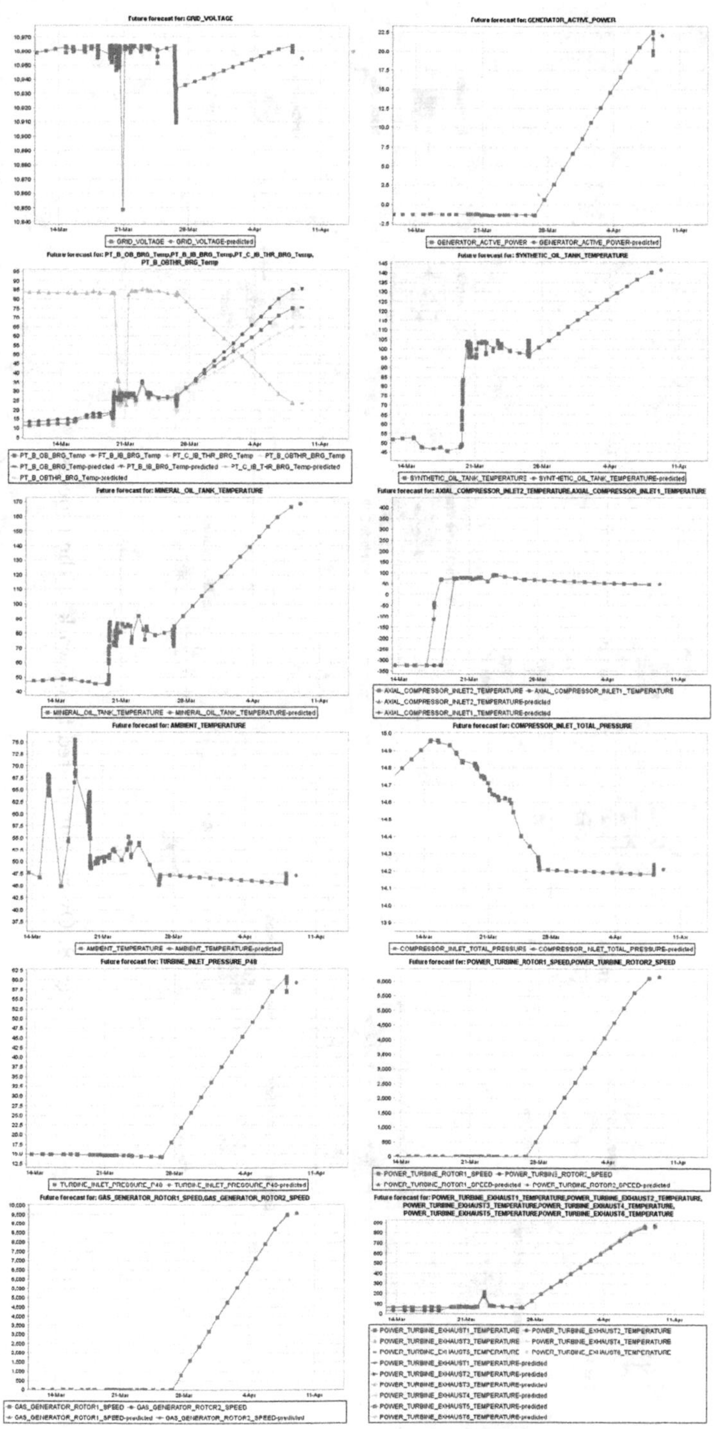

Fig. 7. Predicted Sensor values

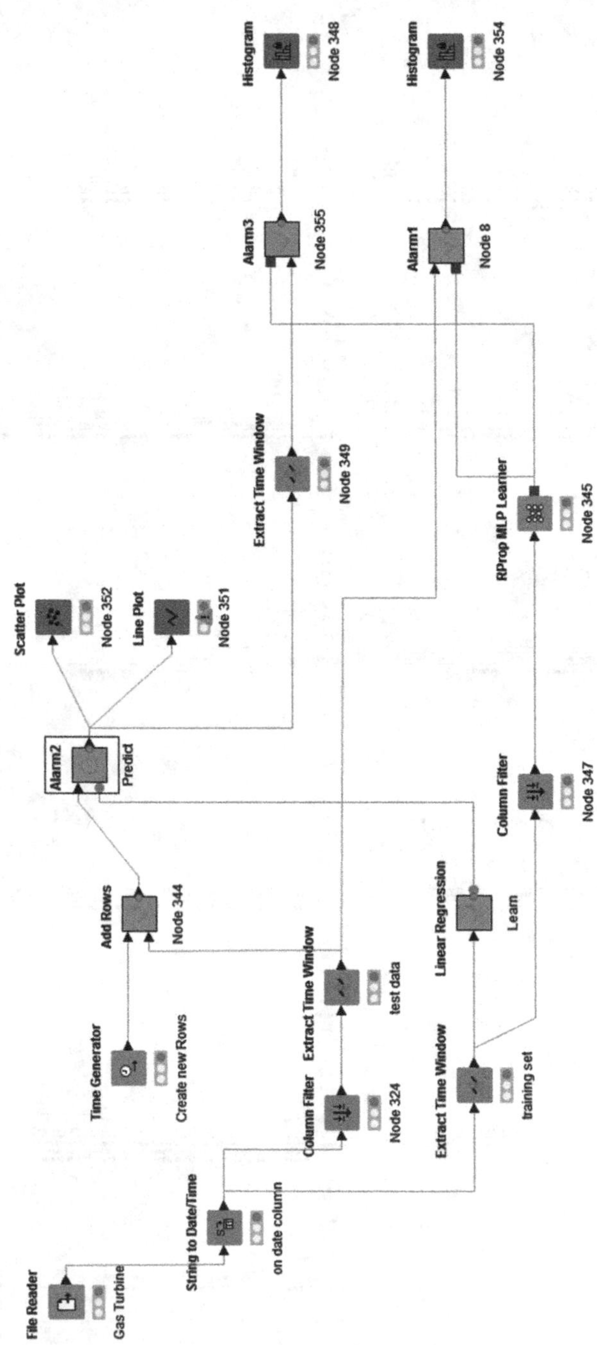

Fig. 8. Overal automated framework of the process

3.6 Evaluation

To test the accuracy and performance of the proposed model, 5 days worth of data was removed from the dataset and the developed model used to predict each of eliminated days hourly. To achieve this, performance of the turbine for the next 1, 3, 6, 9, 12, 14, 16, 18 and 24 h for each days has been predicted. Then the average performance for all these 5 days has been calculated. The average performance shows up until 12 h system could predict the status of the turbine with nearly 99 percent accuracy which is a reasonably high performance. Even for the 16 h period, prediction was around 73 percent which is still considered to be high performance. However after 18 h the prediction performance shows sudden declines and when it gets to prediction of the next 24 h the result is really

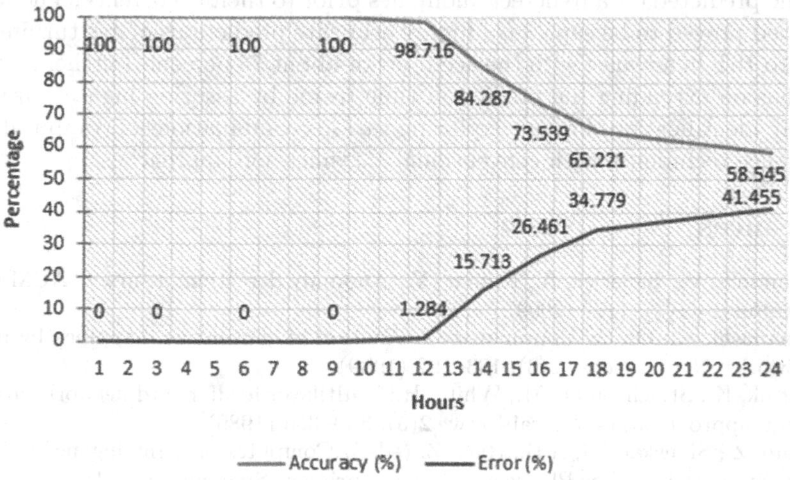

Fig. 9. Hourly performance evaluation

Table 3. Comparison of real-time Status vs. Predicted Status

Hours	Accuracy (%)	Error (%)
	100	0
3	100	0
6	100	0
9	100	0
12	98.716	1.284
14	84.287	15.713
16	73.539	26.461
18	65.221	34.779
24	58.545	41.455

poor by being around 58 percent. Table 3 lists the average value of the result for each prediction.

4 Conclusion

An implementation of a Context-Aware Cyber Physical System using Multilayer Perceptron Artificial Neural Network, to predict the status of a gas turbine on an offshore installation has been successfully developed. In this experiment a three phased model has been proposed. In the processing phase, historical data of 25 sensors was collected from different areas of turbine to train an Artificial Neural Network model as the basis of the prediction model. In the second phase future value of each sensor has been predicted for a certain period of time using linear regression. The final phase makes use of the model developed in phase one to label the predicted data to detect anomalies prior to their occurrence. The model developed proved to be capable of highly accurate predictions of gas turbine status up to 16 h in advance with the accuracy of about 73 percent. Further research will focus on extending the prediction time frame by assuring high accuracy in anomaly identification through exploring various combinations of computational intelligence techniques with conventional classification approaches.

References

1. Chandola, V., Banerjee, A., Kumar, V.: Anomaly detection: a survey. ACM Comput. Surv. **41**(3), 1–72 (2009)
2. Funahashi, K.: On the approximate realization of continuous mappings by neural networks. Neural Netw. **2**(3), 183–192 (1989)
3. Hornik, K., Stinchcombe, M., White, H.: Multilayer feedforward networks are universal approximators. Neural Netw. **2**(5), 359–366 (1989)
4. Khan, Z., Shawkat Ali, A.B., Riaz, Z. (eds.): Computational Intelligence for Decision Support in Cyber-Physical Systems. Springer, Singapore (2014)
5. Lee, E.A.: Cyber physical systems: design challenges. University of California, Berkeley Technical Report No. UCB/EECS-2008-8, January 2008
6. Lun, Y., Cheng, L.: The research on the model of the context-aware for reliable sensing and explanation in cyber-physical system. Procedia Eng. **15**(2011), 1753–1757 (2011)
7. Park, K.J., Zheng, R., Liu, X.: Milestones and research challenges. Comput. Commun. **36**(1), 1–7 (2012)
8. Petrovski, S., Bouchet, F., Petrovski, A.: Data-driven modelling of electromagnetic interferences in motor vehicles using intelligent system approaches. In Proceeding. IEEE Symp. Innovations Intell. Syst. Appl. (INISTA). **2013**, 1–7 (2013)
9. Rattadilok, P., Petrovski, A., Petrovski, S.: Anomaly monitoring framework based on intelligent data analysis. In: Yin, H., Tang, K., Gao, Y., Klawonn, F., Lee, M., Weise, T., Li, B., Yao, X. (eds.) IDEAL 2013. LNCS, vol. 8206, pp. 134–141. Springer, Heidelberg (2013)
10. Pal, S.K., Mitra, S.: Multilayer perceptron, fuzzy sets, and classification. IEEE Trans. Neural Netw. **3**(5), 683–697 (1992)
11. Qiu, M., Song, Y., Akagi, F.: Application of artificial neural network for the prediction of stock market returns: The case of the Japanese stock market. Chaos, Solitons & Fractals **85**, 1–7 (2016)

Time-Series Prediction

Convolutional Radio Modulation Recognition Networks

Timothy J. O'Shea[1(✉)], Johnathan Corgan[2], and T. Charles Clancy[1]

[1] Bradley Department of Electrical and Computer Engineering, Virginia Tech, 900 N Glebe Road, Arlington, VA 22203, USA
oshea@vt.edu

[2] Corgan Labs, 6081 Meridian Ave., Suite 70-111, San Jose, CA 95120, USA
johnathan@corganlabs.com

Abstract. We study the adaptation of convolutional neural networks to the complex-valued temporal radio signal domain. We compare the efficacy of radio modulation classification using naively learned features against using expert feature based methods which are widely used today and e show significant performance improvements. We show that blind temporal learning on large and densely encoded time series using deep convolutional neural networks is viable and a strong candidate approach for this task especially at low signal to noise ratio.

Keywords: Machine learning · Radio · Software radio · Convolutional networks · Deep learning · Modulation recognition · Cognitive radio · Dynamic spectrum access

1 Introduction

Radio communications present a unique signal processing domain with a number of interesting challenges and opportunities for the machine learning community. In this field expert features and decision criterion have been extensively developed, and analyzed for optimality under specific criteria for many years. However in the past few years the trend in machine learning applied to image processing [11] and voice recognition [18] is overwhelmingly that of feature learning from data rather than crafting of expert features, suggesting we should evaluate a similar shift in this domain.

Concurrently wireless data demand is driving a need for improved radio efficiency. High quality spectrum sensing and adaptation to improve spectral allocation and interference mitigation is an important route by which we may achieve this. The FCC in the United States as well as counterparts in Europe are taking seriously and pursuing spectrum policy which leverages some of this ideas from Dynamic Spectrum Access (DSA) [4], making clear the need for improved spectrum sensing and signal identification algorithms allowing sensors and radios to detect and identify spectrum users and interferers at the best possible range and thus signal to noise ratio.

C. Jayne and L. Iliadis (Eds.): EANN 2016, CCIS 629, pp. 213–226, 2016.
DOI: 10.1007/978-3-319-44188-7_16

Ideas such as DSA, opportunistic access and sharing of spectrum, and "Cognitive Radio" (CR) [2], a more broad class of radio optimization through learning, have been widely discussed at the conceptual level. Efforts in these fields however have been constrained to relatively specialized solutions which lack the generality needed to deal with a complex and growing number emitter types, interference types and propagation environments [6,7,9].

This is a significant challenge in the community as expert systems designed to perform well on specialized tasks often lack flexibility and can be expensive and tedious to develop analytically.

Building upon successful strategies from image and voice recognition domains in machine learning, we demonstrate an approach in the radio using Convolutional Neural Networks (CNNs) and Deep Neural Networks (DNNs) which offers flexibility to learn features across a wide range of tasks and demonstrates improved classification accuracy against current day approaches.

2 Modulation Recognition

In Dynamic Spectrum Access (DSA) one of the key sensing performed is that of providing awareness of nearby emitters to avoid radio interference and optimize spectrum allocation. This identifying and differentiating broadcast radio, local and wide area data and voice radios, radar users, and other sources of potential radio interference in the vicinity which each have different behaviors and requirements. Modulation Recognition then is the task of classifying the modulation type of a received radio signal as a step towards understanding what type of communications scheme and emitter is present.

This can be treated as an N-class decision problem where our input is a complex base-band time series representation of the received signal. That is, we sample in-phase and quadrature components of a radio signal at discrete time steps through an analog to digital converted with a carrier frequency roughly centered on the carrier of interest to obtain a $1 \times \mathrm{N}$ complex valued vector. Classically, this is written as in Eq. 1 where $s(t)$ is a time series signal of either a continuous signal or a series of discrete bits modulated onto a sinusoid with either varying frequency, phase, amplitude, trajectory, or some permutation of multiple thereof. c is some path loss or constant gain term on the signal, and $n(t)$ is an additive Gaussian white noise process reflecting thermal noise.

$$r(t) = s(t) * c + n(t) \tag{1}$$

Analytically, this simplified expression is used widely in the development of expert features and decision statistics, but the real world relationship looks much more like that given in Eq. 2 in many systems.

$$r(t) = e^{j * n_{Lo}(t)} \int_{\tau=0}^{\tau_0} s(n_{Clk}(t-\tau))h(\tau) + n_{Add}(t) \tag{2}$$

This considers a number of real world effects which are non-trivial to model: modulation by a residual carrier random walk process, $n_{Lo}(t)$, resampling by a residual clock oscillator random walk, $n_{Clk}(t)$, convolution with a time varying rotating non-constant amplitude channel impulse response $h(t-\tau)$, and the addition of noise which may not be white, $n_{Add}(t)$. Each presents an unknown time varying source of error.

Modeling expert features and decision metrics optimality analytically under each of these harsh realistic assumptions on propagation effects is non-trivial and often forces simplifying assumptions. In this paper we focus on empirical measurement of performance in harsh simulated propagation environments which include all of the above mentioned effects, but do not attempt to analytically trace their performance in closed form.

2.1 Expert Cyclic-Moment Features

Integrated cyclic-moment based features [1] are currently widely popular in performing modulation recognition and for forming analytically derived decision trees to sort modulations into different classes. In general, they take the form given in Eq. 3.

$$s_{nm} = f_m(x^n(t)...x^n(t+T)) \tag{3}$$

By computing the m'th order statistic on the n'th power of the instantaneous or time delayed received signal r(t), we may obtain a set of statistics which uniquely separate it from other modulations given a decision process on the features. For our expert feature set, we compute 32 features. These consist of cyclic time lags of 0 and 8 samples. And the first 2 moments of the first 2 powers of the complex received signal, the amplitude, the phase, and the absolute value of phase for each of these lags.

We train several classifiers on these set of expert features as a benchmark comparison. These leverage scikit-learn and consist of a Decision Tree, K=1-Nearest Neighbor, Gaussian Naive Bayes, and an RBF-SVM. Additionally, we train a 3-layer deep neural network consisting only of fully connected layers of size 512, 256, and 11 neurons. Each of these is measured to provide a performance baseline estimate for how a current day system operating on such a feature set might perform. Best expert-feature performance is obtained from the SVM and DNN based approaches.

2.2 Convolutional Feature Learning

We evaluate several feature learning methods, but our principal method is that of a convolutional neural network (CNN) provided with a windowed input of the raw radio time series $r(t)$. We treat the complex valued input as an input dimension of 2 real valued inputs and use $r(t)$ as a set of $2 \times$ N vectors into a narrow 2D Convolutional Network where the orthogonal synchronously sampled In-Phase and Quadrature (I & Q) samples make up this 2-wide dimension.

3 Evaluation Dataset

While simulation and the use of synthetic data sets for learning is sometimes frowned upon in machine learning, radio communications presents a special case. Training with real data is important and valuable - and will be addressed in future work - but certain properties of the domain allow us to say our simulation is quite meaningful.

Radio communications signals are in reality synthetically generated, and we do so deterministically in a way identical to a real system, introducing modulation, pulse shaping, carried data, and other well characterized transmit parameters identical to a real world signal. We modulate real voice and text data sets onto the signal. In the case of digital modulation we whiten the data using a block randomizer to ensure bits are equiprobable.

Radio channel effects are relatively well characterized. We employ robust models for time varying multi-path fading of the channel impulse response, random walk drifting of carrier frequency oscillator and sample time clocks, and additive Gaussian white noise. We pass our synthetic signal sets through harsh channel models which introduce unknown scale, translation, dilation, and impulsive noise onto our signal.

We model the generation of this dataset in GNU Radio [3] using the GNU Radio channel model [14] blocks and then slice each time series signal up into a test and traning set using a 128 samples rectangular windowing process. The total dataset is roughly 500 MBytes stored as a python pickle file with complex 32 bit floating point samples.

3.1 Dataset Availability

This data will hopefully be of great use to others in the field and may serve as a benchmark for this domain. This dataset is available in pickled python format at http://radioml.com, consisting of time-windowed examples and corresponding modulation class and SNR labels. We hope to grow scope of modulations addressed and the channel realism as interest in this area.

3.2 Dataset Parameters

We focus on a dataset consisting of 11 modulations: 8 digital and 3 analog modulation, all are widely used in wireless communications systems all around us. These consist of BPSK, QPSK, 8PSK, 16QAM, 64QAM, BFSK, CPFSK, and PAM4 for digital modulations, and WB-FM, AM-SSB, and AM-DSB for analog modulations. Data is modulated at a rate of roughly 8 samples per symbol with a normalized average transmit power of 0 dB.

3.3 Dataset Visualization

Inspecting a single example from each class of modulation in the time (Fig. 1) and frequency domain (Fig. 2), we see a number of similarities and differences

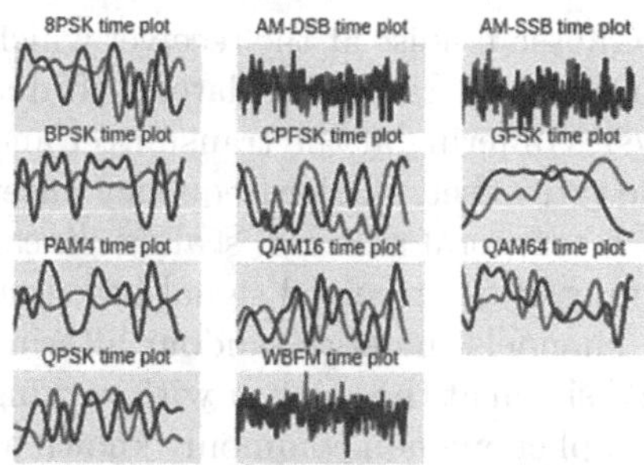

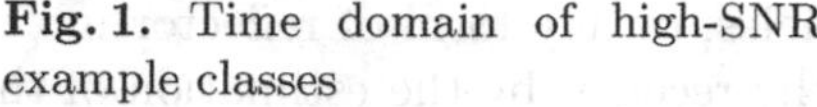

Fig. 1. Time domain of high-SNR example classes

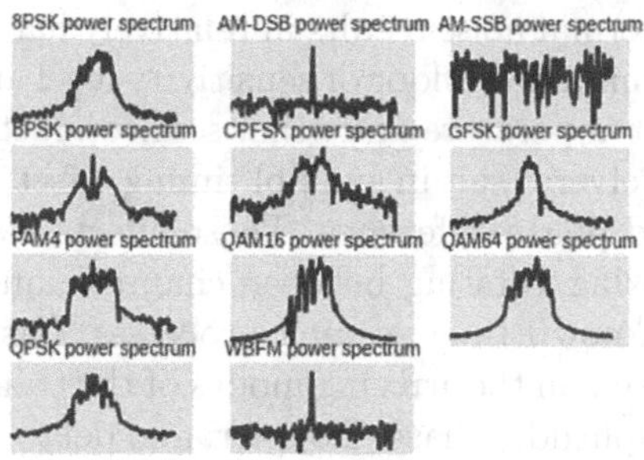

Fig. 2. Power spectrum of high-SNR example classes

between modulations visually, but due to pulse shaping, distortion and other channel effects they are not all readily discernible by a human expert visually.

In the frequency domain, each of the signals follows a similar band-limited power envelope by design whose shape provides some clues as to the modulation, but against poses a difficult noisy task for a human expert to judge visually.

3.4 Modulated Information

In radio communications, signals are typically comprised of a number of modulated data bits on well defined and understood basis functions into discrete modes formed by these bases. Complex baseband representation of a signal decomposes a radio voltage level time-series into its projections onto the sine and cosine functions at a carrier frequency. By manipulating the frequency, amplitude, phase, or sum thereof data bits are then modulated into this space through discrete and separable modes for each distinct symbol period in time in the case of digital, or continuous location in the case of analog modulation. For the case of QPSK this phase-mapping is shown in 4.

$$s(t_i) = e^{j2\pi f_c t + \pi \frac{2c_i+1}{4}}, c_i \in 0,1,2,3 \tag{4}$$

Pulse shaping filters such as root-raised cosine are then typically applied to band-limit the signal in frequency and remove sharp wide-band transients between these distinct modes, resulting in mixing of adjacent symbols' bases at the transmitter in a deterministic and invertible. In our simulated data set we use a root-raised cosine pulse shaping filter with an excess bandwidth of 0.35 for each digital signal.

3.5 Effects on the Modulated Signal

Channel effects in contrast are not deterministic and not completely invertible in a communications system. Real systems experience a number of effects on the transmitted signal, which make recovery and representation thereof challenging.

Thermal noise results in relatively flat white Gaussian noise at the receiver which forms a noise floor or sensitivity level and signal to noise ratio. Oscillator drift due to temperature and other semiconductor physics differing at the transmitter and receiver result in symbol timing offset, sample rate offset, carrier frequency offset and phase difference. These effects result in a temporal shifting, scaling, linear mixing/rotating between channels, and spinning of the received signal based on unknown time varying processes. Last, real channels undergo random filtering based on the arriving modes of the transmitted signal at the receiver with varying amplitude, phase, Doppler, and delay. This is a phenomenon commonly known as multi-path fading or frequency selective fading, which occurs in any environment where signals may reflect off buildings, vehicles, or any form of reflector in the environment. This is typically removed at the receiver by the estimation of the instantaneous value of the time varying channel response and deconvolution of it from the received signal.

3.6 Generating a Dataset

To generate a well characterized dataset, we select a collection of modulations which are used widely in practice and operate on both discrete binary alphabets (digital modulations), and continuous alphabets (analog modulations). We modulate known data over each modem and expose them each to the channel effects described above using GNU Radio. We segment the millions of samples into a dataset consisting of numerous short-time windows in a fashion similar to how a continuous acoustic voice signal is typically windowed for voice recognition tasks. We extract steps of 128 samples with a shift of 64 samples to form our extracted dataset.

After segmentation, examples are roughly 128 μ sec each assuming a sample rate of roughly 1 MSamp/sec. Each contains between 8 and 16 symbols with random time offset, scaling, rotation, phase, channel response, and noise. These examples represent information about the modulated data bits, information about how they were modulated, information about the channel effects the signal passed through during propagation, and information about the state of the transmitted and receiver device states and contained random processes. We focus specifically on recovering the information about how the signal was modulated and thus label the dataset according to a discrete set of 11 class labels corresponding to the modulation scheme.

4 Technical Approach

In a radio communication system, one class of receiver which is commonly considered is a "matched-filter" receiver. That is on the receive side of a communications link, expert designed filters matched with each transmitted symbol representation are convolved with the incoming time signal, and form peaks as the correct symbol slides over the correct symbol time in the received signal. By convolving, we average out the impulsive noise in the receiver in an attempt to optimize signal to noise. Typically, before this convolutional stage, symbol timing and carrier frequency is recovered using an expert envelope or moment based estimators derived analytically for a specific modulation and channel model. The intuition behind the use of a convolutional neural network in this application then is that they will learn to form matched filters for numerous temporal features, each of which will have some filter gain to operate at lower SNR, and which when taken together can form a robust basis for classification.

4.1 Learning Invariance

Many of these recovery processes in radio communications systems can be thought of in terms of invariance to linear mixing, rotation, time shifting, scaling, and convolution through random filters (with well characterized probabilistic envelopes and coherence times). These are analogous to similar learning invariance which is heavily addressed in vision domain learning where matched filters for specific items or features in the image may undergo scaling, shifting, rotation, occlusion, lighting variation, and other forms of noise. We seek to leverage the shift-invariant properties of the convolutional neural network to be able to learn matched filters which may delineate symbol encoding features naively, without expert understanding or estimation of the underlying waveform.

4.2 Evaluation Networks

We train against several candidate neural networks. A 4-layer network utilizing two convolutional layers and two dense fully connected layers (CNN and CNN2). Layers use rectified linear (ReLU) activation functions except for a Softmax activation on the one-hot output layer. We use this network depth as it is roughly equivalent to networks which work well on similar simple datasets in the vision domain such as MNIST.

Regularization is used to prevent over-fitting. CNN uses Dropout, a $\|W\|_2$ norm penalty on the convolutional layer weights, encouraging minimum energy bases, and a $\|\mathbf{h}\|_1$ norm penalty on the first dense layer activation, to encourage

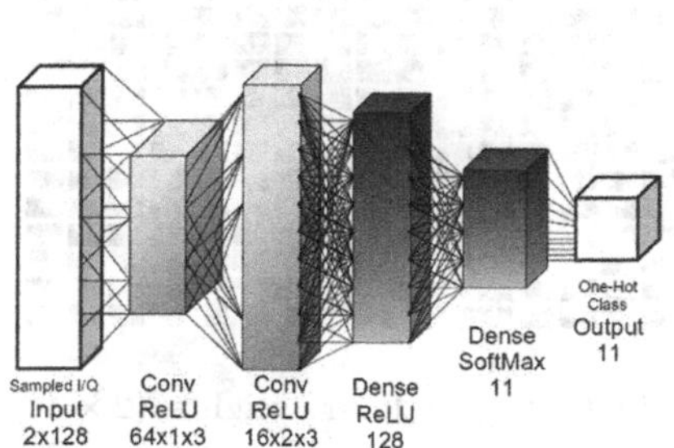

Fig. 3. CNN architecture

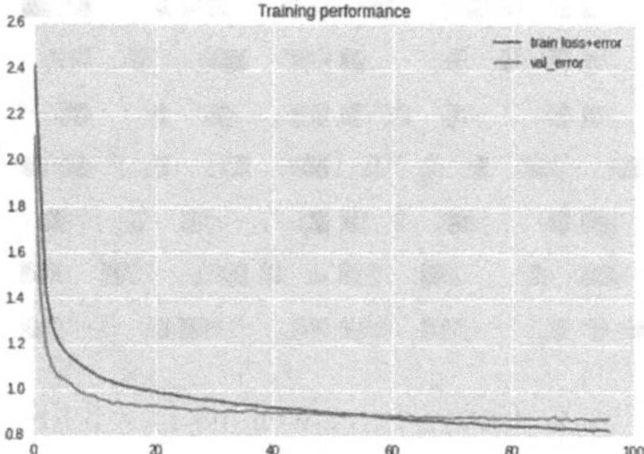

Fig. 4. Loss plots CNN2 60 % dropout

sparsity of solutions [5,10]. CNN2 uses only dropout, and DNN uses only dropout. Training is conducted using a categorical cross entropy loss function and an Adam [15] solver which seems to slightly outperform RMSProp [12] on our dataset. We implement our network training and prediction in Keras [16] running on top of TensorFlow [19] on an NVIDIA Cuda [8] enabled Titan X GPU in a DIGITS Devbox.

An illustration of the CNN architecture is shown in Fig. 3. CNN2 is identical but larger, containing 256 and 80 filters in layers 1 and 2, and 256 neurons in layer 3. The DNN evaluated contains 4 dense layers of size 512, 256, 128, and n-classes neurons.

4.3 Training Complexity

We train our highest complexity model for approximately 23 min with the Adam solver over the $\sim 900,000$ sample training set in batch sizes of 1024. Epochs take roughly 15 s and we do observe some over-fitting despite out regularization, but validation loss does not significantly inflect and we keep the best validation loss model for evaluation Fig. 4.

4.4 Learned Features

Plotting learned features can sometimes give us an intuition as to what the network is learning about the underlying representation. In this case, we plot convolutional layer 1 and convolutional layer 2 filter weights below. In Fig. 5, the first layer, we have 64 filters of 1×3. In this case we simply get a set of 1D edge and gradient detectors which operate across each I and the Q channel.

In convolutional layer 2, weights shown in Fig. 6 we compose this first layer feature map into $64*16 \times 2 \times 3$ larger feature maps, which comprise what is occurring on both I and Q channels simultaneously. These feature maps do not look hugely different than those seen at the lower levels of an image conv-net comprising of 2D learned edge detectors and Gabor-like filters.

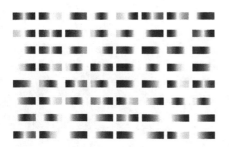

Fig. 5. Conv1 layer weights (1×3 filters)

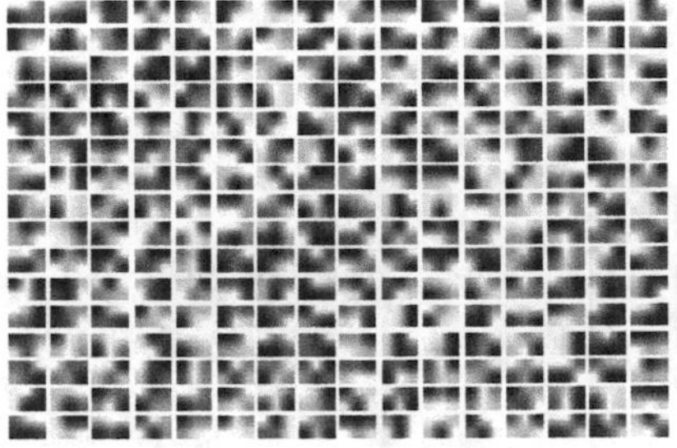

Fig. 6. Conv2 layer weights (2×3 filters)

5 Results

To evaluate the performance of our classifier, we look at classification performance on a test data set. We train on a corpus of approximately 12 million complex samples divided across the 11 modulations. These are divided into training examples of 128 samples in length. We use approximately 96,000 example for training, and 64,000 examples for testing and validation. These samples are uniformly distributed in SNR from −20dB to +20dB and tagged so that we can evaluate performance on specific subsets.

After training, we achieve a validation loss of 0.874 and a classification accuracy of 66.9 % across all signal to noise ratios on the test dataset, but to understand the meaning of this we must inspect how this classification accuracy breaks down across the SNR values of the different training examples, and how it compares to the performance of existing expert feature based based classifiers.

Plot test set modulation classification accuracy as a function of example signal to noise ratio for each classifier in Fig. 7. Solid lines show classifiers trained directly on the radio time series data performing deep feature learning, while dotted lines indicate classifiers using only the expert features previously described as input. This view is a critical way to inspect results as performance at low SNR impacts range and coverage area over which we can effectively use the classifier. We obtain significantly better low-SNR classification accuracy performance from

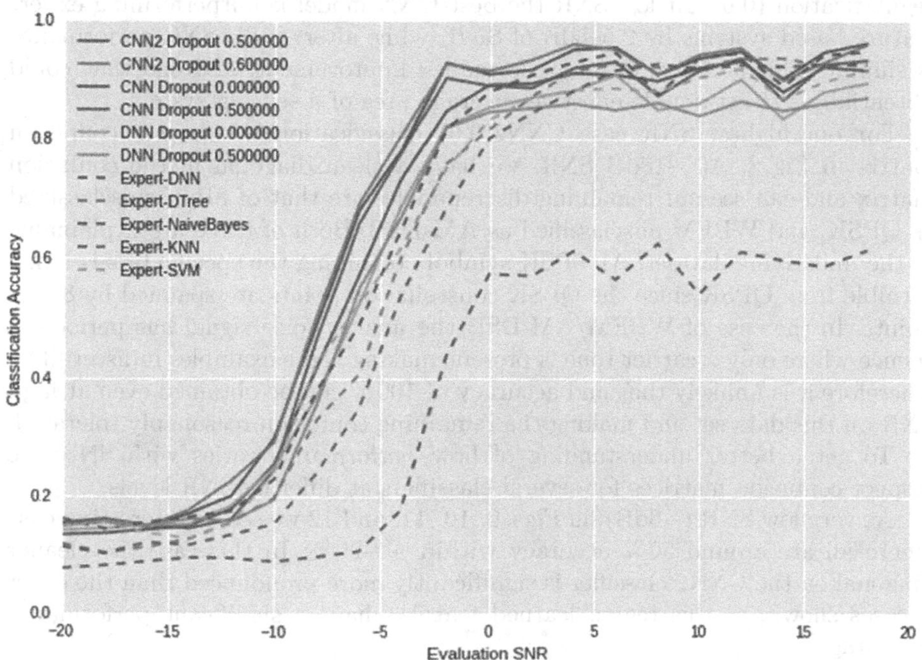

Fig. 7. Classifier Performance vs SNR

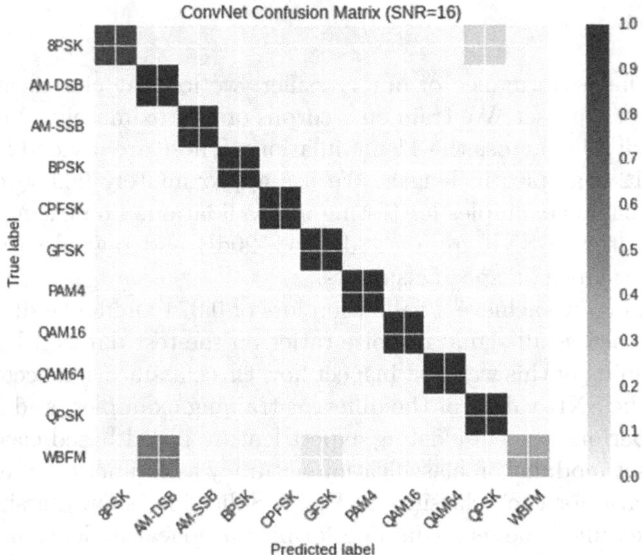

Fig. 8. Conv net confusion matrix at +18dB SNR

large convolutional neural networks (CNN2) with significant amounts of dropout regularization (0.6). At low-SNR the best CNN model is outperforming expert feature based systems by 2.5–5dB of SNR, while after +5dB SNR performance is similar. This is a significant performance improvement, and one that could potentially at least double effective coverage area of a sensing system.

For our highest SNR case CNN2(0.6) classification we show a confusion matrix in Fig. 8. At +18dB SNR, we have a clean diagonal in the confusion matrix and can see our remaining discrepancies are that of 8PSK misclassified as QPSK, and WBFM misclassified as AM-DSB. Both of these are explainable in the underlying dataset. An 8PSK symbol containing the specific bits is indiscernible from QPSK since the QPSK constellation points are spanned by 8PSK points. In the case of WBFM/AM-DSB the analog voice signal has periods of silence where only a carrier tone is present making these examples indiscernible. Therefore it is unlikely that and accuracy of 100 % can be obtained even at high SNR on this data set and making the remaining confusion reasonably tolerated.

To get a better understanding of how performance varies with SNR, we inspect confusion matrices for several classifiers at differing SNR levels.

At very low SNR (−6dB), in Figs. 9, 10, 11, and 12 we see an interesting case where all are around 50 % accuracy within +−20 %. In this case the cleaner diagonal on the CNN2 classifier is significantly more pronounced than the other 3 cases shown, in this region learned features have a significant performance advantage.

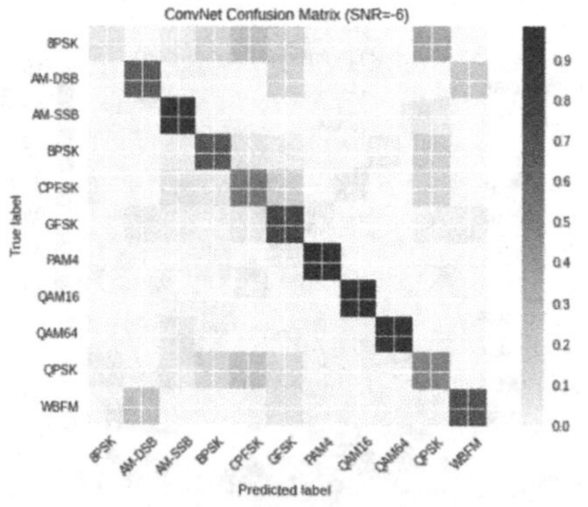

Fig. 9. −6dB performance of CNN2 on raw sample data

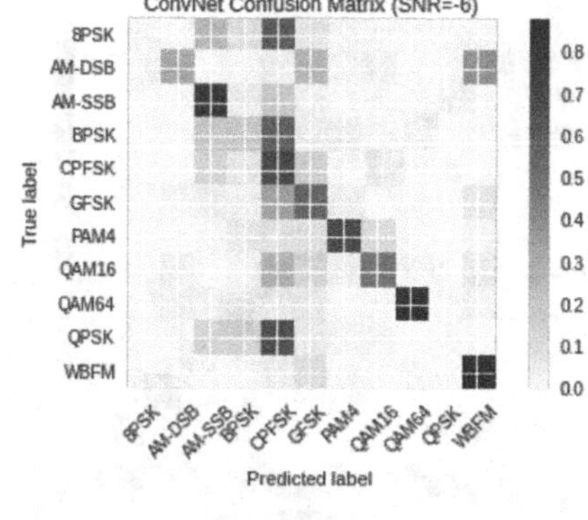

Fig. 10. −6dB performance of DNN on expert features

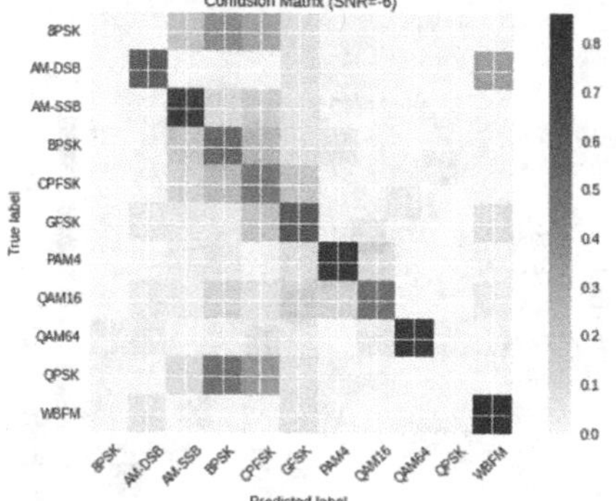

Fig. 11. −6dB performance of SVM on expert features

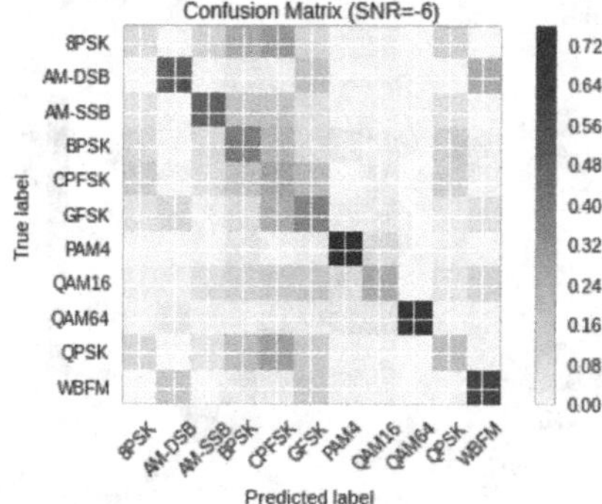

Fig. 12. −6dB performance of decision tree on expert features

At slightly higher, but still low SNR (0dB) performance for all 4 classifiers now has a well defined diagonal, but we see less mis-classifications occurring off-diagonal, especially in the 8PSK case.

6 Model Complexity

An important consideration in many radio system is the training and classification run time due to computational complexity. One common critique of deep learning is its need for large amounts of compute resources, however in this paper our network is relatively compact and the dataset relatively small. We compare the training and classification run times for each of the models below (Figs. 13, 14, 15 and 16).

In Fig. 17 we can see that our CNN model does take a significant amount of time to train, but is lower than the time required for for the SVM training case.

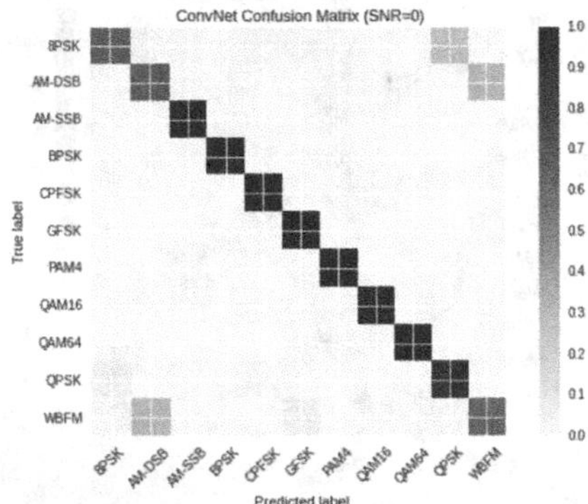

Fig. 13. 0dB performance of CNN2 on raw sample data

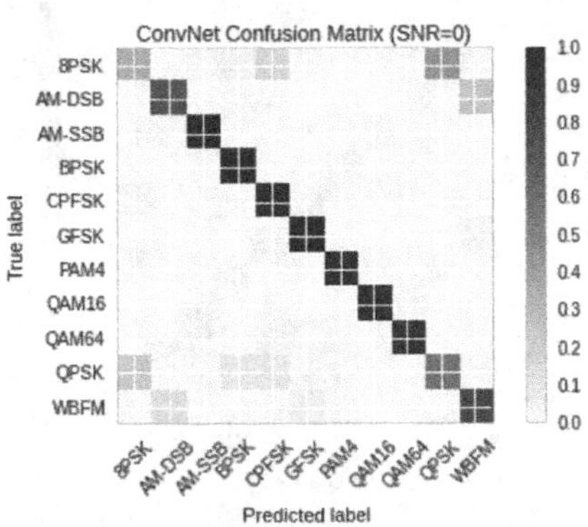

Fig. 14. 0dB performance of DNN on expert features

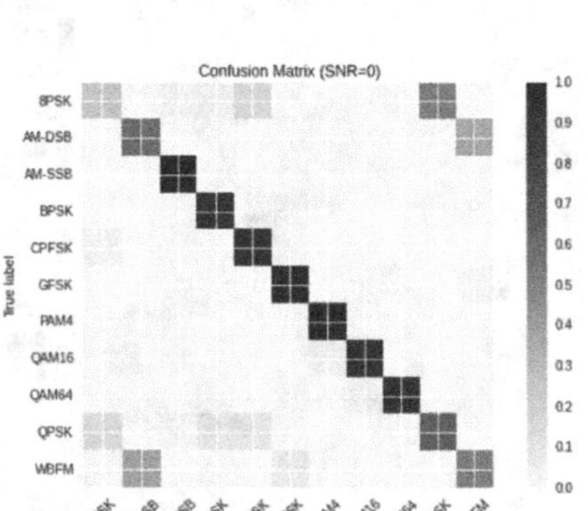

Fig. 15. 0dB performance of SVM on expert features

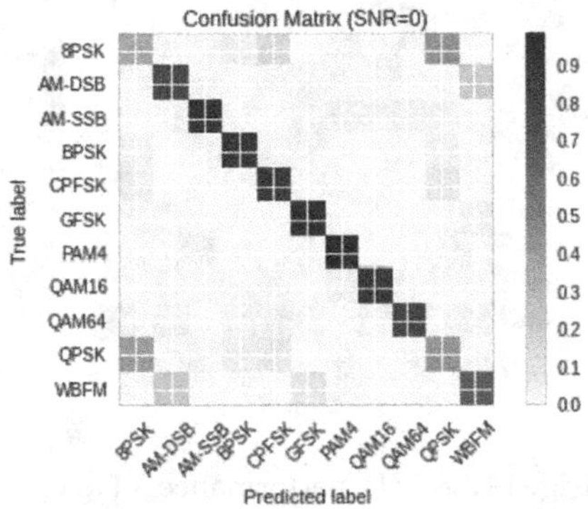

Fig. 16. 0dB performance of decision tree on expert features

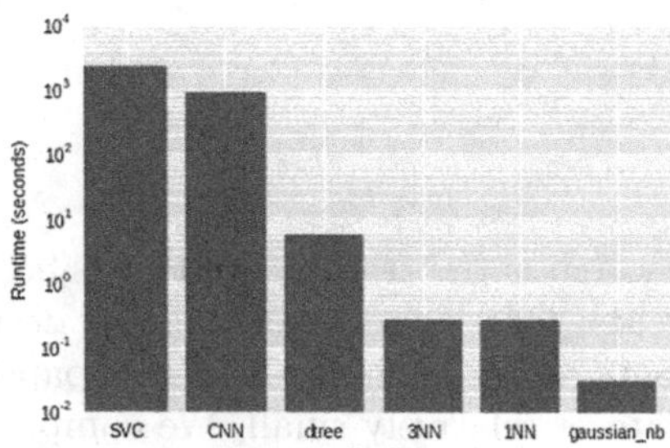

Fig. 17. Model training runtime in seconds

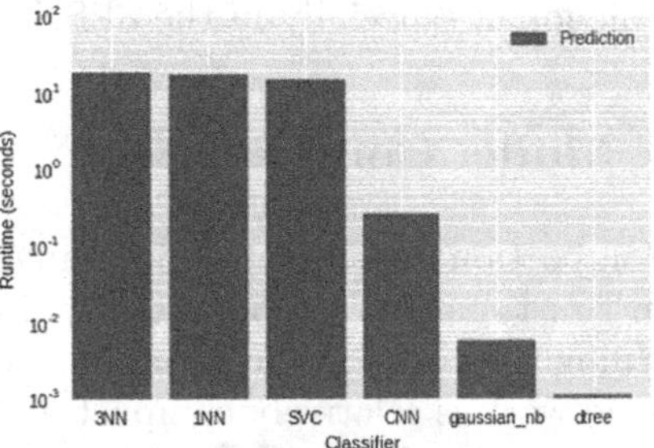

Fig. 18. Signal classification time in seconds (per SNR-batch)

In Fig. 18 it turns out that classification time with this model using Keras compiled python is significantly faster than most of the other models including nearest-neighbor and SVM models using scikit-learn. Only the Decision Tree and GaussianNB models obtain faster classification run times.

In both cases, a ConvNet based classification model of this scale for such a dataset presents an attractive choice for this task when classification performance is considered.

7 Conclusions

While these results are not a comprehensive comparison of existing best case expert feature based modulation classifiers, they do demonstrate that compared to a relatively well expert regarded approach, blind Convolutional Networks on time series radio signal data are viable and work quite well. In Fig. 7, we compared accuracy to SNR for several classifier strategies and believe that for low SNR and short-time examples (128 complex samples), this represents a powerful and likely state of the art accuracy approach to modulation classification. This approach holds the potential to easily scale to additional modulation classes and should be considered as a strong candidate for DSA and CR systems which rely on robust low SNR classification of radio emitters.

8 Future Work

Our results compare to a reasonable approximation of the current best expert system approach, but because no robust competition data sets exist in the emerging field of machine learning in the radio domain, it is difficult to directly compare performance to current state of the art approaches. We hope to further evaluate this in later work, and improve both the feature learning and expert approaches from their current level. Performance refinements are inevitable on the CNN2 network architecture, we expended some effort optimizing it but did not do so exhaustively. Larger filters, differing architecture, and pooling layers all may affect performance significantly, but were not fully investigated for their suitability in this work. Numerous additional techniques could be applied to the problem including the introduction of invariance to additional channel induced effects such as dilation, I/Q imbalance, phase offset and others. Spatial Transformer Networks [17] have demonstrated a powerful ability to learn this type of invariance on image data and may serve as an interesting candidate for enabling improved invariance learning to these effects. Sequence models and recurrent layers [13] may be able to represent a signal sequence embedding and will almost certainly prove valuable in longer time representation, but we have yet to investigate this area fully. This application domain is ripe for a wide array of further investigation and applications which will significantly impact the state of the art in wireless signal processing and cognitive radio domains, shifting them more towards machine learning and data driven approaches.

Acknowledgments. The authors would like to thank the Bradley Department of Electrical and Computer Engineering at the Virginia Polytechnic Institute and State University, the Hume Center, and DARPA all for their generous support in this work.

This research was developed with funding from the Defense Advanced Research Projects Agency's (DARPA) MTO Office under grant HR0011-16-1-0002. The views, opinions, and/or findings expressed are those of the author and should not be interpreted as representing the official views or policies of the Department of Defense or the U.S. Government.

References

1. Gardner, W.A., Spooner, C.M.: Signal interception: Performance advantages of cyclic-feature detectors. IEEE Trans. Commun. **40**(1), 149–159 (1992)
2. Mitola III, J., Maguire Jr., G.Q.: Cognitive radio: Making software radios more personal. IEEE Pers. Commun. **6**(4), 13–18 (1999)
3. Blossom, E.: GNU radio: Tools for exploring the radio frequency spectrum. Linux J. **2004**(122), 4 (2004)
4. Kolodzy, P.J.: Dynamic spectrum policies: Promises and challenges. CommLaw Conspectus **12**, 147 (2004)
5. Lee, H., Battle, A., et al.: Efficient sparse coding algorithms. In: Advances in Neural Information Processing Systems, pp. 801–808 (2006)
6. Clancy, C., Hecker, J., Stuntebeck, E., O'Shea, T.: Applications of machine learning to cognitive radio networks. IEEE Wirel. Commun. **14**(4), 47–52 (2007)
7. Kim, K., Akbar, I.A., et al.: Cyclostationary approaches to signal detection and classification in cognitive radio. In: New Frontiers in Dynamic Spectrum Access Networks, pp. 212–215 (2007)
8. Nvidia, C.: Compute unified device architecture programming guide (2007)
9. Rondeau, T.W.: Application of artificial intelligence to wireless communications. Ph.D. thesis, Virginia Tech (2007)
10. Zeiler, M.D., Krishnan, D., Taylor, G.W., Fergus, R.: Deconvolutional networks. In: 2010 IEEE Conference on Computer Vision and Pattern Recognition (CVPR), pp. 2528–2535. IEEE (2010)
11. Krizhevsky, A., Sutskever, I., Hinton, G.E.: Imagenet classification with deep convolutional neural networks. In: Advances in Neural Information Processing Systems, pp. 1097–1105 (2012)
12. Tieleman, T., Hinton, G.: Lecture 6.5-rmsprop: Divide the gradient by a running average of its recent magnitude. In: COURSERA: Neural Networks for Machine Learning, vol. 4, p. 2 (2012)
13. Graves, A., Mohamed, A., Hinton, G.E.: Speech recognition with deep recurrent neural networks. CoRR, vol. abs/1303.5778 (2013). http://arxiv.org/abs/1303.5778
14. O'Shea, T.: GNU radio channel simulation. In: GNU Radio Conference (2013)
15. Kingma, D.P., Ba, J.: Adam: A method for stochastic optimization. CoRR, vol. abs/1412.6980 (2014). http://arxiv.org/abs/1412.6980
16. Chollet, F.: Keras (2015). https://github.com/fchollet/keras
17. Jaderberg, M., Simonyan, K., Zisserman, A., Kavukcuoglu, K.: Spatial transformer networks. CoRR, vol. abs/1506.02025 (2015). http://arxiv.org/abs/1506.02025
18. Sainath, T.N., et al.: Learning the speech front-end with raw waveform CLDNNS. In: Proceedings of the Interspeech (2015)
19. Abadi, M., Agarwal, A., et al.: Tensorflow: Large-scale machine learningon heterogeneous systems (2015). Software available from https://www.tensorflow.org

Mutual Information with Parameter Determination Approach for Feature Selection in Multivariate Time Series Prediction

Tianhong Liu, Haikun Wei(✉), Chi Zhang, and Kanjian Zhang

Key Laboratory of Measurement and Control of CSE, Ministry of Education,
School of Automation, Southeast University,
Nanjing 210096, People's Republic of China
hkwei@seu.edu.cn

Abstract. For modeling of multivariate time series, input variable selection is a key problem. Feature selection is to select a relevant subset to reduce the dimensionality of the problem without significant loss of information. This paper presents the estimation of mutual information and its application in feature selection problem. Mutual information is one of the most common strategies borrowed from information theory for feature selection. However, the calculation of probability density function (PDF) according to the definition of mutual information is difficult, especially for high dimensional variables. A k-nearest neighbor (k-NN) method based estimator is widely used to estimate the mutual information between two variables directly from the data set. Nevertheless, this estimator depends on smoothing parameter. There is no theoretically method to choose the parameter. This paper purposes to solve two problems: one is to employ resampling methods to help the mutual information estimator to improve feature selection and the other is to apply these methods to a wind power prediction problem.

Keywords: Mutual information · Feature selection · K-nearest neighbor · Permutation test · Wind power prediction

1 Introduction

Multivariate time series are widely exist in the real world. Such as in economics [1], meteorology [2] and many other fields. It has been proved that the forecasting model with multivariate time series can achieve higher accuracy than those with univariate time series [3]. Input data of the multivariate time series are originally high-dimensional. However, many of the features are either irrelevant or redundant to the real problem. In order to avoid the curse of dimensionality caused by learning with large number of features and limited sample size, feature selection is necessary before any further learning steps. Selecting features is important in practice, especially when distance-based methods like k-NN, radial basis function networks (RBF) and support vector machines (SVM) are considered.

C. Jayne and L. Iliadis (Eds.): EANN 2016, CCIS 629, pp. 227–237, 2016.
DOI: 10.1007/978-3-319-44188-7_17

These methods are indeed quite sensitive to irrelevant inputs: their performances tend to decrease when useless variables are added to the data [4].

To tackle the feature selection problem [5], two broad classes: *filter* and *wrapper* methods are concerned. The filter approach consists in a preprocessing of the input data before the model is built. The wrapper approach attempts to design the model at the same time that performs the feature selection. In this paper, both of the two methods are involved.

For the input selection strategy, two different strategies are commonly used: forward selection and backward elimination. The backward elimination procedure starts to build a model from all the initial features. With high-dimensional data, this procedure will be too time consuming. A forward selection procedure starts from an empty set, then the selected feature input is added to the empty set one by one. Mutual information (MI) is unique in its close ties to Shannon entropy [6] and the theoretical advantages derived from this. It has been widely used for feature selection tasks [7–13]. The combination of mutual information and a forward procedure is considered to be an option of feature selection in this paper.

According to the definitions of mutual information, two different methods can be used to estimate MI values. One is to compute the PDFs of the input variables and the other is to estimate the entropies instead. The widespread approaches for estimating MI by computing PDFs are histogram methods, kernel-based estimators. The k-NN approach which was popularized by Kraskov et al. [14] has been widely used because of its ability to estimate the MI directly from the data set. This avoids direct PDF estimation. Nevertheless, choosing an optimal value of k in a practical application is always a problem, due to the fact that only a finite amount of training data is available. This problem is well known as the bias/variance dilemma in the statistical learning community [5]. The result of the feature selection highly depends on the value of the parameter k. By choosing proper values, the algorithm allows to minimize either the statistical or the systematic errors.

This paper proposes to apply resampling methods to help the MI estimators to choose k value. A synthetic data set is to demonstrate the effectivity of the methods. After the k is determined, a proposed procedure is then used to select features. Multilayer Perceptron (MLP) neural network is as the nonlinear model. The application of these methods is illustrated on a real-world wind power prediction problem. The data collected have eighteen inputs and one output. Relevant features are selected and the performance can be improved obviously.

The structure of the rest is organized as follows. Section 2 presents the MI criterion and shows how to extend the MI concept by using nearest neighbor when the dimension of the original space is high. In Sect. 3, the number of neighbors is determined and the selection procedure is given. The application of these methods is outlined in Sect. 4. Finally, conclusions are given in Sect. 5.

2 Feature Selection Based on Mutual Information

2.1 Mutual Information

A powerful formalization of the uncertainty of random variables is Shannon's entropy. Denote the marginal density functions and joint probability density function of two random variables X and Y as $\mu_X(x)$, $\mu_Y(y)$ and $\mu_{XY}(x,y)$. The marginal densities of X and Y are $\mu_X(x) = \int \mu(x,y)dy$ and $\mu_Y(y) = \int \mu(x,y)dx$. According to the formulation of Shannon [6], the uncertainty on X is given by its entropy defined as

$$H(X) = -\int \mu_X(x) log \mu_X(x)dx \tag{1}$$

$$H(Y) = -\int \mu_Y(y) log \mu_Y(y)dy \tag{2}$$

If Y depends on X, the uncertainty on Y is reduced when X is known. This is formalized through the concept of conditional entropy:

$$H(Y|X) = -\int \mu_X(x) \int \mu_Y(y|X=x) log \mu_Y(y|X=x)dydx \tag{3}$$

The joint entropy is used to examine the amount of information among multiple variables. The joint entropy of the two continuous random variables X and Y is as follows:

$$H(X,Y) = -\int\int \mu_{XY}(x,y) log \mu_{XY}(x,y)dxdy \tag{4}$$

The MI is aimed to measure the loss of uncertainty on Y when X is know. Due to the mutual information and entropy properties, the mutual information can be defined as [3]:

$$I(X,Y) = H(X) + H(Y) - H(X,Y) \tag{5}$$

The MI between two random variables X and Y can also be computed as:

$$I(X,Y) = \int\int \mu_{XY}(x,y) log \frac{\mu_{XY}(x,y)}{\mu_X(x)\mu_Y(y)} dxdy \tag{6}$$

It corresponds to the Kullback-Leibler distance between the joint probability density of X and Y, and the product of their respective marginal distributions. Only the estimate of the joint PDF between X and Y is needed to estimate the MI between the two variables.

2.2 Mutual Information Estimation

The k-NN method has been widely used as estimator for the entropy and had been extended to the MI by Alexander Kraskov etc. [14]. The MI means to answer the question whether some knowledge on the value of X may help indentifying what can be the possible values for Y. For a specific data set, if its neighbors in the X and Y spaces correspond to the same data, then knowing X helps in knowing Y, which reflects a high MI.

Assume some metrics to be given on the spaces spanned by X, Y and $Z = (X, Y)$. For each point $z_i = (x_i, y_i)$, its neighbors by distance $d_{i,j} = \|z_i - z_j\|$. x and y can be either a scalar or vector. The maximum norm for any pair of points z and z' is defined by:

$$\left\|z - z^{'}\right\| = max\left\{\left\|x - x^{'}\right\|, \left\|y - y^{'}\right\|\right\} \tag{7}$$

while any norm can be used for $\left\|x - x^{'}\right\|$ and $\left\|y - y^{'}\right\|$. Denote by $\epsilon(i)/2$ the distance from z_i to its k-th neighbors, and by $\epsilon_x(i)/2$ and $\epsilon_y(i)/2$ the distances between the same points projected into the X and Y subspaces. Obviously, $\epsilon(i) = max\{\epsilon_x(i), \epsilon_y(i)\}$.

If $n_x(i)$ is the number of the points x_j whose distance from x_i is strictly less than $\epsilon(i)/2$, and similarly for y instead of x. The estimate of MI is then

$$I(X,Y) = \psi(k) + \psi(N) - \frac{1}{N}\sum_{i=1}^{N}[\psi(n_x(i)+1) + \psi(n_y(i)+1)] \tag{8}$$

Here, $\psi(x)$ is the digamma function,

$$\psi(x) = \Gamma(x)^{-1} d\Gamma(x)/dx \tag{9}$$

with

$$\Gamma(t) = \int_0^\infty u^{t-1}e^{-u}du \tag{10}$$

The digamma function satisfies the recursion $\psi(x+1) = \psi(x) + 1/x$ and $\psi(1) = -C$ where $C = 0.5772156\cdots$ is the Euler-Mascheroni constant. The estimator presented in Eq. (8) is one of the two proposed in Kraskov's literature.

3 Proposed Method

3.1 The Number of Nearest Neighbors

In the estimation of MI, the number of neighbors k acts as a smoothing parameter. With a small value of k, the estimator has a large variance and a small bias, whereas a large value of k leads to a small variance and a large bias. In a word, while the k-NN may be helpful for measuring the MI between variables, it still suffers from the limitations of choosing appropriate parameter k. The aim of this subsection is to investigate this problem.

Since the effectiveness of Eq. (8) depends on careful choose of neighbors k, resampling techniques based on [4,15] are used in this paper. A different evaluation criterion is applied. A cross-validation approach is used to evaluate the variance of the estimator while a permutation method provide some baseline value of the mutual information that can reduce the influence of the bias. Consider X_i is the input feature, a randomized Y_i is denoted as $Y_{i,\pi}$. Two resampling distributions are built for both $MI(X_i, Y)$ and $MI(X_i, Y_\pi)$. $MI(X_i, Y)$ is the estimations of the mutual information between X_i and Y, while $MI(X_i, Y_\pi)$ is the mutual information between a randomized version of Y_π and X_i. The procedure results in two samples of estimates of $MI(X_i, Y)$ and $MI(X_i, Y_\pi)$, where π denotes the permutation operation. This is done by performing several estimations of (i) the mutual information between X_i and Y and (ii) the mutual information between Y_π and X_i, using several non-overlapping subsets of the original sample, in a cross-validation resampling scheme. To evaluate the differences between them, a *z-test* is used to determine the optimal value of k instead of using *t-test* as [4,15] did. Because *z-test* is a commonly used average difference test method in large sample (i.e., sample size greater than 30). The optimal value of k is the one that best separates those two distributions according to the *z-test*:

$$z_{i,k} = \frac{\mu - \mu_\pi}{\sqrt{\frac{\sigma^2}{N} + \frac{\sigma_\pi^2}{N}}} \tag{11}$$

where μ and μ_π denote the empirical mean of $MI(X_i, Y)$ and $MI(X_i, Y_\pi)$, σ and σ_π denote their empirical standard deviation respectively. The optimal k is chosen as the value of k that corresponding to the largest value of $z_{i,k}$.

3.2 The Procedure of Feature Selection

The primary aim of feature selection is to improve the quality of the prediction though the choice of feature inputs. Both the *filter* and *wrapper* methods are involved in this paper. The *filter* method is used at the first two steps and the *wrapper* method is used at the last step. The procedure is shown as the following steps.

Step 1: Form the initial input and output pairs in accordance with the original dataset. The input-output pairs can be expressed as $S_o = \{(X, Y)\}$. Select an optimal value of k according to the resampling methods mentioned in the Sect. 3.1.

Step 2: Calculate the mutual information between the input and output variables $MI(X_i, Y)$ on the basis of Eq. (8), i is the number of the original input series. Maintain the MI values in vector M.

Step 3: Set a threshold α and a span β on the mutual information according to the computed MI values. If $MI_i < \alpha$, the feature is deleted from M. The rest of the features are used as inputs to construct the MLP forecasting model. If $MI_i > \alpha + \beta$, these features are used as model inputs. And by this analogy, if $MI_i > \alpha + n\beta$, $(1 \leqslant n < i$, n is determined according to the specific problem), the selected features are applied to construct forecasting model. Compare the performances of all the groups and select the best one as feature input set (This step will be described in the real world application in detail).

4 Experiments and Discussion

The mutual information, with a nearest neighbor-based estimator, and the forward search combined together present a good compromise between computation time and performances. Two experiments are examined to illustrate the parameter setting in the MI estimation, the effectiveness of feature selection and the application of these methods.

4.1 A Simulation Study

A synthetic data is used in this section to illustrate the effect of the parameter. This data is derived from the following functions. These functions refer to the example in references [16,17]. Some modifications have been made to assess the efficiency of the method in this paper. Five input variables $X_i, 1 \leq i \leq 5$ and one output variable Y are given as follow:

$$\begin{cases} X1 = t^2 - t + e^t + \epsilon_1 \\ X2 = sin(t) + \epsilon_2 \\ X3 = t^3 + t + \epsilon_3 \\ X4 = t^2 + 2 * t + cos(t) + \epsilon_4 \\ X5 = t * sin(t) - t^2 + \epsilon_5 \end{cases} \tag{12}$$

$$Y = X1^2 + X1 * X2 + 4 * cos(X3) + 10 * X4 + \epsilon_6 \tag{13}$$

where t is uniformly distributed over $[-1, 1]$, and $\epsilon_i, 1 \leq i \leq 6$ are uniformly distributed noise over $[-0.1, 0.1]$. $X1, X2, X3, X4$ are relevant input variables while $X5$ are irrelevant variables and have no predictive power. The sample size

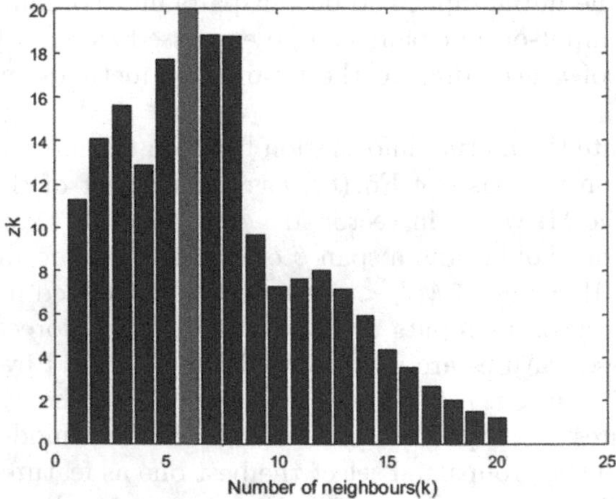

Fig. 1. Values of z_k for $MI(X_4, Y).(k = 6)$.

is 2000. Apparently, the variable X_4 is a relevant feature. When evaluating the MI between Y and X_4, a z_k value is obtained for each value of k, as shown in Fig. 1. The largest z_k value is when $k = 6$ in the $MI(X_4, Y)$. This value corresponds to the parameter k that best separates the relevant variable and its permutation variable. The other k values can be calculated in a similar manner. The optimal k for all features is chosen as the one corresponding to the largest value of z_k over all values of k over all features. This way, features that are useless do not participate in the choice of the optimal value. In this paper, k is chosen as 8. The computed $MI_i = I(X_i, Y)$ values are presented in Table 1, from which it can be seen that the proposed method helps to select four relevant variables and can successfully avoid the influence of irrelevant variable.

Table 1. Mutual information value of the 5 variables

MI	MI_1	MI_2	MI_3	MI_4	MI_5
Value	0.101	0.161	0.435	0.652	0.0001

4.2 Application in the Wind Power Prediction

To evaluate these methods, a wind data collected by the low-wind-speed wind turbine FD-77 from a wind power plant of Jiangsu province in China is used for this study. Wind power is undergoing the fastest rate of growth of any form of electricity generation in the world. As wind power technology has become mature, it can now be considered as a valuable supplement to conventional energy sources. This data were recorded from June 1, 2012 to December 31, 2012 and sampled once every five minutes. It has eighteen inputs and one output. The current moment inputs are used to predict the current moment output. The variables of the dataset are shown in Table 1. The columns represent the average speed, the average wind direction, the standard deviation, wind speed, wind direction respectively. The lines represent the heights of the measuring points. They are 10 m, 50 m, and 70 m. The last two lines are the environment temperature, humidity and pressure. The output is the wind power which is represented as Y. To get a precise prediction, irrelevant inputs should be removed from the feature inputs. Thus, feature selection is an important preprocessing step. To construct experiments, 1000 samples are portioned into two parts: the first 700 samplings for training and the rest 300 samplings for testing.

To evaluate the accuracy of the forecasting model, two different criterions including the root mean squared error (RMSE) and the mean absolute error (MAE) are used in the experiments. These performance indexes can be written as

$$RMSE = \sqrt{\frac{1}{N}\sum_{i=1}^{N}[\hat{y}_i - y_i]^2} \tag{14}$$

$$MAE = \frac{1}{N}\sum_{i=1}^{N}|\hat{y}_i - y_i| \tag{15}$$

Table 2. Variables of the wind data

	avgsp(m/s)	avgdir(°)	standard	sp(m/s)	dir
10 m	x_1	x_2	x_3	x_4	x_5
50 m	x_6	x_7	x_8	x_9	x_{10}
70 m	x_{11}	x_{12}	x_{13}	x_{14}	x_{15}
	temp(°C)	hum	pressure(0.1KPa)	power(W)	
	x_{16}	x_{17}	x_{18}	Y	

where N is the number of observed values, $\hat{y_i}$ is the predicted values and y_i is the actual values (Table 2).

Multilayer perceptron (MLP) neural networks are employed as the prediction model. Data enter the MLP from the input layer and the final response is presented from the output layer. The hidden layer usually plays a filtering and synthesis role for the inputs. The optimal value (on the learning set) of k, searched between 1 and 20, is found to be 15. Figure 2 displays the mutual information between 18 variables and the power output. All relevant features have higher mutual information then non-relevant ones only for well-chosen values. According to the step 3 in Sect. 3.2, the MI threshold $\alpha = 0.1$, the span $\beta = 0.1$ in this problem. If $MI_i < \alpha$, the features are deleted. Compute the RMSE and MAE values to get the model performance. And by this analogy, if $MI_i > \alpha + \beta = 0.2$, $MI_i > \alpha + 2\beta = 0.3$, etc., compute the RMSE and MAE values respectively. Table 3 illustrates that the selection of feature inputs according to the MI values. Table 4 shows the averaged RMSE and MAE of the 10 runs with MLP for wind power prediction. From Table 4 it can be seen that all the selected feature inputs improve the forecasting performance. The mutual

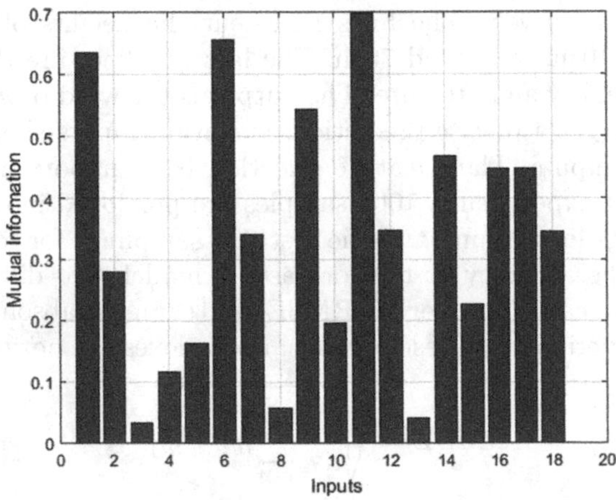

Fig. 2. Mutual information between 18 variables and the wind power output

Table 3. Select feature inputs according to the MI values

MI value	Symbol	Feature inputs
> 0.1	X_1	$X \setminus x_3x_8x_{13}$
> 0.2	X_2	$X \setminus x_3x_4x_5x_8x_{10}x_{13}$
> 0.3	X_3	$x_1x_6x_7x_9x_{11}x_{12}x_{14}x_{16}x_{17}x_{18}$
> 0.4	X_4	$x_1x_6x_9x_{11}x_{14}x_{16}x_{17}$
> 0.5	X_5	$x_1x_6x_9x_{11}$
> 0.6	X_6	$x_1x_6x_{11}$

information of x_3, x_8, x_{13} are much smaller than other values as shown in Fig. 2. These variables are removed from the input set. It means that the standard deviations are irrelevant to the output. The $X_5(x_1, x_6, x_9, x_{11})$ has the smallest values of RMSE and MAE. This can be interpreted as the input set X_5 can represent the relevance between the inputs and the output. Another input set X_6 contains three variables, however, the RMSE and MAE are slightly larger then that of X_5. That is to say, the variable x_9 has some positive effect on the forecasting performance. Thus, the number of feature inputs is 4 in this experiment. The feature inputs are the wind speed variables. The variable selection criteria based on the MI estimator and the selection procedure work well for relative feature selection and removing redundant variables. The variables they select perform well for forecasting models.

Table 4. Select feature inputs according to the MI values

	Original inputs	X_1	X_2	X_3	X_4	X_5	X_6
RMSE(W)	3.94	3.19	3.27	3.79	2.01	1.89	1.90
MAE(W)	3.49	2.66	3.44	3.04	1.57	1.46	1.47

5 Conclusion

Feature selection problem encountered in multivariate time series forecasting is investigated in this paper. It is a fundamental preprocessing step in supervised regression problems. Combining the use of the MI and a forward procedure is a good option for feature selection. The main disadvantage of this method is that the estimation of the MI is often difficult in high-dimensional spaces. Nearest neighbor-based MI estimator is one of the acceptable approaches for such estimation. However, the number of neighbors should be determined before estimating the MI values. The proposed procedure first helps to determine the k value. In this paper, the optimal number of neighbors is chosen by using the permutation resampling methods and the measuring standard is the *z-test*.

Once the number of neighbors is fixed, the feature selection procedure is used to select optimal inputs. The third step of the procedure is in a wrapper manner. A threshold is set to give a criterion criteria of selection.

The methods applied to a wind power prediction indicated that, the k value and the MI effect the feature selection and improve the forecasting performances. The relevant inputs are selected from the original data set. The model uses selected features achieved better accuracy than the model trained with the original set. It has been shown that the quality of the inputs is more significant than the size of the training set. The proposed method can be applied to other multivariate time series forecasting problems.

Acknowledgments. The authors gratefully acknowledge the financial support of this research by the National Natural Science Foundation of China (Grant No. 61374006), the Major Program of National Natural Science Foundation of China (Grant No. 11190015) and the Natural Science Foundation of Jiangsu (Grant No. BK20131300).

References

1. Keynia, F.: A new feature selection algorithm and composite neural network for electricity price forecasting. Eng. Appl. Artif. Intell. **25**(8), 1687–1697 (2012)
2. Wu, C.L., Chau, K.W.: Prediction of rainfall time series using modular soft computing methods. Eng. Appl. Artif. Intell. **26**(3), 997–1007 (2013)
3. Du Preez, J., Witt, S.F.: Univariate versus multivariate time series forecasting: An application to international tourism demand. Int. J. Forecast. **19**(3), 435–451 (2003)
4. Franois, D., Rossi, F., Wertz, V.: Resampling methods for parameter-free and robust feature selection with mutual information. Neurocomputing **70**, 1276–1288 (2007)
5. Geman, S., Bienenstock, E., Doursat, R.: Neural networks and the bias/variance dilemma. Neual. Comput. **4**(1), 1–58 (1992)
6. Shannon, C.E.: A mathematical theory of communication. Bell Syst. Tech. J. **27**, 379–423 (1948)
7. Frnay, B., Doquire, G., Verleysen, M.: Is mutual information adequate for feature selection in regression? Neural Netw. **48**, 1–7 (2013)
8. Han, M., Ren, W.J., Liu, X.X.: Joint mutual information-based input variable selection for multivariate time series modeling. Eng. Appl. Rtif. Intel. **37**, 250–257 (2015)
9. Lin, Y., Hu, Q., Liu, J., Chen, J., Duan, J.: Multi-label feature selection based on neighborhood mutual information. Appl. Soft. Comput. **38**, 244–256 (2016)
10. Battiti, R.: Using mutual information for selecting features in supervised neural net learning. IEEE Trans. Neural Netw. **5**, 537–550 (1994)
11. Van Dijck, G., Van Hulle, M.M.: Speeding up the wrapper feature subset selection in regression by mutual information relevance and redundancy analysis. In: Kollias, S.D., Stafylopatis, A., Duch, W., Oja, E. (eds.) Artificial Neural Networks–ICANN 2006. LNCS, vol. 4131, pp. 31–40. Springer, Heidelberg (2006)
12. Fleuret, F.: Fast binary feature selection with conditional mutual information. J. Mach. Learn. Res. **5**, 1531–1555 (2004)

13. Rossi, F., Franois, D., Wertz, V., Meurens, M., Verleysen, M.: Fast selection of spectral variables with b-spline compression. Chemometr. Intell. Lab. **86**(2), 208–218 (2007)
14. Kraskov, A., Stgbauer, H., Grassberger, P.: Estimating mutual information. Phys. Rev. E **69**, 066–138 (2004)
15. Verleysen, M., Rossi, F., François, D.: Advances in feature selection with mutual information. In: Biehl, M., Hammer, B., Verleysen, M., Villmann, T. (eds.) Similarity-Based Clustering. LNCS, vol. 5400, pp. 52–69. Springer, Heidelberg (2009)
16. Stogbauer, H., Kraskov, A., Astakhov, S.A., Grassberger, P.: Least dependent component analysis based on mutual information. Phys. Rev. E **70**, 066–123 (2004)
17. Peng, H.C., Long, F.H., Ding, C.: Feature selection based on mutual information criteria of max-dependency, max-relevance, and min-redundancy. IEEE Trans. Pattern. Anal. Mach. Intell. **27**(8), 1226–1238 (2005)

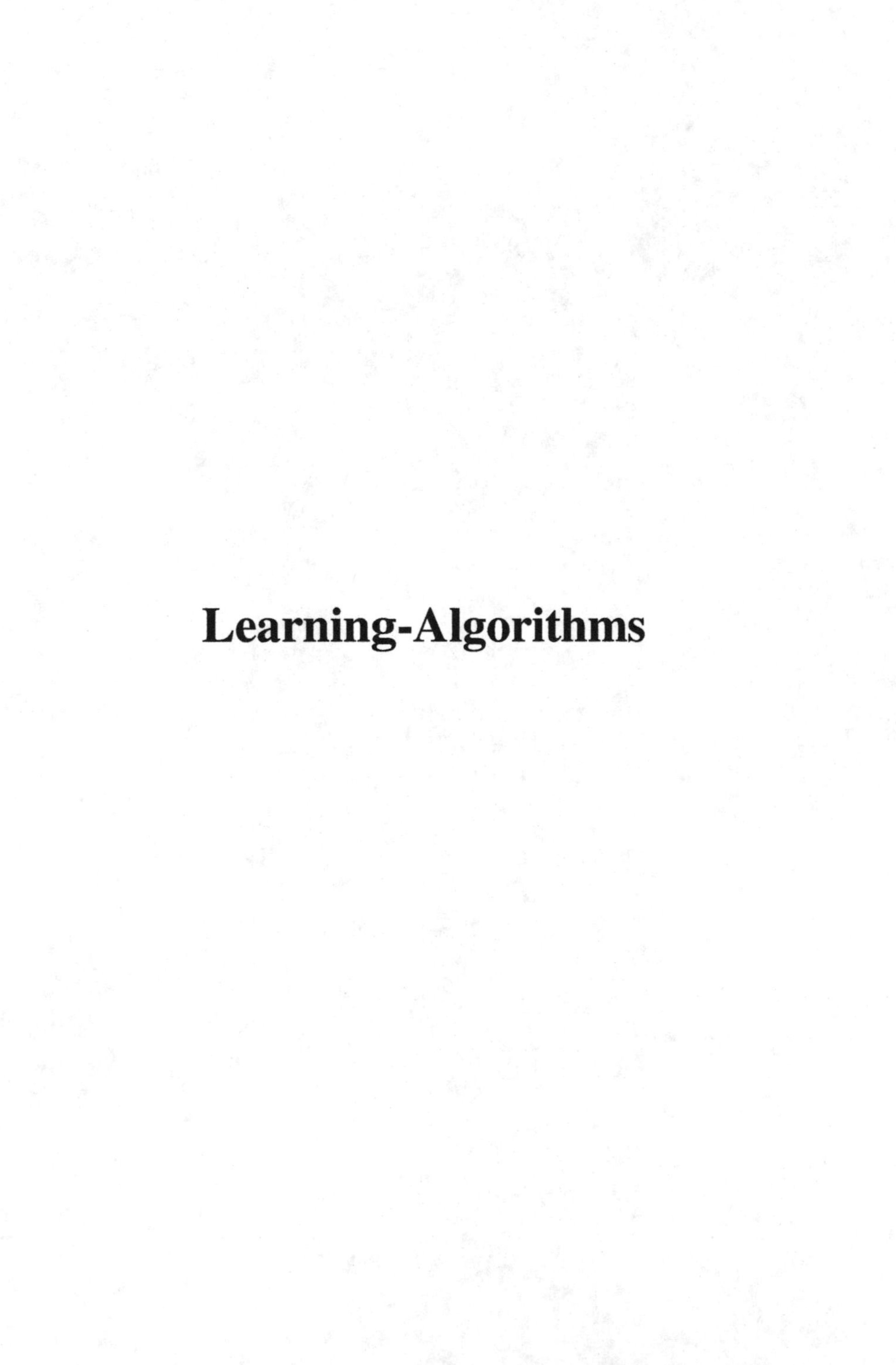

Learning-Algorithms

On Learning Parameters of Incremental Learning in Chaotic Neural Network

Toshinori Deguchi[1(✉)] and Naohiro Ishii[2]

[1] National Institute of Technology, Gifu College, Motosu 501–0495, Japan
deguchi@gifu-nct.ac.jp

[2] Aichi Institute of Technology, Toyota, Aichi 470–0392, Japan
ishii@aitech.ac.jp

Abstract. The incremental learning is a method to compose an associate memory using a chaotic neural network and provides larger capacity than correlative learning in compensation for a large amount of computation. A chaotic neuron has spatio-temporal sum in it and the temporal sum makes the learning stable to input noise. When there is no noise in input, the neuron may not need temporal sum. In this paper, to reduce the computations, a simplified network without temporal sum is introduced and investigated through the computer simulations comparing with the network as in the past. Then, to shorten the learning steps, the learning parameters are changed during the learning along 3 functions.

1 Introduction

The incremental learning proposed by the authors is highly superior to the autocorrelative learning in the ability of pattern memorization [1,2].

The idea of the incremental learning is from the automatic learning [3]. In the incremental learning, the network keeps receiving the external inputs. If the network has already known an input pattern, it recalls the pattern. Otherwise, each neuron in it learns the pattern gradually. Therefore, the weak point of the learning is computational complexity. The network takes steps to learn input patterns.

The neurons used in this learning are the chaotic neurons, and their network is the chaotic neural network, which was developed by Aihara [4].

A chaotic neuron has spatio-temporal sum in it and the temporal sum makes the learning possible with noisy inputs. But, when inputs don't include any noises, the neuron can be more simple without the temporal sum. This simplification reduces the computational complexity. In this paper, a network with spatio-temporal sum is called a usual network and a network without the sum is called a simplifed network.

In this paper, first, we explain the chaotic neural networks and the incremental learning, then simplify the network by eliminating the temporal sum from the chaotic neurons and examine the simplified network comparing with the usual network. Secondly, we investigate the learning, changing the learning parameter along a linear, exponential, and sigmoid function, to shorten the learning steps.

C. Jayne and L. Iliadis (Eds.): EANN 2016, CCIS 629, pp. 241–252, 2016.
DOI: 10.1007/978-3-319-44188-7_18

2 Chaotic Neural Networks and Incremental Learning

The incremental learning was developed by using the chaotic neurons. The chaotic neurons and the chaotic neural networks were proposed by Aihara [4].

We presented the incremental learning that provides an associative memory [1]. The network type is an interconnected network, in which each neuron receives one external input, and is defined as follows [4]:

$$x_i(t+1) = f\big(\xi_i(t+1) + \eta_i(t+1) + \zeta_i(t+1)\big), \tag{1}$$

$$\xi_i(t+1) = k_s \xi_i(t) + \upsilon A_i(t), \tag{2}$$

$$\eta_i(t+1) = k_m \eta_i(t) + \sum_{j=1}^{n} w_{ij} x_j(t), \tag{3}$$

$$\zeta_i(t+1) = k_r \zeta_i(t) - \alpha x_i(t) - \theta_i(1 - k_r), \tag{4}$$

where $x_i(t+1)$ is the output of the i-th neuron at time $t+1$, f is the output sigmoid function described below in (5), k_s, k_m, k_r are the time decay constants, $A_i(t)$ is the input to the i-th neuron at time t, υ is the weight for external inputs, n is the size—the number of the neurons in the network, w_{ij} is the connection weight from the j-th neuron to the i-th neuron, and α is the parameter that specifies the relation between the neuron output and the refractoriness.

$$f(x) = \frac{2}{1 + \exp(\frac{-x}{\varepsilon})} - 1. \tag{5}$$

In the incremental learning, each pattern is inputted to the network for some fixed steps before moving to the next. In this paper, this term is called "input period", and "one set" is defined as a period for which all the patterns are inputted. The patterns are inputted repeatedly for some fixed sets.

During the learning, a neuron which satisfies the condition of (6) changes the connection weights as in (7) [1].

$$\xi_i(t) \times (\eta_i(t) + \zeta_i(t)) < 0. \tag{6}$$

$$w_{ij} = \begin{cases} w_{ij} + \Delta w, & \xi_i(t) \times x_j(t) > 0 \\ w_{ij} - \Delta w, & \xi_i(t) \times x_j(t) \le 0 \end{cases} \quad (i \neq j), \tag{7}$$

where Δw is the learning parameter.

If the network has learned a currently inputted pattern, the mutual interaction $\eta_i(t)$ and the external input $\xi_i(t)$ are both positive or both negative at all the neurons. This means that if the external input and the mutual interaction have different signs at some neurons, a currently inputted pattern has not been learned completely. Therefore, a neuron in this condition changes its connection weights. To make the network memorize the patterns firmly, if the mutual interaction is less than the refractoriness $\zeta_i(t)$ in the absolute value, the neuron also changes its connection weights.

In this learning, the initial values of the connection weights can be 0, because some of the neurons' outputs are changed by their external inputs and this makes the condition establish in some neurons. Therefore, all initial values of the connection weights are set to be 0 in this paper. $\xi_i(0)$, $\eta_i(0)$, and $\zeta_i(0)$ are also set to be 0.

To confirm that the network has learned a pattern after the learning, the pattern is tested on the normal Hopfield's type network which has the same connection weights as the chaotic neural network. That the Hopfield's type network with the connection weights has the pattern in its memory has the same meaning as that the chaotic neural network recalls the pattern quickly when the pattern inputted. Therefore, it is the convenient way to use the Hopfield's type network to check the success of the learning.

3 Simulations with Simplified Network

3.1 Capacity

The simplified network is given by letting all the k-parameters be zero to eliminate the temporal sum, namely, $k_s = k_m = k_r = 0$.

In this paper, a set of patterns are generated and used to be learned by the networks. These patterns are random patterns generated with the method that all elements in a pattern are set to be -1 at first, then the half of the elements are chosen at random to turn to be 1.

For preliminary simulations, we checked how many patterns the networks can learn.

From the result of the former work [5,6], the parameters are assigned in Table 1 for the usual network. Both the input period and the number of sets are set to be 100. These numbers are obtained experimentally.

Table 1. Parameters in incremental learning and chaotic neuron

$\Delta w = 3 \times 10^{-6}$,
$\alpha = 6 \times 10^{-4}$,
$k_s = 0.95$,
$k_m = 0.1$,
$k_r = 0.95$,
$v = 2.0$,
$\theta_i = 0$,
$\varepsilon = 0.015$

The result of the simulation on the usual 100-neuron network is shown in Fig. 1. From the result, until 162 patterns, all the inputted patterns are learned completely. In this paper, this number is called the capacity.

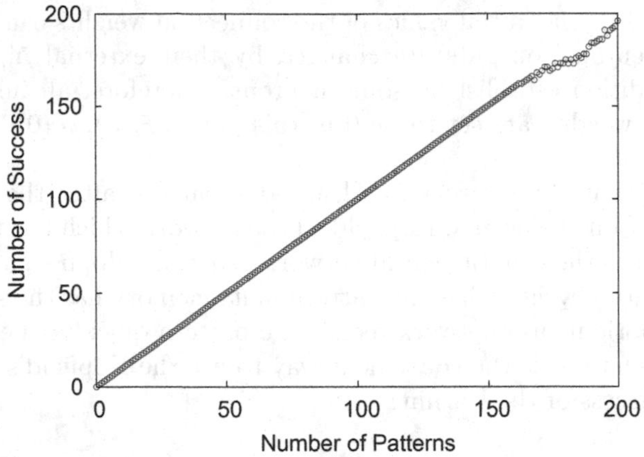

Fig. 1. Capacity of usual network

On the simplified network under the same condition, the network was not able to learn the same number of patterns as the usual network. Through some trial, it was found that setting $\Delta w = 3 \times 10^{-7}$ makes the simplified network learn 162 patterns as in Fig. 2. Thus, the simplified network still learns the same number of patterns without the temporal sums, although the capacity is 158 patterns because when the number of inputs is 159, the network learns 158 patterns.

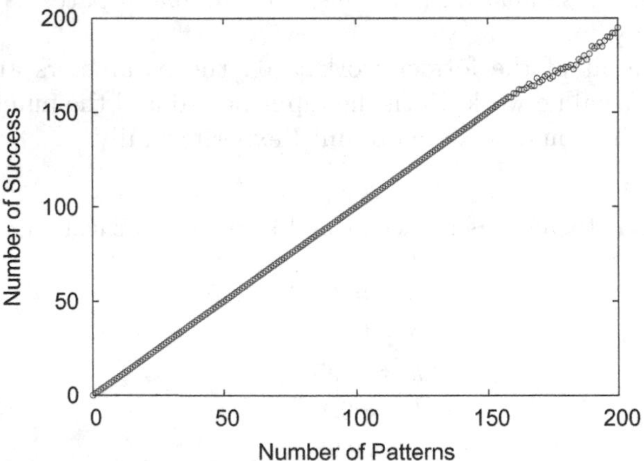

Fig. 2. Capacity of simplified network

3.2 Neurons Which Learn

In the incremental learning, not all the neurons change their weights at the same time but the neurons which learn in that step are decided by the learning condition. To investigate the differences between the simplified and the usual network,

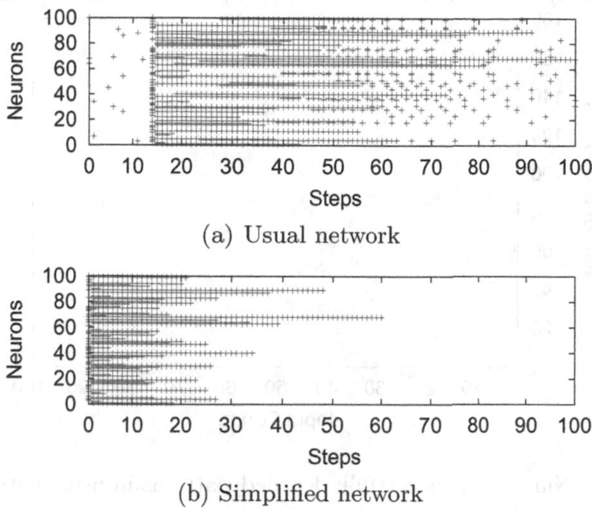

(a) Usual network

(b) Simplified network

Fig. 3. Neurons which learn

when and which neuron learns are inspected. Figure 3 shows the results. All the neurons are arranged vertically, and the horizontal axis shows the steps during the 1st pattern inputted in 10th set. The mark + indicates that the neuron learns at that step.

From Fig. 3(b), in the simplified network without the temporal sum, almost all the neurons learn from step 1, while, from Fig. 3(a), in the usual network, few neurons learn from step 1 and it was after step 14 that almost all the neurons learn. This is because the temporal sum is keeping the previous pattern information for some steps (for 13 steps in this case), and the new pattern information gets overwhelming at step 14 in the usual network. Although it would be unable for the simplified network to learn with noisy inputs without temporal sum which smooths noisy inputs [7], the simplified network may learn patterns faster.

3.3 Input Period

In the former simulations, the input period was kept to be 100 or 50. But, the simplified network may learn patterns with shorter input period. In the next simulations, to investigate the effect of the input period, the input period is changed from 1 to 100 and the number of successfully learned patterns is counted. The set of patterns is the same as in Sect. 3.1 and the number of input patterns is fixed to 162. The simulation results are shown in Figs. 4 and 5.

Figure 4 shows the results on the usual network. The horizontal axis is the input period and the vertical axis is the number of success. When the input period is short, no pattern is stored in the network, because the temporal sum is keeping the previous pattern information. With the input periods longer than 82 steps, the usual network was able to learn all the patterns.

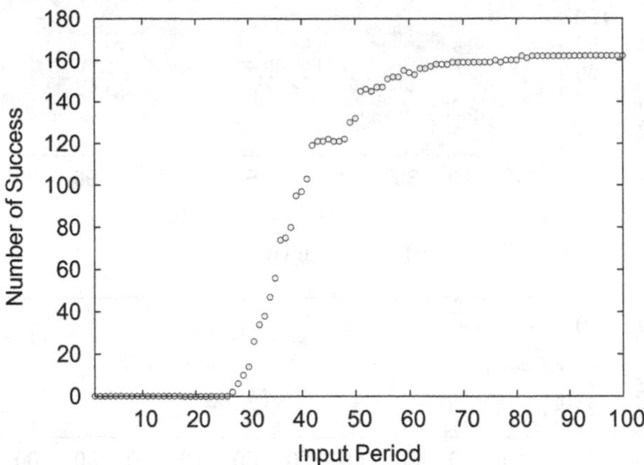

Fig. 4. Number of successfully learned patterns in usual network

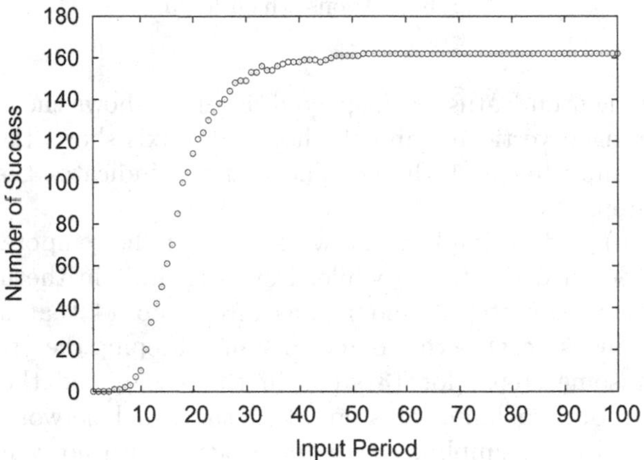

Fig. 5. Number of successfully learned patterns in simplified network

Figure 5 shows the results on the simplified network. The number of success begins to rise from the input period of 5 steps, and reach 162 at the period of 52 which is 63 % of that in the usual network. From these results, the simplified network can learn patterns faster than the usual network.

3.4 Noisy Inputs

The simplified network would lose the ability to learn in noisy inputs without the temporal sums. To verify the handling of noisy inputs, the following simulations investigate this ability.

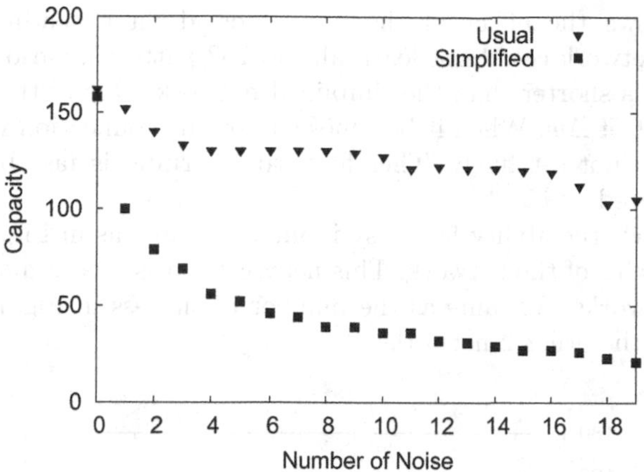

Fig. 6. Learning ability with noisy inputs

In these simulations, to add noises, a fixed number of elements in an input pattern are chosen randomly every step and they reversed before they are inputted to the network. This fixed number is the number of noises. The results are shown in Fig. 6.

The horizontal axis is the number of noises and the vertical axis is the capacity of the network, which is, in this paper, the maximum number of patterns when all the input patterns are stored in the network. As predicted previously, though the capacity of the usual network is above 100, that of the simplified network became below 100 over 2 noises. Thus, the simplified network loses the ability to learn with noisy inputs.

3.5 Adjusted Network

The difference between the usual network and the simplified network is k-parameters (and Δw). In this section, the other parameters are introduced to find the parameters with which the network can learn faster than the usual network and have more ability of learning in noisy inputs than the simplified network. Because the rage of the parameters are 4 dimensional including Δw, the parameters are changed with linear relation as follows:

$$k_s = 0.95r, \tag{8}$$

$$k_m = 0.1r, \tag{9}$$

$$k_r = 0.95r, \tag{10}$$

$$\Delta w = 2.7 \times 10^{-6} r + 3 \times 10^{-7}, \tag{11}$$

where r is the control parameter.

The simulations from $r = 0.1$ to 0.9 are carried out to adjust the parameters. From these simulations, a reasonable set of parameters is found at $r = 0.8$. In this paper, the network with these parameters is called the adjusted network.

Figure 7 shows the effect of the input period on the adjusted network. Although the network could not learn all the 162 patterns, curiously, the input period it needs is shorter than the simplified network. One of the available reasons is the effect of Δw. When it becomes larger, the connection weights change quickly, but are not set finely. Therefore, the learning is fast but not all the patterns are stored.

Figure 8 shows the ability for noisy inputs. As same as in Fig. 6, the vertical axis is the capacity of the network. This network shows medium ability between the other 2 networks. As same as the number of success in Fig. 7, the capacity is smaller than the other 2 networks.

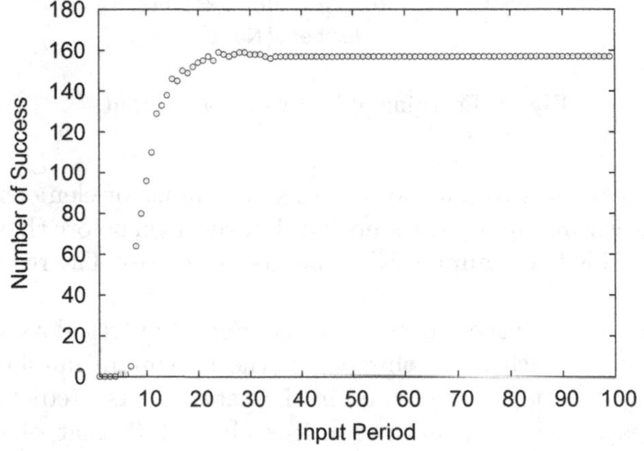

Fig. 7. Number of successfully learned patterns in adjusted network

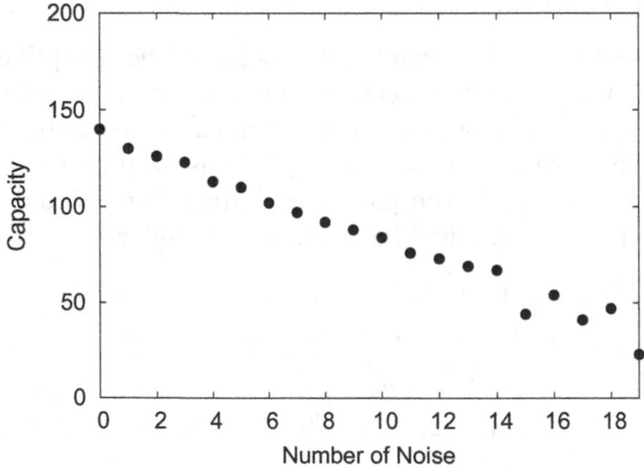

Fig. 8. Learning ability with noisy inputs in adjusted network

3.6 Varying the Learning Parameter

Described above, the adjusted network learns in shorter steps using Δw at $r = 0.8$. At the simplified network with the same Δw, there is a possibility that the network also learns in short steps.

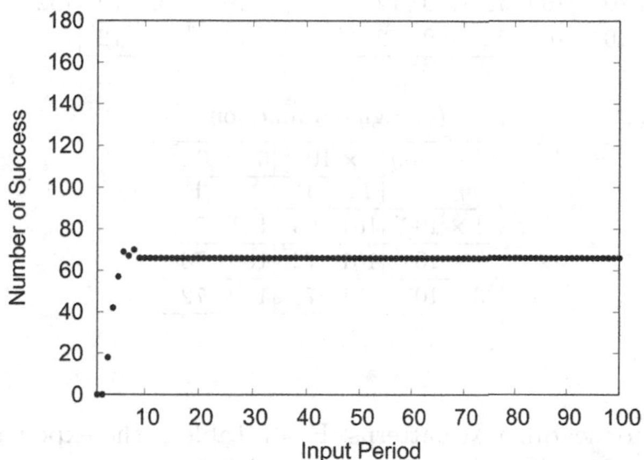

Fig. 9. Number of successfully learned patterns with $\Delta w = 2.46 \times 10^{-6}$

Figure 9 shows the result of the simulations in which $\Delta w = 2.46 \times 10^{-6}$. This value makes the network learn more quickly but less patterns. The network learned only 66 patterns out of 162. When Δw is large, the network learns more quickly, and when Δw is small, the network learns more patterns.

Then, there arises the question that whether varying Δw from large value to small value during the learning makes the network learn all the patterns quickly at the simplified network and the usual network.

In our former work, Δw is varied linearly [8]. In this paper, Δw is varied according to the 3 functions which are the linear function (Eq. (12)), the exponential function (Eq. (13)), and the sigmoid function (Eq. (14)).

$$\Delta w = w_H - (w_H - w_L)\frac{s-1}{99} \tag{12}$$

$$\Delta w = w_H \exp\left(\log\left(\frac{w_L}{w_H}\right) \cdot \frac{s-1}{99}\right) \tag{13}$$

$$\Delta w = \frac{w_H - w_L}{1 + \frac{1}{\exp(k(\frac{s-1}{99} - 0.5))}} + w_L \tag{14}$$

where w_H is the large value, w_L is the small value, s is the number of learning set and $k = -10$.

For the usual network, we chose w_H and w_L as in Table 2, and the results are summarized. M is the maximum number of learned patters and P is the input

Table 2. Summary of learning varying Δw in usual network

(a) Linear function

w_L \ w_H	3×10^{-5}		5×10^{-5}	
	M	P	M	P
1×10^{-6}	160	94	158	84
2×10^{-6}	160	72	160	72
3×10^{-6}	160	72	160	72

(b) Exponential function

w_L \ w_H	3×10^{-5}		5×10^{-5}	
	M	P	M	P
1×10^{-6}	159	90	159	97
2×10^{-6}	162	72	162	72
3×10^{-6}	161	72	161	72

(c) Sigmoid function

w_L \ w_H	3×10^{-5}		5×10^{-5}	
	M	P	M	P
1×10^{-6}	161	95	161	85
2×10^{-6}	161	72	160	73
3×10^{-6}	161	72	160	72

period needed for learning M patterns. From Table 2, the exponential function with $w_L = 2\times10^{-6}$ gives the best result—all 162 patterns are stored and the input period is 72, that is 89 % of that of the constant $\Delta w = 3\times10^{-6}$.

For the simplified network, we chose w_H and w_L as in Table 3, and the results are summarized. When w_L is large, the input period to store maximum number of patterns is small with all three functions.

Although the number of success did not reach 162 as seen in Table 3, the sigmoid function with $w_H = 4\times10^{-7}$ and $w_L = 2\times10^{-6}$ gives the best result between the simulations. In Fig. 5, the input period needs 52 to reach 162, and 46 to reach 160. Using this sigmoid function, the input period becomes 38 %.

Table 3. Summary of learning varying Δw in simplified network

(a) Linear function

w_L \ w_H	3×10^{-5}		5×10^{-5}	
	M	P	M	P
2×10^{-6}	160	32	156	36
3×10^{-6}	159	23	154	23
4×10^{-6}	156	19	152	19

(b) Exponential function

w_L \ w_H	3×10^{-5}		5×10^{-5}	
	M	P	M	P
2×10^{-6}	160	33	160	32
3×10^{-6}	160	23	160	22
4×10^{-6}	159	18	159	18

(c) Sigmoid function

w_L \ w_H	3×10^{-5}		5×10^{-5}	
	M	P	M	P
2×10^{-6}	160	32	160	28
3×10^{-6}	160	23	160	21
4×10^{-6}	160	18	159	17

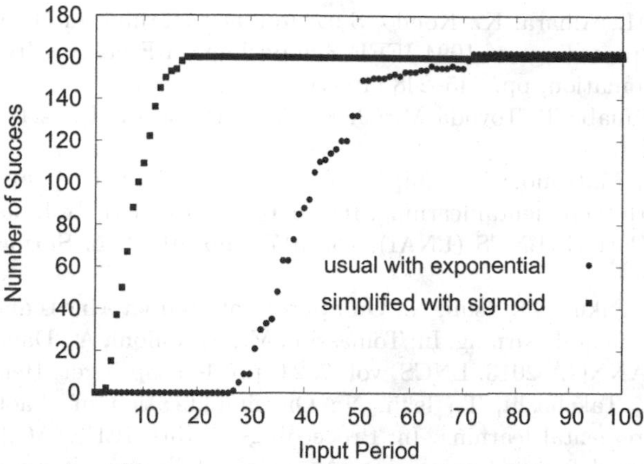

Fig. 10. Number of successfully learned patterns in usual network varying Δw with exponential function and in simplified network varying Δw with sigmoid function

To compare with Figs. 4 and 5, the learning of the usual network with the exponential change of Δw and that of the simplified network with the sigmoid are shown in Fig. 10. While the obvious difference was not seen except for the input period at which the network learns all the patterns in the usual network, varying Δw affects the number of success for the input period in the simplified network.

4 Conclusion

To reduce the amount of computation in the incremental learning, the simplified network was introduced and the behaviour of this network was investigated comparing with the usual network. The simplified network was able to learn patterns in shorter steps than the usual network and had almost the same capacity as the usual network, although the simplifed network loses its learning ability when the input includes noise. Furthermore, varying the learning parameters during the learning, the amount of steps for the learning can be reduced in the simplified network.

References

1. Asakawa, S., Deguchi, T., Ishii, N.: On-demand learning in neural network. In: Proceedings of the ACIS 2nd International Conference on Software Engineering, Artificial Intelligence, Networking & Parallel/Distributed Computing, pp. 84–89 (2001)
2. Deguchi, T., Ishii, N.: On refractory parameter of chaotic neurons in incremental learning. In: Negoita, M.G., Howlett, R.J., Jain, L.C. (eds.) KES 2004. LNCS (LNAI), vol. 3214, pp. 103–109. Springer, Heidelberg (2004)

3. Watanabe, M., Aihara, K., Kondo, S.: Automatic learning in chaotic neural networks. In: Proceedings of 1994 IEEE Symposium on Emerging Technologies and Factory Automation, pp. 245–248 (1994)
4. Aihara, K., Tanabe, T., Toyoda, M.: Chaotic neural networks. Phys. Lett. A **144**(6–7), 333–340 (1990)
5. Deguchi, T., Matsuno, K., Ishii, N.: On capacity of memory in chaotic neural networks with incremental learning. In: Lovrek, I., Howlett, R.J., Jain, L.C. (eds.) KES 2008, Part II. LNCS (LNAI), vol. 5178, pp. 919–925. Springer, Heidelberg (2008)
6. Deguchi, T., Fukuta, J., Ishii, N.: On appropriate refractoriness and weight increment in incremental learning. In: Tomassini, M., Antonioni, A., Daolio, F., Buesser, P. (eds.) ICANNGA 2013. LNCS, vol. 7824, pp. 1–9. Springer, Heidelberg (2013)
7. Deguchi, T., Takahashi, T., Ishii, N.: On simplification of chaotic neural network on incremental learning. In: Proceedings of 15th IEEE/ACIS International Conference on Software Engineering, Artificial Intelligence, Networking and Parallel/Distributed Computing (SNPD), pp. 1–4 (2014)
8. Deguchi, T., Takahashi, T., Ishii, N.: On acceleration of incremental learning in chaotic neural network. In: Rojas, I., Joya, G., Catala, A. (eds.) IWANN 2015. LNCS, vol. 9095, pp. 370–379. Springer, Heidelberg (2015)

Accelerated Optimal Topology Search for Two-Hidden-Layer Feedforward Neural Networks

Alan J. Thomas(✉), Simon D. Walters, Miltos Petridis, Saeed Malekshahi Gheytassi, and Robert E. Morgan

School of Computing Engineering and Mathematics, University of Brighton, Brighton, UK
{a.j.thomas,s.d.walters,m.petridis, m.s.malekshahi,r.morgan2}@brighton.ac.uk

Abstract. Two-hidden-layer feedforward neural networks are investigated for the existence of an optimal hidden node ratio. In the experiments, the heuristic $n_1 = int(0.5n_h + 1)$, where n_1 is the number of nodes in the first hidden layer and n_h is the total number of hidden nodes, found networks with generalisation errors, on average, just 0.023 %–0.056 % greater than those found by exhaustive search. This reduced the complexity of an exhaustive search from quadratic, to linear in n_h, with very little penalty. Further reductions in search complexity to logarithmic could be possible using existing methods developed by the Authors.

Keywords: Two-hidden-layer feedforward · ANN · Exhaustive search · Optimal topology · Optimal node ratio · Heurix · Universal function approximation

1 Introduction

Function approximators are an important class of artificial neural networks. Since it was shown that multilayer feedforward neural networks with as few as a single hidden layer are universal function approximators [1], they have enjoyed an upsurge of popularity in diverse domains. In the automotive arena, they are used increasingly to predict engine emissions, and typically involve an exhaustive or 'trial and error' search through one or two hidden layers to find the optimal topology - though the former is by far the most common [2–5]. This could well be because of the prohibitive time required to conduct an exhaustive quadratic search through two hidden layers. This paper addresses the question: 'Does there exist an optimal ratio of nodes between the first and second hidden layers of a two-hidden-layer neural network (TLFN)?' If so, this could be combined with existing network topology optimisation techniques to reduce their complexity. For example, the complexity of an exhaustive search for a TLFN would be reduced from a quadratic search $O(n^2)$ to a linear search $O(n)$, diagonally along the optimal ratio line.

C. Jayne and L. Iliadis (Eds.): EANN 2016, CCIS 629, pp. 253–266, 2016.
DOI: 10.1007/978-3-319-44188-7_19

In this paper, a heuristic relationship between the total number of hidden nodes, n_h and the number of nodes in the first hidden layer n_1 is proposed. TLFNs created using this heuristic are compared with the best of those found by searching all possible combinations of nodes in the first and second hidden layers (n_1 and n_2 respectively) such that $n_h = n_1 + n_2$. Although this heuristic is not guaranteed to produce the best node ratio, in our experiments the generalisation error is only 0.023 %–0.056 % greater than the best of any other node combination.

2 Problem Description

When designing a feedforward neural network, the number of inputs and outputs are easily selected as these are determined by the application. The number of hidden layers required depends on the complexity of the function. For functions which are linearly separable, no hidden layers are required at all. Given a sufficiently large number of hidden units, a single layer will suffice [1], however two hidden layers can often achieve better result than a single layer [6]. In the Authors' own experience, node for node, a TLFN will give a better generalisation capability than an single-hidden-layer feedforward neural network (SLFN) in many cases.

The most challenging and time consuming aspect of the design is choosing the optimal number of hidden nodes. It is assumed here that 'optimal' means 'yielding the best generalisation capability'. Too few hidden nodes, and the network simply will not have the capacity to solve the problem. Conversely, too many, and the network will memorise noise within the training data, leading to poor generalisation capability. Thus the challenge is finding a network which achieves the best balance, ideally in a reasonable time.

3 Related Work

Many optimisation techniques for feedforward neural networks have been proposed in the literature. These can be broadly summarised as:

3.1 Rules of Thumb

These are generally associated with guessing the best number of hidden nodes for single-hidden-layer feedforward networks (SLFNs). There do not appear to be any that pertain to TLFNs.

3.2 Trial and Error

This is a very primitive approach likely to yield extremely sub-optimal results. However, this term is occasionally applied to an exhaustive search between certain bounds. In [5], for example, the term is used to describe the search for a TLFN which varies the number of nodes in each hidden layer between 1 and 20, with a resulting search space of 400 different topologies.

3.3 Exhaustive Search

This involves training networks with every possible combination of hidden nodes between 1 and some upper bound, N_h, and choosing the network with the best generalisation performance. Huang and Babri rigorously proved that an SLFN with at most N_h hidden neurons can learn N_s distinct samples with zero error [7]. Though this gives us an upper bound on the number of hidden neurons, it also means that at this bound the network will also overfit by exactly learning the noise within the training set. Thus an exhaustive search for an SLFN should vary the number of hidden nodes from 1 up to an absolute maximum of $N_h = N_s$.

Huang later proved that the upper bound on the number of hidden nodes N_h for TLFNs with sigmoid activation function is given by $N_h = 2\sqrt{(n_3+2)N_s}$, where n_3 is the number of outputs. These can learn at least N_s distinct samples with any degree of precision [8]. Interestingly, Huang also demonstrated that the storage capacity can be increased by reducing the number of outputs, which is probably the best argument for limiting the number of outputs for function approximation to a single output. With that in mind, substituting $n_3 = 1$ we have $N_h = \sqrt{12N_s}$. This is $\sqrt{N_s/12}$ times lower than that of an SLFN. This means that an exhaustive search for a TLFN with for example 10,000 training samples, would have an upper bound which is 29 times lower than for an SLFN.

With exhaustive searches, several networks of each topology need to be trained to filter out networks where the initial random weight allocation might cause the training to get trapped in local minima. Because of this effect, it is unlikely that the actual global optimum will be found. Since this depends on all the weights being exactly correct, the probability of finding the global optimum will increase with the number of weights. The result is that the 'optimal' topology returned by an exhaustive search will be different on successive searches. For SLFNs, the complexity of an exhaustive search is linear $O(n)$, whereas for TLFNs, it is quadratic $O(n^2)$.

3.4 Growing Algorithms

At their simplest, these are similar to exhaustive searches. They generally start with a single hidden node, and increase the number of hidden nodes one by one until the improvement in generalisation error is negligible. Using this approach with TLFNs is problematic because the sudden variation of node ratio on new rows will result in spikes on the generalisation landscape resulting in premature termination. Other types of growing algorithms combine simultaneous growing and training. These can be classified as non-evolutionary [9], and evolutionary [10]. The latter are notoriously time consuming.

3.5 Pruning Algorithms

With this approach, an oversized network is trained and the relative importance of the weights subsequently analysed. The least important weights are removed and the network retrained. The problem with these is determining what constitutes an oversized network in the first instance, and their time complexity. Brute force approaches which set each weight in turn to zero and eliminates it if it has a negligible effect on the generalisation error. These have a complexity of $O(N_s w^3)$, where N_s is the number of samples in the training set, and w is the number of weights in the original oversized network [11].

3.6 Heuristic Algorithms

These estimate the optimal number of hidden nodes by sampling a sub-set of topologies and using curve fitting techniques to predict the optimum topology. A system previously developed by the Authors [12] can create SLFNs with a generalisation error of as little as 0.4 % greater than those found by exhaustive search with a complexity of $O(log_2(n))$.

3.7 Proposed Method

This is not a separate method *per se*, but rather a heuristic to be used in conjunction with another optimisation method. If there exists an 'optimal' node ratio for a TLFN, then it effectively reduces its complexity to that of an SLFN.

4 Experiments

All experiments were carried out using the Matlab R2014b environment. The networks were created using the Neural Network Toolbox 'fitnet' function to generate the SLFNs and TLFNs where appropriate. Two separate datasets were used, with different numbers of inputs. These were trained the Levenberg-Marquardt training function, 'trainlm' which is commonly used for function approximation as it has often been found to yield the best results [2–5]. For comparison, the second dataset was also trained with the Scaled Conjugate Gradient training function 'trainscg'.

4.1 Data Preparation

The datasets were chosen because of their availability in the public domain, allowing the findings to be independently verified. In all cases, the data is split into three subsets: Training (80 %), Validation (10 %) and Test (10 %). The Validation set is used to stop the training process when the validation error starts to rise, and the Test set is used exclusively as an estimate of the generalisation error.

For any given dataset, exactly the same subsets were used for every single network created in the experiment. By eliminating any bias in the error surface that may have resulted from a different random split for each network, it was ensured that they were all competing on the same playing field. The only random element at play was thus the initial randomisation of the weights. This initial starting point determines which local minimum in the error surface the training might get stuck in and thus has a direct impact on the generalisation error. For complex error surfaces, it is extremely unlikely that the global minimum will be found.

Dataset 1. The 'engine_data' (available in Matlab), consists of 1199 samples organised as two inputs (fuel and speed) and two targets (torque and NOx). These were reorganised to use torque as a third input, with a single output, NOx. They were subsequently split into Training, Validation and Test subsets (959, 120, and 120 samples, respectively).

Dataset 2. The NASA Airfoil Self-Noise dataset, available from the UCI Machine Learning Repository [13]. This consists of 1503 samples with five inputs: Frequency (Hz), Angle of attack (°), Chord length (m), Free-stream velocity (m/s), and Suction side displacement thickness (m). It has a single output, scaled sound pressure level (dB). These were split into Training, Validation and Test subsets (1201, 151 and 151 samples, respectively).

4.2 Training Algorithms

In all cases, data preprocessing was 'mapminmax' for both inputs and outputs, the transfer function was 'tansig' and the error function for training was 'mse'. However, the generalisation error in the experiments was reported as the normalised root mean squared error (NRMSE), which is given by:

$$NRMSE_y = \frac{1}{\hat{y}_{max} - \hat{y}_{\min}} \sqrt{\frac{\sum_{i=1}^{n} (y_i - \hat{y}_i)^2}{N_s}} \tag{1}$$

where N_s represents the number of samples, $\hat{y}_i$ is the target value, and y_i is the actual value.

Training Algorithm 1. Levenberg-Marquardt training algorithm 'trainlm' using default training parameters: 1000 epochs, training goal of 0, minimum gradient of 10^{-7}, 6 validation failures, $\mu = 0.001$, $\mu_{dec} = 0.1$, $\mu_{inc} = 10$ and $\mu_{max} = 10^{-10}$.

Training Algorithm 2. Scaled Conjugate Gradient training algorithm 'trainscg' using default parameters: 1000 epochs, training goal of 0, minimum gradient of 10^{-6}, 6 validation failures, $\sigma = 5 \times 10^{-5}$, and $\lambda = 5 \times 10^{-7}$.

5 Experimental Method

Three separate domains were tested:

- **Domain 1** - Dataset 1 using Training Algorithm 1, with the generalisation error averaged over 100 rounds of Fig. 1.
- **Domain 2** – Dataset 2 using Training Algorithm 1, with the generalisation error averaged over 100 rounds of Fig. 1.
- **Domain 3** – Dataset 2 using Training Algorithm 2, with the generalisation error averaged over 300 rounds of Fig. 1. The number of rounds were increased here because of the higher variance in generalisation error when using Algorithm 2.

```
function e = singleRound(nh)
  for n1 = 1 to nh-1 do
    n2 = nh - n1              % Calculate n2

    % create and train 100 networks recording NRMSE
    for run = 1 to 100 do
      net = createNetwork(n1,n2)
      nrmse[run] = trainNetwork(net)
    end do

    e[n1] = min(nrmse)        % Calculate winner`s error
  end do
  return e
end function
```

Fig. 1. Pseudo-code for a single round

Within these domains, a number of experiments were carried out each with a constant total number of hidden nodes:

$$n_h = n_1 + n_2 \tag{2}$$

where n_1 and n_2 are the number of nodes in hidden layers 1 and 2 respectively. The values of n_h chosen for these experiments were given by the set:

$$n_h = \{34, 20, 16, 14, 13, 12, 11, 10, 9, 8, 7, 6, 5, 4, 3\}. \tag{3}$$

For each value of n_h, TLFNs were created using all possible combinations of n_1 and n_2 satisfying (2). For example if $n_h = 4$, $n_1 = \{1,2,3\}$ and $n_2 = \{3,2,1\}$. This yielded 3 possible networks with topologies $n_0 : 1 : 3 : n_3, n_0 : 2 : 2 : n_3$ and $n_0 : 3 : 1 : n_3$. To reduce the effect that the random initial weights have on the generalisation error, 100 networks of each topology were created, and the NRMSE of each calculated from (1). The generalisation errors of the best generalisers (those with the minimum NRMSE on the test set) formed the results of a single round. This is shown in the pseudo-code in Fig. 1, which from any given n_h, returns an array, e, length $n_h - 1$, indexed by the number of nodes in the first hidden layer.

The array, e, was then averaged over 100, 100, and 300 of these rounds for domains 1, 2 and 3, respectively. This was repeated for every value of n_h within the set defined by (3).

6 Results and Discussion

6.1 Optimal Node Ratio Investigation

The results were not at all as expected. In the preliminary investigative experiments with Domain 1, the median NRMSE was used instead of the minimum to determine the winner, and 200 rounds were used. The contents of the arrays, e, were displayed along the y-axis, and their indices (representing the values of n_1) were displayed as an offset from $0.5n_h$ along the x-axis. In other words, the x-axis = $n_1 - 0.5n_h$ as this was where the optimum, if it existed, was expected to lie. However, there seemed a marked symmetry in the region bounded by $n_1 = 4$ and $n_2 = 2$ (narrow dash) and centred on $0.5n_h + 1$ (wide dash) as shown in Fig. 2. From left to centre, this shows a series of contour lines of decreasing constant values of n_h represented by the set in (3). The 'sweet spot' seemed to imply that the number of nodes in the first hidden layer should be greater than 3 and those in the second layer should be greater than 1. Since the dataset had 3 inputs and 1 output, it was suspected at this stage that the sweet spot might be governed by the number of inputs and outputs.

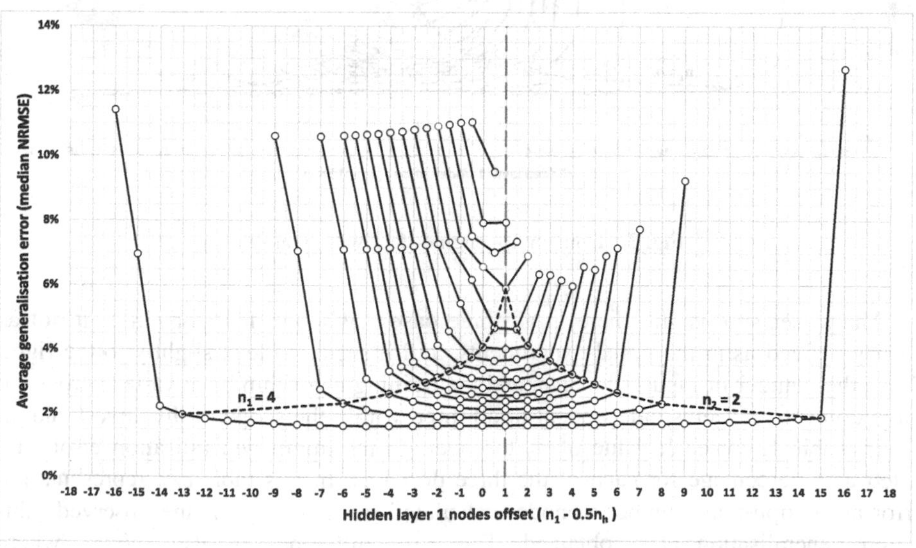

Fig. 2. Initial Domain 1 experiments using average median NRMSE

The main investigation tested whether $n_1 = 0.5n_h + 1$ could also be used to describe the optimum for other datasets and training algorithms. Since n_h can be either even or odd, rounding down was used as a heuristic for the optimal value of n_1, i.e.:

$$n_{1(\mathrm{opt})} = int(0.5n_h + 1) \tag{4}$$

As a measure of the accuracy of this prediction, the root mean square difference (rmsd) between the observed minimum generalisation errors, and those obtained using node ratio (4) were calculated. In the preliminary case above, this is less than 0.011 %.

In the main body of experiments, each round searched for the networks with the minimum NRMSE (as described in Fig. 1). In this respect, a single round was more representative of an actual exhaustive search for the best generaliser, and multiple rounds could be considered as multiple exhaustive searches. The results, which are shown graphically in Figs. 3, 4 and 5, show the averages over multiple exhaustive searches (100 for Domains 1–2, and 300 for Domain 3).

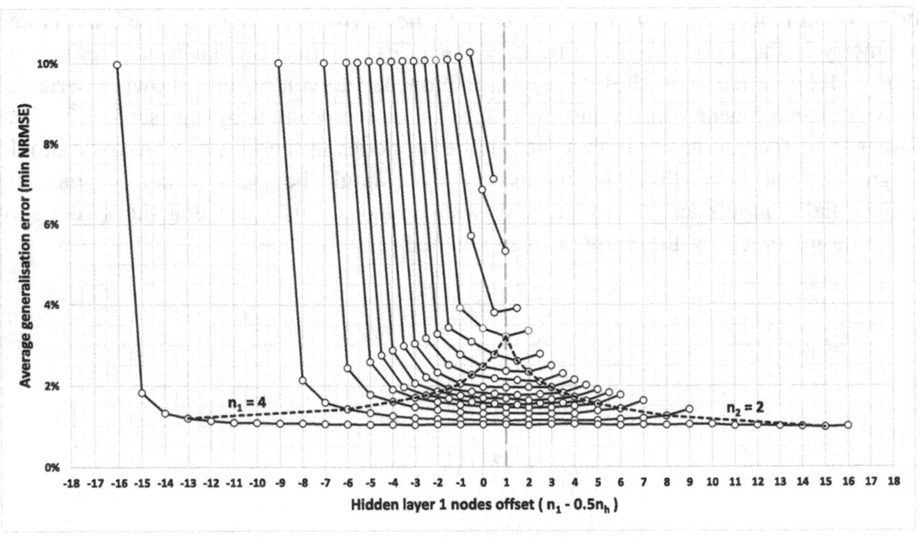

Fig. 3. Domain 1 - engine data with Trainlm

The sweet spot is still there in all three cases, although for Domain 1, it not as clearly defined as in the initial experiments, and it seems to lean slightly to the right. Since the number of inputs, the data, and the training algorithm have varied across the three domains, it seems independent of all three within the scope of this investigation.

In Table 1, for each value of n_h the average minimum generalisation errors are listed as a percentage for each of the three domains. In this table e_{opt} represents the error at the optimum number of nodes calculated from (4), e_{min} is the observed minimum generalisation error obtained. The error difference is also shown, where $\delta e = e_{opt} - e_{min}$. The root mean square difference (rmsd) between these are shown in

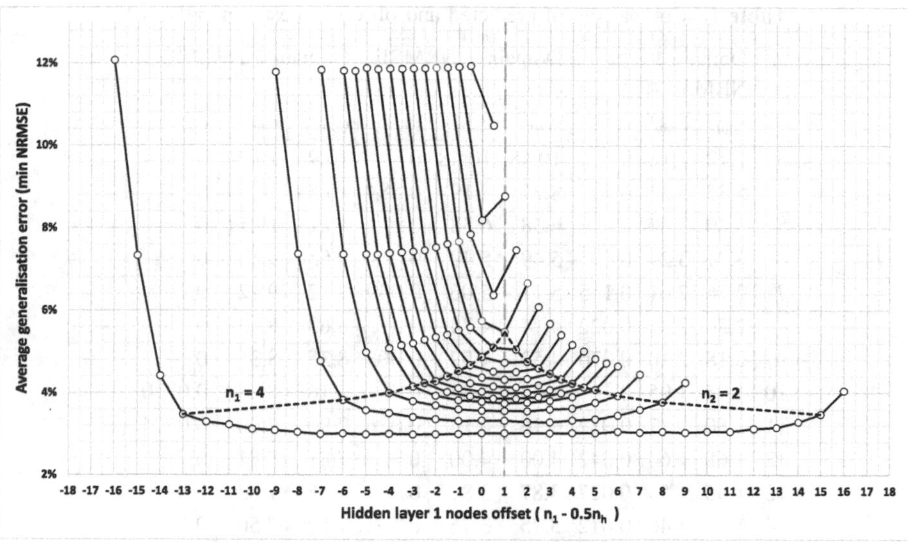

Fig. 4. Domain 2 - airfoil self-noise with Trainlm

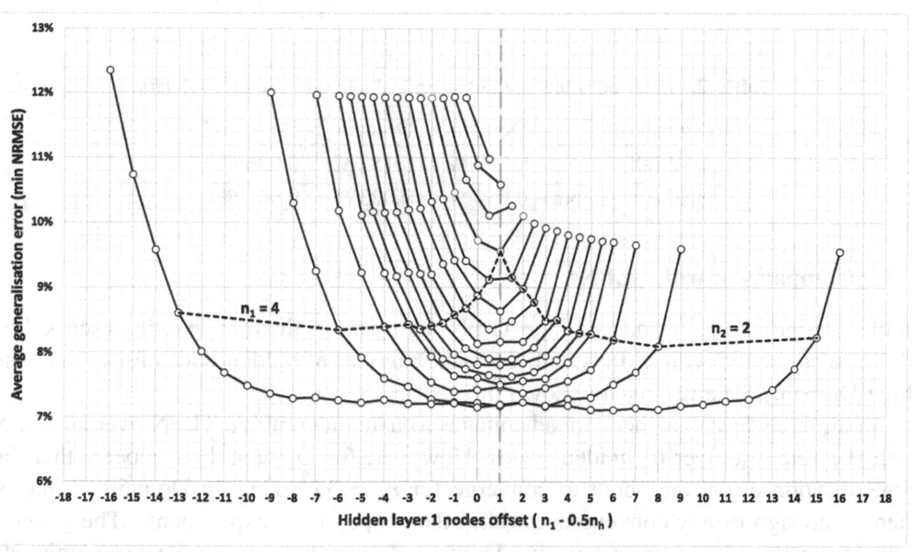

Fig. 5. Domain 3 – airfoil self-noise with Trainscg

Table 2. Since there is an outlier in Domain 2 for $n_h = 4$, which is outside the sweet spot, the rmsd solely within the sweet spot is also included.

These results show that although a linear search along (4) is not guaranteed to find the best generalisers (but then neither is an exhaustive quadratic search), it will find networks within 0.023 % – 0.056 % of these on average.

Table 1. Comparison of predicted and observed best generalisers.

n_h	Domain 1 NRMSE (%)			Domain 2 NRMSE (%)			Domain 3 NRMSE (%)		
	e_{opt}	e_{min}	δ_e	e_{opt}	e_{min}	δ_e	e_{opt}	e_{min}	δ_e
3	7.12	7.12	0	10.48	10.48	0	10.98	10.98	0
4	5.33	5.33	0	8.77	8.19	0.584	10.59	10.59	0
5	3.80	3.80	0	6.38	6.38	0	10.11	10.11	0
6	3.21	3.21	0	5.49	5.49	0	9.55	9.55	0
7	2.76	2.61	0.155	5.11	5.06	0.049	9.12	9.12	0
8	2.36	2.34	0.022	4.74	4.74	0	8.63	8.63	0
9	2.18	2.10	0.070	4.53	4.52	0.008	8.35	8.35	0
10	1.95	1.95	0	4.33	4.33	0	8.16	8.13	0.0270
11	1.80	1.77	0.032	4.16	4.14	0.026	7.90	7.90	0.0074
12	1.68	1.64	0.042	4.00	4.00	0	7.81	7.81	0
13	1.55	1.53	0.027	3.87	3.87	0	7.65	7.63	0.0224
14	1.47	1.46	0.012	3.75	3.75	0	7.50	7.50	0
16	1.32	1.31	0.014	3.57	3.56	0.008	7.47	7.37	0.0993
20	1.13	1.13	0	3.33	3.29	0.035	7.17	7.16	0.0049
34	1.04	0.98	0.067	3.03	2.98	0.042	7.19	7.10	0.0894

Table 2. rmsd between predicted and observed best generalisers

Domain	Domain 1	Domain 2	Domain 3
rmsd (all)	0.050 %	0.152 %	0.036 %
rmsd (sweet spot)	0.056 %	0.023 %	0.040 %

6.2 Comparison with SLFNs

In this section, the performance of the TLFNs using the optimal node ratio described by (4) were compared with SLFNs with the same number of hidden nodes for each of the three Domains. The results are shown in Fig. 6.

In all three domains, there are advantages to using an optimal TLFN over an SLFN with the same number of hidden nodes. However, for Domain 1, it appears that the generalisation errors are about to converge for $n_h > 34$ nodes. In Domains 2 and 3, there is no sign of any convergence within the scope of the experiments. The greatest gain in generalisation error was for Domain 3, which uses the Scaled Conjugate Gradient algorithm. This is a popular training algorithm for larger numbers of hidden nodes as it is much faster than the Levenberg-Marquardt algorithm, since the latter's training time scales exponentially with the number of hidden nodes. An interesting feature is that for Domains 1 and 2, which both use the Levenberg-Marquardt algorithm, the generalisation errors cross over at $n_h = 5$, above which TLFNs outperform SLFNs. Coincidentally, $n_h = 6$ is the apex of the perceived sweet spot in these experiments. It is unclear at this stage whether the two are related.

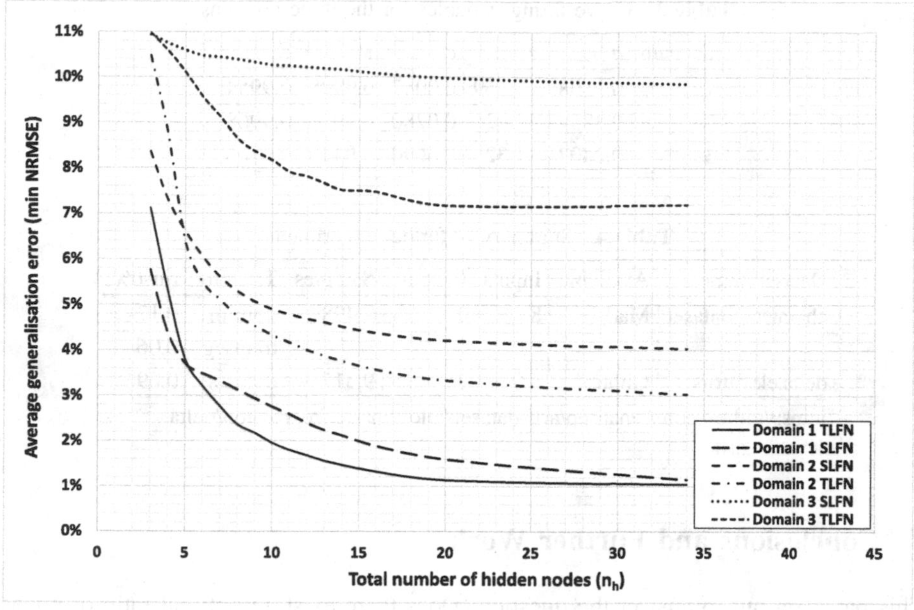

Fig. 6. Node for node comparison of TLFNs and SLFNs.

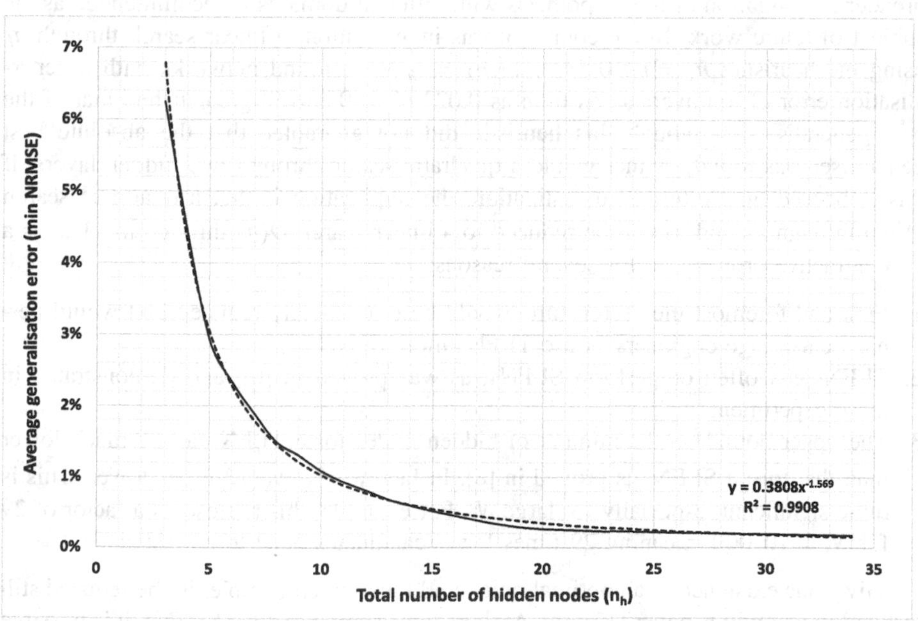

Fig. 7. Domain 1 TLFN with −0.87 % offset

Table 3. Curve fitting variables for the three domains

Domain	a	b	c	R^2	R
1	0.3808	−1.569	0.0087	0.9908	0.9954
2	0.3854	−1.424	0.0280	0.9936	0.9968
3	0.2407	−1.329	0.0690	0.9420	0.9706

Table 4. Summary of further experiments

Dataset name	Available	Inputs	Outputs	Samples	Training	rmsd%
chemical_dataset	Matlab	8	1	498	trainlm	0.17
					trainscg	0.06
delta.elevators	Github[a]	6	1	9,517	trainscg	0.09

[a]https://github.com/renatopp/arff-datasets/blob/master/regression/delta.elevators.arff

7 Conclusions and Further Work

This paper set out to answer the question 'Does there exist an optimal ratio of nodes between the first and second hidden layers of a two-hidden-layer neural network (TLFN)?' Based on the domains tested in the investigation, for $n_h > 5$, this can be described by the relationship $n_1 = 0.5n_h + 1$, or alternatively $n_1 = n_2 + 2$. However, a broader investigation of this hypothesis with different domains is recommended as the subject of future work. In the course of this investigation, a linear search through n_h using the heuristic: $n_1 = int(0.5n_h + 1), n_2 = n_h - n_1$, found networks with a generalisation error of, on average, as little as 0.023 % to 0.056 % greater than that of the best generalisers. Although this heuristic did not guarantee that the absolute best generaliser was found, neither would a quadratic search through two hidden layers. If this is backed up by further investigation, the implication is that a quadratic search $O(n^2)$ through n_1 and n_2 can be reduced to a linear search $O(n)$ through n_h. This is a very attractive proposition for several reasons:

1. First and foremost the search time would be dramatically reduced and would perhaps encourage engineers to use TLFNs more often.
2. TLFNs can often outperform SLFNs, as was proved in [6], and demonstrated in these experiments.
3. The upper bound on the number of hidden nodes for a TLFN can be much lower than that for an SLFN, as proved in [8]. In fact, it is $\sqrt{N_s/12}$ times lower. This is quite significant, especially for large N_s. For example, this represents a factor of 29 for $N_s = 10,000$, meaning 29 times fewer candidates need to be tested.

Given the existence of an optimal ratio, could the search complexity be reduced still further? In a previous paper [12], the Authors have shown that for SLFNs, it is possible to reduce a linear search $O(n)$ to a logarithmic search $O(log_2(n))$. This is achieved by sampling the generalisation error at node values $n_k = 2^k, 0 \leq k \leq N_s$, fitting an error curve of the form $e(n_h) = an_h^{-b} + c$ to these samples, and calculating the optimal number of hidden nodes from its gradient. The choice of gradient determines whether the network is optimised for speed, accuracy or both.

In order for this method to be suitable for TLFNs, their generalisation error must also follow similar power law curves. The easiest way to check this is to subtract an offset from the generalisation error and use the trend line feature of the spreadsheet to fit a power law curve. The offset is adjusted to achieve the best value of R^2. Figure 7 shows the end of this process for Domain 1. This shows that the generalisation error can be described by $e_{opt} = 0.3808n_h^{-1.569} + 0.0087$, with a Pearson's correlation coefficient of $R^2 = 0.9908$ or $R = 0.9954$. A similar process was carried out on Domains 2 and 3. The results are summarised in Table 3.

The results are excellent for Domains 1 and 2, which use the Levenberg-Marquardt training algorithm. This is good news, since this training algorithm yields the best generalisation error. Based on the experiments carried out in this paper, the Authors are quite confident that the Heurix system they previously developed [12] will also be suitable for TLFNs. Subject to further work, this method could also be used to find near-optimal TLFNs automatically, with a search complexity of as little as $O(log_2(n))$.

8 Addendum

Since the initial submission of the paper for review, three further domains have been tested using n_h = {1 to14, 16, 20 and 34}, for 100 rounds each. The results are summarised in Table 4.

These datasets are quite interesting. With the former, over the range tested, SLFNs outperform TLFNs with respect to their genaralisation capability. In the case of the latter, there is little or no advantage over a network with no hidden nodes at all. Whilst the heuristic does still yield reasonable results, this does tend to suggest that cases like these ought to be tested for in order to obtain efficient network response times from stimulus to output. This will be the subject of further work.

Acknowledgements. We thank Prof. Martin T. Hagan of Oklahoma State University for kindly donating the Engine Data Set used in this paper to Matlab. We would also like to thank Dr. Roberto Lopez of Intelnics (robertolopez@intelnics.com) for donating the Airfoil Self-Noise dataset; also the dataset's creators: Thomas F. Brooks, D. Stuart Pope and Michael A. Marcolini of NASA.

References

1. Hornik, K., Stinchcombe, M., White, H.: Multilayer feedforward networks are universal approximators. Neural Netw. **2**, 359–366 (1989)
2. Mocanu, F.: On-board fuel identification using artificial neural networks. Int. J. Engines **7**, 937–946 (2014)
3. Yap, W.K., Ho, T., Karri, V.: Exhaust emissions control and engine parameters optimization using artificial neural network virtual sensors for a hydrogen-powered vehicle. Int. J. Hydrog. Ener. **37**, 8704–8715 (2012)

4. Roy, S., Banerjee, R., Das, A.K., Bose, P.K.: Development of an ANN based system identification tool to estimate the performance-emission characteristics of a CRDI assisted CNG dual fuel diesel engine. J. Nat. Gas Sci. Eng. **21**, 147–158 (2014)
5. Taghavifar, H., Taghavifar, H., Mardani, A., Mohebbi, A.: Modeling the impact of in-cylinder combustion parameters of DI engines on soot and NOx emissions at rated EGR levels using ANN approach. Ener. Convers. Manag. **87**, 1–9 (2014)
6. Chester, D.L.: Why two hidden layers are better than one. In: International Joint Conference on Neural Networks, vol. 1, pp. 265–268. Laurence Erlbaum, New Jersey (1990)
7. Huang, G.-B., Babri, H.A.: Upper bounds on the number of hidden neurons in feedforward networks with arbitrary bounded nonlinear activation functions. IEEE Trans. Neural Netw. **9**, 224–229 (1989)
8. Huang, G.-B.: Learning capability and storage capacity of two-hidden-layer feedforward networks. IEEE Trans. Neural Netw. **14**, 274–281 (2003)
9. Kwok, T.-Y., Yeung, D.-Y.: Constructive algorithms for structure learning in feedforward neural networks for regression problems. IEEE Trans. Neural Netw. **8**, 630–645 (1997)
10. Azzini, A., Tettamanzi, A.G.B.: Evolutionary ANNs: a state of the art survey. Intell. Artif. **5**, 19–35 (2011)
11. Reed, R.: Pruning algorithms – a survey. IEEE Trans. Neural Netw. **4**, 740–747 (1993)
12. Thomas, A.J., Petridis, M., Walters, S.D., Malekshahi Gheytassi, S., Morgan, R.E.: On predicting the optimal number of hidden nodes. In: 2015 International Conference on Computational Science and Computational Intelligence, pp. 565–570. IEEE CPS (2015)
13. UCI Machine Learning Repository. http://archive.ics.uci.edu/ml

An Outlier Ranking Tree Selection Approach to Extreme Pruning of Random Forests

Khaled Fawagreh, Mohamed Medhat Gaber(✉), and Eyad Elyan

School of Computing Science and Digital Medial, Robert Gordon University, Garthdee Road, Aberdeen AB10 7GJ, UK
mohamed.m.gaber@gmail.com

Abstract. Random Forest (RF) is an ensemble classification technique that was developed by Breiman over a decade ago. Compared with other ensemble techniques, it has proved its accuracy and superiority. Many researchers, however, believe that there is still room for enhancing and improving its performance in terms of predictive accuracy. This explains why, over the past decade, there have been many extensions of RF where each extension employed a variety of techniques and strategies to improve certain aspect(s) of RF. Since it has been proven empirically that ensembles tend to yield better results when there is a significant diversity among the constituent models, the objective of this paper is twofold. First, it investigates how an unsupervised learning technique, namely, Local Outlier Factor (LOF) can be used to identify diverse trees in the RF. Second, trees with the highest LOF scores are then used to create a new RF termed *LOFB-DRF* that is much smaller in size than RF, and yet performs at least as good as RF, but mostly exhibits higher performance in terms of accuracy. The latter refers to a known technique called ensemble pruning. Experimental results on 10 real datasets prove the superiority of our proposed method over the traditional RF. Unprecedented pruning levels reaching as high as 99 % have been achieved at the time of boosting the predictive accuracy of the ensemble. The notably extreme pruning level makes the technique a good candidate for real-time applications.

1 Introduction

Ensemble classification is an application of ensemble learning to boost the accuracy of classification. Ensemble learning is a supervised machine learning paradigm where multiple models are used to solve the same problem [20,28,29]. Since single classifier systems have limited predictive performance [21,28,29,38], ensemble classification was developed to yield better predictive performance [21,28,29]. In such an ensemble, multiple classifiers are used. In its basic mechanism, majority voting is then used to determine the class label for unlabeled instances where each classifier in the ensemble is asked to predict the class label of the instance being considered. Once all the classifiers have been queried, the class that receives the greatest number of votes is returned as the final decision of the ensemble.

C. Jayne and L. Iliadis (Eds.): EANN 2016, CCIS 629, pp. 267–282, 2016.
DOI: 10.1007/978-3-319-44188-7_20

Three widely used ensemble approaches could be identified, namely, boosting, bagging, and stacking. Boosting is an incremental process of building a sequence of classifiers, where each classifier works on the incorrectly classified instances of the previous one in the sequence. AdaBoost [13] is the representative of this class of techniques. However, AdaBoost is proned to overfitting. The other class of ensemble approaches is the Bootstrap Aggregating (Bagging) [5]. Bagging involves building each classifier in the ensemble using a randomly drawn sample of the data with replacement, having each classifier give an equal vote when labeling unlabeled instances. Bagging is known to be more robust than boosting against model overfitting. Random Forest (RF) is the main representative of bagging [7]. Stacking (sometimes called stacked generalization) extends the cross-validation technique that partitions the data set into a held-in data set and a held-out data set; training the models on the held-in data; and then choosing whichever of those trained models performs best on the held-out data. Instead of choosing among the models, stacking combines them, thereby typically getting performance better than any single one of the trained models [37]. Stacking has been successfully used in both supervised learning tasks (regression) [6], and unsupervised learning (density estimation) [33].

The ensemble method that is relevant to our work in this paper is RF. RF has been proved to be the state-of-the-art ensemble classification technique. In a recent evaluation study made by [11] where 179 classifiers arising from 17 families were evaluated, RF has proven to be the best family of classifiers. Since RF algorithms typically build between 100 and 500 trees [36], it would be useful to reduce the number of trees participating in majority voting and yet achieve better performance both in terms of accuracy and speed. In this paper, we propose an unsupervised learning approach to improve speed and accuracy of RF. For speed, our approach avoids having all trees participate in majority voting as only a small subset of the trees is selected. For accuracy, since it has been proven empirically that ensembles tend to yield better results when there is a significant diversity among the models [1,9,20,34], our approach ensures that diverse trees in the ensemble are selected.

We will utilize the Local Outlier Factor (LOF) [8] for the first time ever to extreme prune RF ensembles by assigning each tree an LOF value and then selecting the top k (where k is a predefined integer) trees with the highest LOF scores as shown in Fig. 1. In this figure, an 80 % pruning level has been achieved since the top 4 trees were picked from a total of 20 trees in the initial ensemble to form the pruned ensemble.

This paper is organized as follows. First we discuss related work in Sect. 2. Section 3 covers preliminaries related to motivation and introduction to RF. Section 4 describes the Local Outlier Factor that will be utilized in our proposed extension of RF. Section 5 formalizes our proposed method and corresponding algorithm. Experimental study demonstrating the superiority of the proposed technique over the traditional RF is detailed in Sect. 6. The paper is then concluded with a summary and pointers to future directions in Sect. 7.

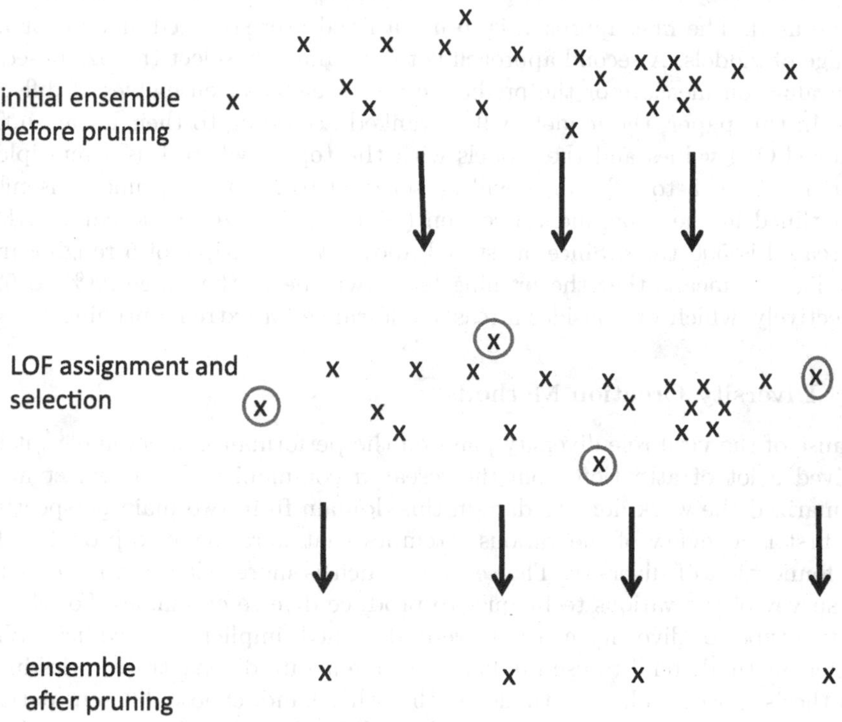

Fig. 1. Extreme pruning via local outlier factor

2 Related Work

Several attempts have been made in recent years in order to produce a subset of an ensemble that performs as well as, or better than, the original ensemble. The purpose of ensemble pruning is to search for such a good subset. This is particularly useful for large ensembles that require extra memory usage, computational costs, and occasional decreases in effectiveness. Grigorios et al. [35] recently amalgamated a survey of ensemble pruning techniques where they classified such techniques into four categories: ranking based, clustering based, optimization based, and others. Ranking based methods, that are relevant to us in this paper, are conceptually the simplest. Since using the predictive performance to rank models is too simplistic and does not yield satisfying results [27,39], ranking based methods employ an evaluation measure to rank models. Kappa statistic measure κ was used in [22] for pruning AdaBoost ensembles. For bagging ensembles, however, kappa has proven to be non-competitive [25]. For bagging ensembles, [24] developed an efficient and effective pruning method based on orientation ordering where the classifiers obtained from bagging are reordered and a subset is selected for aggregation.

An interesting issue that remains after ranking the models is to determine the models that will be chosen to form the pruned ensemble. For this, two approaches can be used. The first approach is to use a fixed user-specified amount or percentage of models. A second approach is to dynamically select the size based on the evaluation measure or the predictive performance of ensembles of different sizes. In this paper, the models will be ranked according to their Local Outlier Factor (LOF) values and the models with the top k (where k is a multiple of 5 ranging from 5 to 40) values will be selected to form the pruned ensemble. As outlined in the experimental section (Sect. 6), the size of the parent RF to be created is 500 trees. Since, as stated above, k is multiple of 5 ranging from 5 to 40, this means that the pruning levels will be in the range 99 % to 92 % respectively, which we consider a reasonable range for extreme pruning.

2.1 Diversity Creation Methods

Because of the vital role diversity plays on the performance of ensembles, it had received a lot of attention from the research community. G. Brown et al. [9] summarized the work done to date in this domain from two main perspectives. The first is a review of the various attempts that were made to provide a formal foundation of diversity. The second, which is more relevant to this paper, is a survey of the various techniques to produce diverse ensembles. For the latter, two types of diversity methods were identified: implicit and explicit. While implicit methods tend to use randomness to generate diverse trajectories in the hypothesis space, explicit methods, on the other hand, choose different paths in the space deterministically. In light of these definitions, bagging and boosting in the previous section are classified as implicit and explicit respectively.

G. Brown et al. [9] also categorized ensemble diversity techniques into three categories: starting point in hypothesis space, set of accessible hypotheses, and manipulation of training data. Methods in the first category use different starting points in the hypothesis space, therefore, influencing the convergence place within the space. Because of their poor performance of achieving diversity, such methods are used by many authors as a default benchmark for their own methods [21]. Methods in the second category vary the set of hypotheses that are available and accessible by the ensemble. For different ensembles, these methods vary either the training data used or the architecture employed. In the third category, the methods alter the way space is traversed. Occupying any point in the search space, gives a particular hypothesis. The type of the ensemble obtained will be determined by how the space of the possible hypotheses is traversed.

In this paper, we propose a new diversity creation method based on unsupervised learning. The method utilizes an existing unsupervised learning technique that, to the best of our knowledge, has not been used before in the production of pruned ensembles.

2.2 Diversity Measures

Regardless of the diversity creation technique used, diversity measures were developed to measure the diversity of a certain technique or perhaps to compare the diversity of two techniques. Tang et al. [34] presented a theoretical analysis on six existing diversity measures: disagreement measure [32], double fault measure [14], KW variance [19], inter-rater agreement [12], generalized diversity [26], and measure of difficulty [12]. The goal was not only to show the underlying relationships between them, but also to relate them to the concept of margin, which is one of the contributing factors to the success of ensemble learning algorithms.

We suffice to describe the first two measures as the others are outside the scope of this paper. The disagreement measure is used to measure the diversity between two base classifiers h_j and h_k, and is calculated as follows:

$$dis_{j,k} = \frac{N^{10} + N^{01}}{N^{11} + N^{10} + N^{01} + N^{00}}$$

where

- N^{10}: means number of training instances that were correctly classified by h_j, but are incorrectly classified by h_k
- N^{01}: means number of training instances that were incorrectly classified by h_j, but are correctly classified by h_k
- N^{11}: means number of training instances that were correctly classified by h_j and h_k
- N^{00}: means number of training instances that were incorrectly classified by h_j and h_k

The higher the disagreement measure, the more diverse the classifiers are. The double fault measure uses a slightly different approach where the diversity between two classifiers is calculated as:

$$DF_{j,k} = \frac{N^{00}}{N^{11} + N^{10} + N^{01} + N^{00}}$$

The above two diversity measures work only for binary classification (AKA binomial) where there are only two possible values (like Yes/No) for the class label, hence, the objects are classified into exactly two groups. They do not work for multiclass (AKA multinomial) classification where the objects are classified into more than two groups.

3 Preliminaries

3.1 Motivation

As mentioned before, RF algorithms tend to build between 100 and 500 trees [36]. Our research aims at producing child RFs that are significantly smaller in size and yet, have accuracy performance that is at least as good as that of the parent RF from which they were derived. The classification speed of each child is guaranteed to be much faster than that of the parent RF because (1) it has much fewer trees and (2) any tree used in the child is also in the parent (i.e., no new trees were introduced in the child).

3.2 Random Forest

RF is an ensemble learning method used for classification and regression. Developed by Breiman [7], the method combines Breiman's bagging sampling approach [5], and the random selection of features, introduced independently by Ho [15,16] and Amit and Geman [2], in order to construct a collection of decision trees with controlled variation. Using bagging, each decision tree in the ensemble is constructed using a sample with replacement from the training data. Statistically, the sample is likely to have about 64 % of instances appearing at least once in the sample. Instances in the sample are referred to as in-bag-instances, and the remaining instances (about 36 %), are referred to as out-of-bag instances. Each tree in the ensemble acts as a base classifier to determine the class label of an unlabeled instance. This is done via majority voting where each classifier casts one vote for its predicted class label, then the class label with the most votes is used to classify the instance. Algorithm 1 below depicts the RF algorithm [7] where N is the number of training samples and S is the number of features in data set.

Algorithm 1. Random Forest Algorithm

```
{User Settings}
input N, S
{Process}
Create an empty vector RF→
for i = 1 → N do
   Create an empty tree T_i
   repeat
      Sample S out of all features F using Bootstrap sampling
      Create a vector of the S features F_S→
      Find Best Split Feature B(F_S→)
      Create A New Node using B(F_S→) in T_i
   until No More Instances To Split On
   Add T_i to the RF→
end for
{Output}
A vector of trees RF→
```

4 Local Outlier Factor

The Local Outlier Factor (LOF) algorithm was developed by Breunig et al. [8] to measure the outlierness of an object. The higher the LOF value assigned to an object, the more isolated the object is with respect to its neighbors. It is considered a very powerful anomaly detection technique in machine learning and classification. Earlier work on outlier detection was investigated in [3,17,18,30], however, the work was limited by treating an outlier as a binary property to classify an object as an outlier or not, without assigning it a value to measure its outlierness as was done in [8].

The LOF can be used as a method to achieve diversity. It was one of 3 strategies used to obtain diversity when constructing an ensemble for the KDDCup 1999 dataset [10]. Schubert et al. [31] proposed methods for measuring similarity and diversity of methods for building advanced outlier detection ensembles using LOF variants and other algorithms.

Formally, Breunig et al. [8] introduced the concept of reachability distance in order to calculate the LOF. If the distance of object A to the k nearest neighbor is denoted by k-distance(A), where the k nearest neighbors is denoted by N_k(A), the following equation defines the reachability distance (rd):

$$rd_k(A,B) = max\{k{-}distance(B), d(A,B)\} \tag{1}$$

where $d(A,B)$ is the distance between objects A and B. The local reachability density of object A is then defined by

$$lrd(A) = \frac{\sum_{B \in N_k(A)} rd_k(A,B)}{|N_k(A)|} \tag{2}$$

Using the local reachability density of object A as defined in the previous equation, the LOF for object A is given by:

$$LOF_k(A) = \frac{\sum_{B \in N_k(A)} \frac{lrd(B)}{lrd(A)}}{|N_k(A)|} \tag{3}$$

5 LOF-Based Diverse Random Forest (LOFB-DRF)

In this section, we propose an extension of RF called *LOFB-DRF* that spawns a child RF that is (1) much smaller in size than the parent RF and (2) has an accuracy that is at least as good as that of the parent RF. In this extension, we use the LOF discussed in Sect. 4. As shown in Fig. 2, each tree predictions on the training dataset (denoted by the vector $C(t_i,T)$) is assigned an LOF value that indicates the degree of its outlierness. The top k (k=5,10,...,40) trees corresponding to these predictions with the highest weighted LOF values (to be discussed next) are then selected to become members of the resulting *LOFB-DRF*. In the remainder of this paper, we will refer to the parent/original traditional Random Forest as *RF*, and refer to the resulting child RF based on our method as *LOFB-DRF*.

Based on Fig. 2, we formalize the *LOFB-DRF* algorithm as shown in Algorithm 2 where T is the training set, and N refers to the number of training samples. The constant k refers to the number of trees that will have the highest weighted LOF values as will be discussed later. As aforementioned, the domain of this constant is multiple of 5 in the range 5 to 40. This way and as we shall see in the experiments section, we can compare the performance *RF* with an *LOFB-DRF* of different sizes.

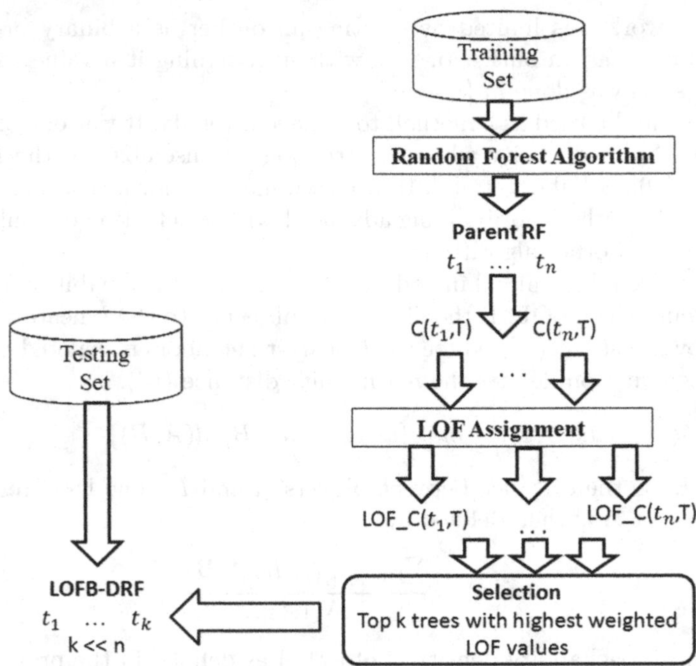

Fig. 2. LOFB-DRF approach

Algorithm 2. LOFB-DRF Algorithm

{User Settings}
input T, N, k
{Process}
Create an empty vector $\overrightarrow{treesPredictions}$
Create an empty vector $\overrightarrow{LOFB-RF}$
Using N, call Random Forest Algorithm 1 above to create $\overrightarrow{RF}$
for $i = 1 \rightarrow RF.size()$ **do**
  $\overrightarrow{treesPredictions} = \overrightarrow{treesPredictions} \cup$ C(RF.tree(i), T)
end for
for $i = 1 \rightarrow \overrightarrow{treesPredictions}.size()$ **do**
  assignNormalizedLOF($\overrightarrow{treesPredictions}$.element(i))
end for
for $i = 1 \rightarrow \overrightarrow{treesPredictions}.size()$ **do**
  assignWeight($\overrightarrow{treesPredictions}$.element(i))
end for
Select the top k instances in $\overrightarrow{treesPredictions}$ with highest weighted LOF values
Select the corresponding trees from RF and add them to $\overrightarrow{LOFB-DRF}$
{Output}
A vector of trees $\overrightarrow{LOFB-DRF}$

5.1 Selection of Trees

With reference to Algorithm 2, the selection of trees in *RF* that will become members of *LOFB-DRF* proceeds as follows. First, predictions of each tree on the training dataset T is computed as a vector and added to the vector $\overrightarrow{treesPredictions}$. At the conclusion of the first **for** loop, $\overrightarrow{treesPredictions}$ becomes a super vector containing vectors where each vector stores the predictions of each tree. By the second **for** loop, each instance in $\overrightarrow{treesPredictions}$ is then assigned a normalized LOF value between 0 and 1. This way, each normalized value describes the probability of the instance being an outlier [10]. By the third **for** loop, each instance is assigned a weight that is the product of the normalized LOF value and the accuracy rate of the corresponding tree on the training data. Formally, let c_i be an instance in the super vector $\overrightarrow{treesPredictions}$, NormalizedLOF($c_i$) be the normalized LOF value assigned to this instance, and AccuracyRate(Tree(c_i),T) be the accuracy rate of Tree(c_i) on the training dataset T where Tree(c_i) is the tree that corresponds to the instance c_i. The weight assigned to this instance is given by:

$$weight = NormalizedLOF(c_i) \times AccuracyRate(Tree(c_i), T) \quad (4)$$

The instances are then sorted in descending order by this weight and the corresponding top k trees are then selected.

6 Experiments

For our experiments, we have used 10 real datasets with varying characteristics from the UCI repository [4]. To use the holdout testing method, each dataset was divided into 2 sets: training and testing. Two thirds (66 %) were reserved for training and the rest (34 %) for testing. Each dataset consists of input variables (features) and an output variable. The latter refers to the class label whose value will be predicted in each experiment. In Fig. 2, the initial *RF* to produce *LOFB-DRF* had a size of 500 trees, a typical upper limit setting for RF ensembles [36].

The *LOFB-DRF* algorithm described above was implemented using the Java programming language utilizing the API of Waikato Environment for Knowledge Analysis (WEKA) [23]. We ran this algorithm 10 times on each dataset where a new *RF* was created in each run. We then calculated the average of the 10 runs for each resulting *LOFB-DRF* to produce the average for a variety of metrics including accuracy rate, minimum accuracy rate, maximum accuracy rate, standard deviation, FMeasure, and AUC as shown in Table 3. For *RF*, we just calculated the average accuracy rate, FMeasure, and AUC as shown in the last 3 columns of the table.

6.1 Results

Table 3 compares the performance of *LOFB-DRF* and *RF* on the 10 datasets used in the experiment. To show the superiority of *LOFB-DRF*, we have highlighted in boldface the average accuracy rate of *LOFB-DRF* when it is greater than that of *RF*. With the exception of the *audit* and *vote* datasets (last 2 datasets), we find that *LOFB-DRF* performed at least as good as *RF*. Interestingly enough, of the 10 datasets, *LOFB-DRF*, regardless of its size, completely outperformed *RF* on 3 of the datasets, namely, *squash-stored*, *eucalyptus*, and *sonar*.

6.2 Pruning Level

In ensemble pruning, a pruning level refers to the reduction ratio between the original ensemble and the pruned one. For example, if the size of the original ensemble is 500 trees and the pruned one is of size 50, then $100\,\% - \frac{50}{500} \times 100\,\% = 90\,\%$ is the pruning level that was achieved in the pruned ensemble. This means that the pruned ensemble is 90 % smaller than the original one. Table 1 shows the pruning levels where the first column shows the maximum possible pruning level for an *LOFB-DRF* that has outperformed *RF*, and the second column shows the pruning level of the best performer *LOFB-DRF*. We can see that with extremely healthy pruning levels ranging from 95 % to 99 %, our technique outperformed *RF*. This makes *LOFB-DRF* a natural choice for real-time applications, where fast classification is an important desideratum. In most cases, 100 times faster classification can be achieved with a 99 % pruning level, as shown in the table. In the worst case scenario, only 16.67 times faster classification with 95 % pruning level in the *squash-unstored* dataset. Such estimates are based on the fact that the number of trees traversed in the RF is the dominant factor in the classification response time. This is especially true, given that RF trees are unpruned bushy trees.

Table 1. Maximum pruning level with best possible performance

Dataset	Maximum pruning level	Best performance pruning level
breast-cancer	97 %	95 %
squash-unstored	95 %	93 %
squash-stored	99 %	98 %
eucalyptus	99 %	99 %
soybean	98 %	97 %
diabetes	96 %	96 %
car	99 %	99 %
sonar	99 %	99 %

6.3 Analysis

For each dataset, Fig. 3 shows the number of *LOFB-DRF*s outperforming *RF*. As shown in the figure, with the exception of the *audit* and *vote* datasets, we have at least one *LOFB-DRF* outperformer for each dataset.

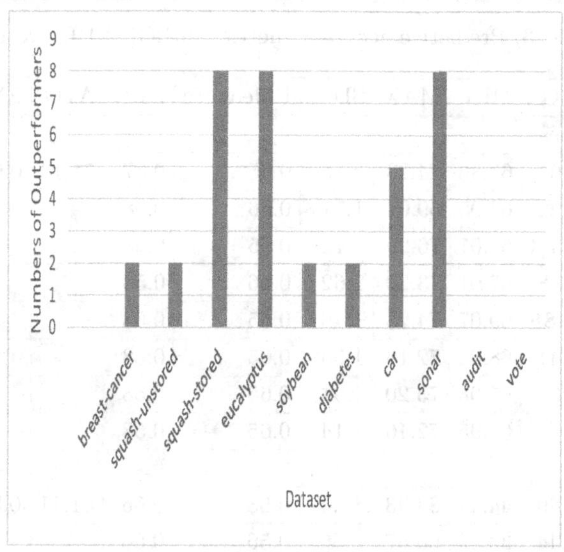

Fig. 3. Number of LOFB-DRFs outperforming RF

6.4 Outperformance Range

Tables 2 below depicts the outperformance range of *LOFB-DRF* over *RF*. A negative number indicates that *RF* was superior to *LOFB-DRF* and the absolute value of this number refers to the performance difference between *RF* and best performer *LOFB-DRF*. Taking a closer look at this table, we can see that *LOFB-DRF* outperformed *RF* on 8/10 datasets with a maximum outperformance range of 12.11 %.

Table 2. Outperformance range of LOFB-DRF over RF

Dataset	Range
breast-cancer	0.21 % - 0.73 %
squash-unstored	2.22 % - 6.11 %
squash-stored	0.55 % - 3.88 %
eucalyptus	1.08 % - 5.88 %
soybean	0.86 % - 1.98 %
diabetes	0.12 % - 0.16 %
car	0.09 % - 1.91 %
sonar	1.69 % - 12.11 %
audit	-0.05 %
vote	-0.27 %

Table 3. Predictive accuracy metrics of LOFB-DRF & RF

LOFB-DRF	AVG	MIN	MAX	SD	FMeasure	AUC	AVG	FMeasure	AUC
breast-cancer									
5	67.01	61.86	74.23	3.16	0.65	0.57	71.13	0.65	0.58
10	67.22	64.95	69.07	1.71	0.66	0.58			
15	**71.34**	67.01	76.29	3.12	0.65	0.58			
20	69.48	67.01	73.20	2.62	0.66	0.58			
25	**71.86**	69.07	74.23	1.46	0.65	0.58			
30	70.41	68.04	72.16	1.53	0.65	0.58			
35	70.62	65.98	73.20	1.91	0.65	0.58			
40	69.18	64.95	72.16	2.14	0.65	0.58			
squash-unstored									
5	58.89	44.44	83.33	12.47	0.58	0.66	61.11	0.52	0.64
10	54.44	33.33	66.67	9.56	0.56	0.66			
15	60.56	50.00	83.33	8.77	0.55	0.65			
20	60.00	50.00	66.67	5.98	0.54	0.66			
25	**63.33**	55.56	77.78	7.93	0.54	0.65			
30	58.33	44.44	77.78	8.70	0.53	0.65			
35	**67.22**	50.00	83.33	10.08	0.54	0.66			
40	57.78	50.00	66.67	6.19	0.53	0.65			
squash-stored									
5	**56.67**	38.89	66.67	9.56	0.57	0.59	55.56	0.51	0.56
10	**59.44**	44.44	66.67	7.05	0.54	0.58			
15	**58.33**	50.00	66.67	4.48	0.54	0.58			
20	**58.33**	50.00	61.11	3.73	0.55	0.58			
25	**58.33**	50.00	66.67	5.12	0.53	0.57			
30	**56.67**	55.56	61.11	2.22	0.52	0.56			
35	**56.11**	55.56	61.11	1.67	0.52	0.57			
40	**56.11**	55.56	61.11	1.67	0.52	0.56			
eucalyptus									
5	**25.80**	11.20	40.40	8.73	0.26	0.60	19.92	0.21	0.57
10	**21.00**	12.40	28.40	4.70	0.24	0.59			
15	**24.32**	14.80	32.00	5.01	0.24	0.58			
20	**24.48**	15.60	29.60	4.55	0.23	0.58			
25	**24.68**	21.20	29.60	2.35	0.23	0.58			
30	**24.80**	14.80	33.60	5.13	0.23	0.58			
35	**23.96**	20.00	34.40	4.20	0.23	0.58			
40	**21.16**	15.20	28.00	3.69	0.22	0.57			

Table 3. *(Continued)*

LOFB-DRF	AVG	MIN	MAX	SD	FMeasure	AUC	AVG	FMeasure	AUC
soybean									
5	77.28	60.78	85.78	6.80	0.79	0.88	77.59	0.73	0.85
10	**78.45**	70.69	85.34	5.46	0.75	0.87			
15	**79.57**	72.84	83.62	3.50	0.76	0.87			
20	76.85	74.57	78.88	1.26	0.74	0.86			
25	76.90	74.14	79.31	1.88	0.74	0.86			
30	76.85	72.41	81.47	2.43	0.74	0.86			
35	77.33	71.98	82.33	3.66	0.73	0.86			
40	76.59	71.98	81.03	2.59	0.73	0.85			
diabetes									
5	80.80	74.71	84.29	3.53	0.72	0.68	81.26	0.71	0.67
10	81.15	74.71	84.29	3.56	0.71	0.68			
15	79.85	77.39	83.14	1.96	0.71	0.67			
20	**81.42**	79.31	83.14	1.24	0.71	0.67			
25	80.96	78.93	82.76	1.31	0.71	0.67			
30	80.88	78.54	82.76	1.14	0.71	0.67			
35	79.81	77.39	81.99	1.40	0.71	0.67			
40	**81.38**	80.08	83.14	0.94	0.71	0.67			
car									
5	**64.17**	62.41	67.52	1.33	0.56	0.78	62.26	0.56	0.78
10	**63.01**	61.56	64.29	0.75	0.56	0.78			
15	**62.36**	60.71	64.29	1.12	0.56	0.78			
20	**62.35**	61.22	63.78	0.82	0.56	0.78			
25	**62.69**	60.88	63.95	0.85	0.56	0.78			
30	62.18	61.05	63.10	0.82	0.56	0.78			
35	61.96	60.88	63.61	0.72	0.56	0.78			
40	61.99	61.05	62.59	0.54	0.55	0.78			
sonar									
5	**12.25**	7.04	18.31	3.34	0.26	0.00	0.14	0.29	0.00
10	**9.15**	0.00	16.90	5.20	0.28	0.00			
15	**6.34**	0.00	14.08	4.47	0.29	0.00			
20	**3.38**	0.00	8.45	2.76	0.29	0.00			
25	**3.10**	0.00	7.04	2.42	0.28	0.00			
30	**1.83**	0.00	4.23	1.27	0.28	0.00			
35	**3.38**	0.00	4.23	1.29	0.28	0.00			
40	**3.38**	0.00	9.86	2.69	0.28	0.00			

Table 3. *(Continued)*

LOFB-DRF	AVG	MIN	MAX	SD	FMeasure	AUC	AVG	FMeasure	AUC
audit									
5	95.63	94.26	96.47	0.72	0.91	0.89	96.31	0.90	0.88
10	95.74	95.00	96.18	0.35	0.90	0.88			
15	95.99	95.29	96.47	0.35	0.90	0.88			
20	96.06	95.29	96.76	0.39	0.90	0.88			
25	96.22	95.88	96.47	0.25	0.91	0.89			
30	96.03	95.59	96.47	0.25	0.90	0.88			
35	96.26	95.88	96.47	0.18	0.90	0.88			
40	96.00	95.59	96.47	0.27	0.90	0.87			
vote									
5	96.82	95.27	97.97	0.80	0.96	0.98	97.97	0.95	0.97
10	97.09	95.27	97.97	0.86	0.96	0.97			
15	97.57	96.62	97.97	0.45	0.95	0.97			
20	97.43	96.62	97.97	0.51	0.95	0.97			
25	97.57	96.62	97.97	0.45	0.95	0.97			
30	97.70	97.30	97.97	0.33	0.95	0.97			
35	97.64	96.62	97.97	0.45	0.95	0.97			
40	97.64	96.62	97.97	0.45	0.95	0.97			

7 Conclusion and Future Directions

Research conducted in this paper was based on how diversity in ensembles tends to yield better results [1,9,20,34]. We have adopted the Local Outlier Factor method to select diverse trees in an RF and then used these trees to form a pruned ensemble of the original one. The selection was based on both the LOF value and the predictive accuracy of each tree. Experimental results have shown the potential of this method, with extreme pruning levels of Random Forests that can outperform the original population of trees, reaching as high as 99 %. This makes the pruned ensemble a suitable candidate for real-time applications.

We have selected trees that correspond to the instances with the top k weighted LOF values. Another interesting variation would be to use a hybrid approach that combines LOF with clustering to boost diversity up. Using this approach, we first create clusters of trees then from each cluster, we select a representative that corresponds to the instance with the highest weighted LOF value. The current implementation also gives equal importance to the peculiarity of the tree as measured by the LOF score and the predictive accuracy, represented by the percentage of correctly classified instances for the tree. However, tuning this significance can play an important role in enhancing the classifier. At one hand, choosing trees with higher predictive accuracy can lead to model overfitting, and on the other hand, using LOF only can lead to leaving out trees that are most representative of the dataset. Balancing between the two can result in an ensemble that is diverse enough to boost the accuracy.

References

1. Garcıa Adeva, J.J., Beresi, U., Calvo, R.: Accuracy and diversity in ensembles of text categorisers. CLEI Electron. J. **9**(1), 1–12 (2005)
2. Amit, Y., Geman, D.: Shape quantization and recognition with randomized trees. Neural Comput. **9**(7), 1545–1588 (1997)
3. Arning, A., Agrawal, R., Raghavan, P.: A linear method for deviation detection in large databases. In: KDD, pp. 164–169 (1996)
4. Bache, K., Lichman, M.: Uci machine learning repository (2013)
5. Breiman, L.: Bagging predictors. Mach. Learn. **24**(2), 123–140 (1996)
6. Breiman, L.: Stacked regressions. Mach. Learn. **24**(1), 49–64 (1996)
7. Breiman, L.: Random forests. Mach. Learn. **45**(1), 5–32 (2001)
8. Breunig, M.M., Kriegel, H.-P., Ng, R.T., Sander, J.: Lof: identifying density-based local outliers. In: ACM Sigmod Record, vol. 29, pp. 93–104. ACM (2000)
9. Brown, G., Wyatt, J., Harris, R., Yao, X.: Diversity creation methods: a survey and categorisation. Inf. Fusion **6**(1), 5–20 (2005)
10. Kriegel, H.-P., Kroger, P., Schubert, E., Zimek, A.: Interpreting and unifying outlier scores. In: 11th SIAM International Conference on Data Mining (SDM), Mesa, AZ (2011)
11. Fernández-Delgado, M., Cernadas, E., Barro, S., Amorim, D.: Do we need hundreds of classifiers to solve real world classification problems? J. Mach. Learn. Res. **15**(1), 3133–3181 (2014)
12. Fleiss, J.L., Levin, B., Paik, M.C.: Statistical Methods for Rates and Proportions. Wiley, New York (2013)
13. Freund, Y., Schapire, R.E.: A decision-theoretic generalization of on-line learning and an application to boosting. J. Comput. Syst. Sci. **55**(1), 119–139 (1997)
14. Giacinto, G., Roli, F.: Design of effective neural network ensembles for image classification purposes. Image Vis. Comput. **19**(9), 699–707 (2001)
15. Ho, T.K.: Random decision forests. In: 1995 Proceedings of the Third International Conference on Document Analysis and Recognition, vol. 1, pp. 278–282. IEEE (1995)
16. Ho, T.K.: The random subspace method for constructing decision forests. IEEE Trans. Pattern Anal. Mach. Intell. **20**(8), 832–844 (1998)
17. Knorr, E.M., Ng, R.T.: Finding intensional knowledge of distancebased outliers. VLDB **99**, 211–222 (1999)
18. Knox, E.M., Ng, R.T.: Algorithms for mining distance-based outliers in large datasets. In: Proceedings of the International Conference on Very Large Data Bases. Citeseer (1998)
19. Kohavi, R., Wolpert, D.H., et al.: Bias plus variance decomposition for zero-one loss functions. In: ICML, pp. 275–283 (1996)
20. Kuncheva, L.I., Whitaker, C.J.: Measures of diversity in classifier ensembles and their relationship with the ensemble accuracy. Mach. Learn. **51**(2), 181–207 (2003)
21. Maclin, R., Opitz, D.: Popular ensemble methods: an empirical study. J. Artif. Intell. Res. **11**(1–2), 169–198 (1999)

22. Margineantu, D.D., Dietterich, T.G.: Pruning adaptive boosting. In: ICML, vol. 97, pp. 211–218. Citeseer (1997)
23. Holmes, G., Pfahringer, B., Reutemann, P., Witten, I.H., Hall, M., Frank, E.: The WEKA data mining software: an update. SIGKDD Explor. Newslett. **11**(1), 10–18 (2009)
24. Martínez-Muñoz, G., Suárez, A.: Pruning in ordered bagging ensembles. In: Proceedings of the 23rd International Conference on Machine Learning, pp. 609–616. ACM (2006)
25. Martinez-Muoz, G., Hernández-Lobato, D., Suárez, A.: An analysis of ensemble pruning techniques based on ordered aggregation. IEEE Trans. Pattern Anal. Mach. Intell. **31**(2), 245–259 (2009)
26. Partridge, D., Krzanowski, W.: Software diversity: practical statistics for its measurement and exploitation. Inf. Softw. Technol. **39**(10), 707–717 (1997)
27. Partridge, D., Yates, W.B.: Engineering multiversion neural-net systems. Neural Comput. **8**(4), 869–893 (1996)
28. Polikar, R.: Ensemble based systems in decision making. IEEE Circ. Syst. Mag. **6**(3), 21–45 (2006)
29. Rokach, L.: Ensemble-based classifiers. Artif. Intell. Rev. **33**(1–2), 1–39 (2010)
30. Ruts, I., Rousseeuw, P.J.: Computing depth contours of bivariate point clouds. Comput. Stat. Data Anal. **23**(1), 153–168 (1996)
31. Schubert, E., Wojdanowski, R., Zimek, A., Kriegel, H.-P.: On evaluation of outlier rankings and outlier scores. In: 2012 Proceedings of the 12th SIAM International Conference on Data Mining (SDM), Anaheim, CA, pp. 1047–1058 (2012)
32. Skalak, D.B.: The sources of increased accuracy for two proposed boosting algorithms. In: Proceedings of American Association for Artificial Intelligence, AAAI-96, Integrating Multiple Learned Models Workshop, vol. 1129, p. 1133. Citeseer (1996)
33. Smyth, P., Wolpert, D.: Linearly combining density estimators via stacking. Mach. Learn. **36**(1–2), 59–83 (1999)
34. Tang, E.K., Suganthan, P.N., Yao, X.: An analysis of diversity measures. Mach. Learn. **65**(1), 247–271 (2006)
35. Tsoumakas, G., Partalas, I., Vlahavas, I.: An ensemble pruning primer. In: Okun, O., Valentini, G. (eds.) Applications of Supervised and Unsupervised Ensemble Methods. SCI, pp. 1–13. Springer, Heidelberg (2009)
36. Williams, G.: Use R: Data Mining with Rattle and R: The Art of Excavating Data for Knowledge Discovery. Springer, New York (2011)
37. Wolpert, D.H.: Stacked generalization. Neural Netw. **5**(2), 241–259 (1992)
38. Yan, W., Goebel, K.F.: Designing classifier ensembles with constrained performance requirements. In: Defense and Security. International Society for Optics and Photonics, pp. 59–68 (2004)
39. Yang, Y., Korb, K.B., Ting, K.M., Webb, G.I.: Ensemble selection for superparent-one-dependence estimators. In: Zhang, S., Jarvis, R.A. (eds.) AI 2005. LNCS (LNAI), vol. 3809, pp. 102–112. Springer, Heidelberg (2005)

Lower Bounds on Complexity of Shallow Perceptron Networks

Věra Kůrková(✉)

Institute of Computer Science, Czech Academy of Sciences,
Pod Vodárenskou věží 2, 18207 Prague, Czech Republic
vera@cs.cas.cz

Abstract. Model complexity of shallow (one-hidden-layer) perceptron networks computing multivariable functions on finite domains is investigated. Lower bounds are derived on growth of the number of network units or sizes of output weights in terms of variations of functions to be computed. A concrete construction of a class of functions which cannot be computed by percetron networks with considerably smaller numbers of units and output weights than the sizes of the function's domains is presented. In particular, functions on Boolean d-dimensional cubes are constructed which cannot be computed by shallow perceptron networks with numbers of hidden units and sizes of output weights depending on d polynomially. A subclass of these functions is described whose elements can be computed by two-hidden-layer networks with the number of units depending on d linearly.

Keywords: Shallow feedforward networks · Signum perceptrons · Finite mappings · Model complexity · Hadamard matrices

1 Introduction

Originally, biologically inspired neural networks were introduced as multilayer computational models, but later one-hidden-layer architectures became dominant in applications (see, e.g., [1,2] and the references therein). Networks composed from several layers of convolutional units were tested already in 1990s [3], but their training has been inefficient till the advent of fast graphic processing units. Recently, network architectures with several hidden layers became called deep [4,5] to distinguish them from networks with merely one hidden layer called shallow. Deep networks with several convolutional and pooling layers achieved excellent performance on computer vision and speech recognition tasks (see the survey article [6] and the references therein). However recently, some reservations about overall superiority of deep networks over shallow ones appeared. An empirical study demonstrated that shallow networks can learn some functions previously learned by deep ones using the same numbers of parameters as the original deep networks [7].

Thus it is desirable to develop a theoretical analysis comparing model complexities of shallow and deep networks. Characterization of tasks, which can be

C. Jayne and L. Iliadis (Eds.): EANN 2016, CCIS 629, pp. 283–294, 2016.
DOI: 10.1007/978-3-319-44188-7_21

computed by deep networks of smaller model complexities than shallow ones can be derived by comparing lower bounds on numbers of units in shallow networks with upper bounds on numbers of units in deep ones. Generally, derivation of lower bounds is much more difficult than derivation of upper ones. For shallow networks, a variety of upper bounds on numbers of units in dependence on types of units, input dimensions, and types of functions to be computed is known (see, e.g., [8] and the references therein), but only few lower bounds is available. Some lower bounds hold merely for types of computational units that are not commonly used such as perceptrons with specially designed activation functions [9]. Bianchini and Scarselli [10] initiated a theoretical research comparing numbers of units in shallow and deep networks in terms of topological properties of input-output functions.

Bengio et al. [11] suggested that a cause of large model complexities of shallow networks might be in the "amount of variations" of functions to be computed. As an example of a highly varying function they presented the parity function on the Boolean cube. They proved that classification of points from the d-dimensional Boolean cube by Gaussian SVM requires at least 2^{d-1} support vectors.

In [12] we showed that the effect of "high variations" of a function depends on a type of computational units. We proposed to use a concept of variational norm tailored to a type of computational units as a measure of variations of a function influencing model complexity of networks with units of the given type. Using a probabilistic argument, we proved that almost any uniformly randomly chosen function on a sufficiently large domain is highly varying with respect to Heaviside or signum perceptrons and thus it cannot be represented by perceptron networks with a reasonably small number of units and sizes of output weights.

However, our argument proving existence of large sets of functions whose implementations by shallow perceptron networks require large numbers of units or large sizes of output weights is non constructive [12]. It is based on probabilistic Chernoff bound related to the law of large numbers. In this paper, we present a concrete construction of classes of such functions. We construct classes of multivariable functions on finite domains in $\mathbb{R}^d$ in the form of $n \times n$ rectangles. We show that these functions cannot be computed by shallow Heaviside or signum perceptron networks having both number of units and sizes of output weights larger than $\frac{\sqrt{n}}{\lceil \log_2 n \rceil}$. In particular, for domains of sizes $2^k \times 2^k$, such functions cannot be computed by shallow perceptron networks with numbers or units or sizes of output weights depending on k polynomially. Our construction is based on properties of square matrices inducing functions which are not correlated to any function computable by a signum perceptron. We describe a subclass of these functions which can be computed by two-hidden-layer perceptron networks with the numbers of units depending on k linearly.

The paper is organized as follows. Section 2 contains basic concepts on shallow networks and dictionaries of computational units. Section 3 reviews variational norms as tools for investigation of network complexity. Section 4 presents existential probabilistic results on functions which cannot be computed by shallow signum perceptron networks with "small" numbers of units and sizes of output weights. In Sect. 5, construction of concrete classes of such functions is presented. In Sect. 6 model complexities of one and two-hidden-layer networks computing a subclass of functions used in our construction are compared. Section 7 is a brief discussion.

2 Preliminaries

The most widespread type of a feedforward neural network architecture is a *one-hidden-layer network with a single linear output.* For a given type of computational units from a set of functions called *dictionary* these networks compute input-output functions belonging to sets of the form

$$\mathrm{span}_n\, G := \left\{ \sum_{i=1}^{n} w_i g_i \,|\, w_i \in \mathbb{R},\, g_i \in G \right\},$$

where the coefficients w_i are called *output weights* and n denotes the number of network units. Recently, one-hidden-layer networks became called *shallow networks* to distinguish them from deep networks with several hidden layers of computational units.

A common type of a computational unit is *perceptron*, which computes functions of the form $\sigma(v \cdot . + b) : X \to \mathbb{R}$, where $\sigma : \mathbb{R} \to \mathbb{R}$ is an *activation function.* It is called *sigmoid* when it is monotonic increasing and $\lim_{t\to-\infty} \sigma(t) = 0$ and $\lim_{t\to\infty} \sigma(t) = 1$. Important types of activation functions are the *Heaviside function* defined as

$$\vartheta(t) := 0 \text{ for } t < 0 \quad \text{and} \quad \vartheta(t) := 1 \text{ for } t \geq 0$$

and the *signum function* $\mathrm{sgn} : \mathbb{R} \to \{-1, 1\}$, defined as

$$\mathrm{sgn}(t) := -1 \text{ for } t < 0 \quad \text{and } \mathrm{sgn}(t) := 1 \quad \text{for } t \geq 0.$$

We denote by $H_d(X)$ the dictionary of functions on $X \subset \mathbb{R}^d$ computable by *Heaviside perceptrons*, i.e.,

$$H_d(X) := \{\vartheta(v \cdot . + b) : X \to \{0,1\} \,|\, v \in \mathbb{R}^d, b \in \mathbb{R}\}, \tag{1}$$

end by $P_d(X)$ the dictionary of functions on X computable by *signum perceptrons*, i.e.,

$$P_d(X) := \{\mathrm{sgn}(v \cdot . + b) : X \to \{-1,1\} \,|\, v \in \mathbb{R}^d, b \in \mathbb{R}\}. \tag{2}$$

Note that from the point of view of number of network units, there is only a minor difference between networks with signum and Heaviside perceptrons as

$$\mathrm{sgn}(t) = 2\vartheta(t) - 1 \quad \text{and} \quad \vartheta(t) = \frac{\mathrm{sgn}(t) + 1}{2}. \tag{3}$$

For a domain $X \subset \mathbb{R}^d$ we denote by

$$\mathcal{F}(X) := \{f \,|\, f : X \to \mathbb{R}\}$$

the *set of all real-valued functions on* X and by

$$\mathcal{B}(X) := \{f \,|\, f : X \to \{-1, 1\}\}$$

the *set of all functions on* X *with values in* $\{-1, 1\}$.

In practical applications, domains $X \subset \mathbb{R}^d$ are finite, but their sizes $\operatorname{card} X$ and/or input dimensions d can be quite large. It is easy to see that when $\operatorname{card} X = m$ and $X = \{x_1, \ldots, x_m\}$ is a linear ordering of X, then the mapping $\iota : \mathcal{F}(X) \to \mathbb{R}^m$ defined as $\iota(f) := (f(x_1), \ldots, f(x_m))$ is an isomorphism. So, on $\mathcal{F}(X)$ we have the Euclidean inner product defined as

$$\langle f, g \rangle := \sum_{u \in X} f(u)g(u)$$

and the Euclidean norm $\|f\| := \sqrt{\langle f, f \rangle}$. In contrast to the inner product $\langle ., . \rangle$ on $\mathcal{F}(X)$, we denote by $\cdot$ the inner product on $X \subset \mathbb{R}^d$, i.e., for $u, v \in X$,

$$u \cdot v := \sum_{i=1}^{d} u_i v_i.$$

3 Variational Norms as Measures of Sparsity

It is desirable that dictionaries of computational units are chosen in such a way that input-output functions representing optimal solutions (or reasonably suboptimal ones) of given tasks can be computed by sufficiently sparse networks. The basic measure of sparsity of a shallow network is the number of computational units in the hidden layer.

Shallow networks with many types of computational units (including perceptrons and positive definite kernel units) can exactly compute any function on any finite domain [13]. However, arguments proving universal representation capabilities of shallow networks provide representations of functions on finite domains by networks with numbers of hidden units equal to sizes of domains of the functions to be computed. For large domains, this number can be too large for efficient implementations.

As minimization of the number of non-zero output weights in a shallow network with units from a dictionary G computing function f is a difficult non convex task, minimization of l_1 and l_2-norms of output weights have been used in weight-decay regularization techniques (see, e.g., [1, p. 220]). A small l_1-norm of output weights implies that a function has both small number of units and small absolute values of output weights. Thus an alternative concept of sparsity in terms of l_1-norm of output weight vector can be considered. When a dictionary is linearly independent, its value is unique for a given function. For other dictionaries, minimum over all representations of a given function as input-output functions has to be considered.

The concept of l_1-sparsity is related to the concept of *variation of a function with respect to a dictionary of computational units* from nonlinear approximation theory. Variation with respect to Heaviside perceptrons (called *variation with respect to half-spaces*) was introduced by Barron [14] and extended to general dictionaries by Kůrková [15]. Here we use this concept for dictionaries on finite domains. Such dictionaries are subsets of finite dimensional Hilbert spaces $\mathcal{F}(X)$. For a bounded subset G of $\mathcal{F}(X)$, *G-variation (variation with respect to the dictionary G)*, denoted by $\|.\|_G$, is defined as

$$\|f\|_G := \inf \left\{ c \in \mathbb{R}_+ \,\middle|\, \frac{f}{c} \in \mathrm{cl}_{\mathcal{X}}\, \mathrm{conv}\,(G \cup -G) \right\}, \tag{4}$$

where $-G := \{-g \mid g \in G\}$, $\mathrm{cl}_{\mathcal{X}}$ denotes the closure with respect to the topology induced by the norm $\|\cdot\|_{\mathcal{F}}$, and conv is the convex hull. For properties of variational norm and its role in estimates of rates of approximation, see [8,16–20].

The next proposition, which follows easily from the definition, shows the role of G-variation in estimates of complexity and sparsity of networks representing a function f.

Proposition 1. *Let G be a finite subset of a normed linear space $(\mathcal{X}, \|.\|)$ with $\mathrm{card}\, G = k$. Then, for every $f \in \mathcal{X}$*

$$\|f\|_G = \min \left\{ \sum_{i=1}^{k} |w_i| \,\middle|\, f = \sum_{i=1}^{k} w_i\, g_i\,,\; w_i \in \mathbb{R},\; g_i \in G \right\}.$$

Thus any representation of a function with "large" G-variation by a shallow network with units from a dictionary G must have "large" number of units and/or absolute values of some output weights must be "large". On the other hand, functions with "small" G-variations can be represented by networks with "small" numbers of units from the dictionary G and "small" output weights. Classes of d-variable functions with G-variations growing with d polynomially are of particular interest as they can be represented by shallow networks with numbers of units from the dictionary G and sizes of output weights growing with d polynomially and thus avoid the "curse of dimensionality".

The following theorem from [21] (see also [19]) shows that lower bounds on G-variation of a function f can be obtained by estimating correlations of f with functions from the dictionary G. By $G^{\perp}$ is denoted the *orthogonal complement of G* in the Hilbert space $\mathcal{F}(X)$ which is isomorphic to the finite dimensional Euclidean space $\mathbb{R}^{\mathrm{card} X}$.

Theorem 1. *Let X be a finite subset of $\mathbb{R}^d$ and G be a bounded subset of $\mathcal{F}(X)$. Then, for every $f \in \mathcal{F}(X) \setminus G^{\perp}$ one has*

$$\|f\|_G \geq \frac{\|f\|^2}{\sup_{g \in G} |\langle g, f \rangle|}.$$

Theorem 1 shows that functions which are almost orthogonal to all elements of a dictionary G have large variations with respect to G.

4 Shallow Networks with Signum Perceptrons

The dictionary of signum perceptrons

$$P_d(X) := \{\mathrm{sgn}(v \cdot . + b) : X \to \{-1,1\} \,|\, v \in \mathbb{R}^d, b \in \mathbb{R}\}$$

occupies a relatively small subset of the set $\mathcal{B}(X)$ of all functions on X with values in $\{-1,1\}$. The size of $P_d(X)$ grows with increasing size m of the domain X only polynomially (the degree of the polynomial is the dimension d of the space $\mathbb{R}^d$ where X is embedded), while the size 2^m of the set $\mathcal{B}(X)$ of all functions from X to $\{-1,1\}$ grows exponentially. The following upper bound is a direct consequence of an upper bound on the number of linearly separable dichotomies of m points in $\mathbb{R}^d$ from [22] combined with an upper bound on partial sum of binomials (see [12]).

Theorem 2. *For every d and every $X \subset \mathbb{R}^d$ such that* $\mathrm{card}\, X = m$*,*

$$\mathrm{card}\, P_d(X) \leq 2\frac{m^d}{d!}.$$

In [12] we showed that for large domains X, almost any uniformly randomly chosen function from X to $\{-1,1\}$ has large variation with respect to signum perceptrons. We proved the following theorem combining a probabilistic Chernoff bound, the geometric lower bound on variational norm from Theorem 1, and the relatively small size of the dictionary $P_d(X)$.

Theorem 3. *Let d be a positive integer, $X \subset \mathbb{R}^d$ with* $\mathrm{card}\, X = m$*, f uniformly randomly chosen in $\mathcal{B}(X)$, and $b > 0$. Then*

$$\Pr\left(\|f\|_{P_d(X)} \geq b\right) \geq 1 - 4\frac{m^d}{d!}e^{-\frac{m}{2b^2}}.$$

Thus for large X, almost any uniformly randomly chosen function cannot be l_1-sparsely represented by a shallow network with signum perceptrons. In particular for $\mathrm{card}\, X = 2^d$ and $b = 2^{\frac{d}{4}}$, Theorem 3 implies a lower bound

$$1 - 4\frac{2^{d^2}}{d!}e^{-(2^{\frac{d}{2}}-1)} \tag{5}$$

on probability that a uniformly randomly chosen function from $\mathcal{B}(\{0,1\}^d)$ has variation with respect to signum perceptrons greater or equal to $2^{\frac{d}{4}}$. So almost any uniformly randomly chosen function on the d-dimensional Boolean cube $\{0,1\}^d$ cannot be computed by a shallow network with the number of signum perceptrons and absolute values of output weights depending on d polynomially.

Theorem 3 is existential. It proves that there exists a lot of functions which cannot be l_1-sparsely represented by shallow signum perceptron networks, but it does not suggest how to construct such functions.

5 Construction of Functions with Large Variations with Respect to Signum Perceptrons

In this section, we present a concrete construction of a class of functions on square domains with "large" variations with respect to signum perceptrons. Such functions are concrete examples of functions whose existence is guaranteed by Theorem 3. They cannot be computed by one-hidden-layer networks with "small" numbers of signum perceptrons and "small" sizes of output weights.

We provide a method of construction of such functions on domains in the form of squares $X = \{x_1, \ldots, x_n\} \times \{y_1, \ldots, y_n\} \subset \mathbb{R}^d$. Functions on such domains can be represented by square matrices. A function f on X can be described by a matrix $M(f)$ defined as $M(f)_{i,j} = f(x_i, y_j)$. An $n \times n$ matrix M induces a function f_M on X such that $f_M(x_i, y_j) = M_{i,j}$. In particular, functions with values in $\{-1, 1\}$ induce matrices with entries equal to -1 or $+1$.

The next theorem gives a lower bound on variation with respect to signum perceptrons of functions on square domains induced by Hadamard matrices. Recall that a *Hadamard matrix* of order n is an $n \times n$ square matrix M with entries in $\{-1, 1\}$ such that any two distinct rows (or equivalently columns) of M are orthogonal. In the proof of our theorem we show that functions induced by these matrices have "small" inner products with all elements of the dictionary of signum perceptrons and thus by Theorem 1, they have large variations with respect to signum perceptrons.

Theorem 4. *Let $d = d_1 + d_2$, $\{x_i \,|\, i = 1, \ldots, n\} \subset \mathbb{R}^{d_1}$, $\{y_j \,|\, j = 1, \ldots, n\} \subset \mathbb{R}^{d_2}$, $X = \{x_i \,|\, i = 1, \ldots, m\} \times \{y_j \,|\, j = 1, \ldots, m\} \subset \mathbb{R}^d$, and $f_M : X \to \{-1, 1\}$ be defined as $f_M(x_i, y_j) = M_{i,j}$, where M is an $n \times n$ Hadamard matrix. Then $\|f_M\|_{P_d(X)} \geq \frac{\sqrt{n}}{\lceil \log_2 n \rceil}$.*

Proof. By Theorem 1,

$$\|f_M\|_{P_d(X)} \geq \frac{\|f_M\|^2}{\sup_{g \in P_d(X)} |\langle f_M, g \rangle|} = \frac{n^2}{\sup_{g \in P_d(X)} |\langle f_M, g \rangle|}. \tag{6}$$

The inner product of f_M with g is equal to the sum of entries of the matrices M and $M(g)$, i.e., $\langle f_M, g \rangle = \sum_{i,j}^{n} M_{i,j} M(g)_{i,j}$ and thus it is invariant under permutations of rows and columns performed jointly on both matrices M and $M(g)$.

Without loss of generality, we can assume that each row and each column of $M(g)$ starts with a (possibly empty) initial segment of -1's followed by a (possibly empty) segment of $+1$'s. Otherwise, we reorder rows and columns in both matrices $M(g)$ and M.

To estimate $\langle f_M, g \rangle = \sum_{i,j=1}^{n} M_{i,j} M(g)_{i,j}$ we define a partition of the matrices M and $M(g)$ into families of submatrices such that each submatrix from the partition of $M(g)$ has all entries either equal to -1 or equal to $+1$. We define the partition of $M(g)$ recursively as a sequence of families of matrices (possibly some of them empty) $\mathcal{P}(g, k) = \{P(g, k, 1), \ldots, P(g, k, 2^k)\}$, $k = 1, \ldots, \lceil \log_2 n \rceil$.

Let $\mathcal{P}(k) = \{P(k,1), \ldots, P(k,2^k)\}$ be a family of submatrices of M formed by the entries from the same rows and columns as corresponding submatrices of $M(g)$ from the family $\mathcal{P}(g,k) = \{P(g,k,1), \ldots, P(g,k,2^k)\}$. Then

$$\left|\langle f_M, g\rangle\right| = \left|\sum_{i,j}^{n} M_{i,j} M(g)_{i,j}\right| = \left|\sum_{k=1}^{\lceil \log_2 n\rceil} \sum_{t=1}^{2^k} P(k,t)_{i,j}\, P(g,k,t)_{i,j}\right|. \qquad (7)$$

As the matrices $P(k,t)$ are submatrices of the Hadamard matrix M, by the Lindsay lemma [23, p. 88], we have $\left|\langle f_M, g\rangle\right| \leq n\sqrt{n}\lceil \log_2 n\rceil$. Thus $\|f_M\|_{P_d(X)} \geq \frac{\sqrt{n}}{\lceil \log_2 n\rceil}$. □

By Proposition 1, functions with large variations with respect to signum perceptrons cannot be l_1-sparsely represented by shallow sigmoidal perceptron networks. Combining this proposition with Theorem 4 we obtain the next corollary.

Corollary 1. *Let $d = d_1 + d_2$, $\{x_i \,|\, i = 1, \ldots, n\} \subset \mathbb{R}^{d_1}$, $\{y_j \,|\, j = 1, \ldots, n\} \subset \mathbb{R}^{d_2}$, $X = \{x_i \,|\, i = 1, \ldots, n\} \times \{y_j \,|\, j = 1, \ldots, n\} \subset \mathbb{R}^d$, and $f_M : X \to \{-1,1\}$ be defined as $f_M(x_i, y_j) = M_{i,j}$, where M is an $n \times n$ Hadamard matrix. Then f_M cannot be computed by a shallow signum perceptron network having both the number of units and absolute values of all output weights smaller than $\frac{\sqrt{n}}{\lceil \log_2 n\rceil}$.*

Theorem 4 provides a method of construction of functions with large variations with respect to signum perceptrons. It can be applied to domains containing sufficiently large squares, for example two-dimensional squares with $2^k \times 2^k$ pixels or $2k$-dimensional Boolean cubes $\{0,1\}^{2k}$.

Corollary 2. *Let k be a positive integer and $f_M : \{0,1\}^k \times \{0,1\}^k \to \{-1,1\}$ be defined as $f_M(x_i, y_j) = M_{i,j}$, where M is a $2^k \times 2^k$ Hadamard matrix. Then $\|f_M\|_{P_d(\{0,1\}^{2k})} \geq \frac{2^{k/2}}{k}$.*

The Corollary 2 shows that functions defined in terms of Hadamard matrices on $2k$-dimensional Boolean cubes has variations with respect to signum perceptrons bounded from below by $\frac{2^{k/2}}{k}$. Such functions cannot be computed by shallow signum perceptron networks with numbers of units and absolute values of output weights bounded by any polynomial of k. So they cannot be sparsely represented by shallow perceptron networks with considerably smaller numbers of units and sizes of output weights than the sizes $2^k \times 2^k$ of their domains.

6 Comparison of Representations by One and Two-Hidden-Layer Networks

Examples of functions for which lower bounds in the previous section were derived can be obtained from a variety of types of Hadamard matrices. Their listings can be found at the Neil Sloane Library of Hadamard matrices [24].

Various constructions of Hadamard matrices are known, e.g., Sylvester's recursive construction of $2^k \times 2^k$ matrices, Paley's construction based on quadratic residues, constructions based on Latin squares, and on Steiner's triples.

Our results can be illustrated by an example of Sylvester-Hadamard matrices. A $2^k \times 2^k$ matrix is called *Sylvester-Hadamard* if it is constructed recursively starting from the matrix

$$S(2) = \begin{vmatrix} 1 & 1 \\ 1 & -1 \end{vmatrix}$$

and iterating the Kronecker product

$$S(l+1) = S(2) \bigotimes S(l) = \begin{vmatrix} S(l) & S(l) \\ S(l) & -S(l) \end{vmatrix}.$$

Sylvester-Hadamard matrices are of orders 2^k and so they induce functions on Boolean $2k$-dimensional cubes. The following theorem together with Corollary 2 provides a comparison of model complexities of one and two-hidden-layer networks representing functions induced by Sylvester-Hadamard matrices. It shows that such functions can be represented by two-hidden-layer Heaviside perceptron networks with k units in each of hidden layers.

Theorem 5. *Let $S(k)$ be a $2^k \times 2^k$ Sylvester-Hadamard matrix, $h_k : \{0,1\}^k \times \{0,1\}^k \to \{-1,1\}$ be defined as $h_k(u,v) = S(k)_{u,v}$. Then h_k can be represented by a network with one linear output and two hidden layers with k Heaviside perceptrons in each one.*

Proof. It is well-known that any $2^k \times 2^k$ Sylvester-Hadamard matrix is equivalent to the matrix with rows formed by generalized parities $p_u(v) : \{0,1\}^k \to \{-1,1\}$ defined as $p_u(v) = -1^{u \cdot v}$ (see, e.g., [25]). Thus we can assume that $S(k)_{u,v} = -1^{u \cdot v}$ (otherwise we permute rows and columns). For any $b \in (1,2)$, define k perceptrons with $2k$ inputs in the first hidden layer as $\vartheta(c^i \cdot x - b)$, where we let $c^i_i = 1$, $c^i_{k+i} = 1$, and all other weights are equal to 0.

Let $w = (w_1, \ldots, w_k)$ be such that $w_j = 1$ for all $j = 1, \ldots, k$. In the second hidden layer, define k perceptrons by $z_j(y) := \vartheta(w \cdot y - j + 1/2)$. Finally, for all $j = 1, \ldots, k$ let the j-th unit from the second hidden layer be connected with one linear output unit with the weight $(-1)^j$.

The two-hidden-layer network obtained in this way computes the function $\sum_{j=1}^{k} (-1)^j \vartheta(\sum_{i=1}^{d/2} \vartheta(c^i \cdot x - b) - j + 1/2) = h_k(x) = h_k(u,v) = -1^{u \cdot v}$. □

Combining Theorems 5 and 4 and the Eq. (3) we obtain the next corollary.

Corollary 3. *Let $S(k)$ be a $2^k \times 2^k$ Sylvester-Hadamard matrix, $h_k : \{0,1\}^k \times \{0,1\}^k \to \{-1,1\}$ be defined as $h_k(u,v) = S(k)_{u,v}$. Then h_k can be represented by a two-hidden-layer network with k Heaviside perceptrons in each hidden layer, but every representation of h_k by one-hidden-layer Heaviside perceptron network has at least $\frac{2^k}{k}$ units or some of absolute values of output weights are greater or equal to $\frac{2^k}{k}$.*

Corollary 3 gives an example of a class of functions on $\{0,1\}^{2k}$ which can be "l_0"-sparsely represented by two-hidden-layer perceptron networks but cannot be l_1-sparsely represented by perceptron networks with merely one hidden layer.

7 Discussion

As estimation of minimal numbers of network units needed for a representation of a given function is a difficult non convex problem, we focused on investigation of sparsity of shallow neural networks in terms of l_1-norms of their output weights. These norms has been used in weight-decay regularization techniques and are related to the concept of a variational norm tailored to a type of computational units which plays an important role in estimates of upper bounds on rates of approximation by shallow networks with various types of computational units [8,19]. Thus variational norms provide a framework for derivation of both upper and lower bounds of model complexities of shallow networks.

For signum perceptrons we derived a lower bound on variational norms of functions defined on rectangular domains in terms of Hadamard matrices. In particular, we proved that such functions on d-dimensional Boolean cubes $\{0,1\}^d$, with d even, cannot be computed by shallow perceptron networks with numbers of units and output weights depending on d polynomially. Our results complement an existential probabilistic argument from [12] which shows that almost any uniformly randomly chosen function on a large domain cannot be computed by a sparse shallow perceptron network.

We also showed that functions from a subclass of the class of functions which we used in our construction can be computed by two-hidden-layer perceptron networks of much smaller model complexities than by shallow ones.

Although for large domains, almost any uniformly randomly chosen function has a large variation with respect to perceptrons, it is not easy to find examples of such functions. We constructed a class of such functions generated by matrices with rather extreme properties. However, it is quite likely that many practical tasks can be represented by functions which can be computed by reasonably small shallow networks. Deep networks seem to be more efficient than shallow ones in tasks which can be naturally described in terms of compositional functions. Such functions can be suitable for description of visual recognition tasks.

Our lower bounds on model complexity were derived merely for perceptron networks. Investigation of lower bounds for other types of computational units is subject of our future work.

Acknowledgments. This work was partially supported by the Czech Grant Agency grant 15-18108S and institutional support of the Institute of Computer Science RVO 67985807.

References

1. Fine, T.L.: Feedforward Neural Network Methodology. Springer, Heidelberg (1999)
2. Kecman, V.: Learning and Soft Computing. MIT Press, Cambridge (2001)
3. LeCun, Y., Bottou, L., Bengio, Y., Haffner, P.: Gradient-based learning applied to document recognition. Proc. IEEE **86**, 2278–2324 (1998)
4. Hinton, G.E., Osindero, S., Teh, Y.W.: A fast learning algorithm for deep belief nets. Neural Comput. **18**, 1527–1554 (2006)
5. Bengio, Y.: Learning deep architectures for AI. Found. Trends Mach. Learn. **2**, 1–127 (2009)
6. LeCunn, Y., Bengio, Y., Hinton, G.: Deep learning. Nature **521**, 436–444 (2015)
7. Ba, L.J., Caruana, R.: Do deep networks really need to be deep?. In: Ghahrani, Z., et al. (eds.) Advances in Neural Information Processing Systems, vol. 27, pp. 1–9 (2014)
8. Kainen, P.C., Kůrková, V., Sanguineti, M.: Dependence of computational models on input dimension: tractability of approximation and optimization tasks. IEEE Trans. Inf. Theory **58**, 1203–1214 (2012)
9. Maiorov, V., Pinkus, A.: Lower bounds for approximation by MLP neural networks. Neurocomputing **25**, 81–91 (1999)
10. Bianchini, M., Scarselli, F.: On the complexity of neural network classifiers: a comparison between shallow and deep architectures. IEEE Trans. Neural Netw. Learn. Syst. **25**(8), 1553–1565 (2014)
11. Bengio, Y., Delalleau, O., Roux, N.L.: The curse of highly variable functions for local kernel machines. In: Advances in Neural Information Processing Systems, vol. 18, pp. 107–114. MIT Press (2006)
12. Kůrková, V., Sanguineti, M.: Model complexities of shallow networks representing highly varying functions. Neurocomputing **171**, 598–604 (2016)
13. Ito, Y.: Finite mapping by neural networks and truth functions. Math. Sci. **17**, 69–77 (1992)
14. Barron, A.R.: Neural net approximation. In: Narendra, K. (ed.) Proceedings of the 7th Yale Workshop on Adaptive and Learning Systems, pp. 69–72. Yale University Press (1992)
15. Kůrková, V.: Dimension-independent rates of approximation by neural networks. In: Warwick, K., Kárný, M. (eds.) Computer-Intensive Methods in Control and Signal Processing: The Curse of Dimensionality, pp. 261–270. Birkhäuser, Boston (1997)
16. Kůrková, V., Sanguineti, M.: Comparison of worst-case errors in linear and neural network approximation. IEEE Trans. Inf. Theory **48**, 264–275 (2002)
17. Kainen, P.C., Kůrková, V., Vogt, A.: A Sobolev-type upper bound for rates of approximation by linear combinations of heaviside plane waves. J. Approximation Theory **147**, 1–10 (2007)
18. Kůrková, V.: Minimization of error functionals over perceptron networks. Neural Comput. **20**, 250–270 (2008)
19. Kůrková, V.: Complexity estimates based on integral transforms induced by computational units. Neural Netw. **33**, 160–167 (2012)
20. Gnecco, G., Sanguineti, M.: On a variational norm tailored to variable-basis approximation schemes. IEEE Trans. Inf. Theory **57**, 549–558 (2011)
21. Kůrková, V., Savický, P., Hlaváčková, K.: Representations and rates of approximation of real-valued Boolean functions by neural networks. Neural Netw. **11**, 651–659 (1998)

22. Cover, T.: Geometrical and statistical properties of systems of linear inequalities with applications in pattern recognition. IEEE Trans. Electron. Comput. **14**, 326–334 (1965)
23. Erdös, P., Spencer, J.H.: Probabilistic Methods in Combinatorics. Academic Press, New York (1974)
24. Sloane, N.J.A.: A library of Hadamard matrices. http://www.research.att.com/~njas/hadamard/
25. MacWilliams, F., Sloane, N.J.A.: The Theory of Error-Correcting Codes. North-Holland, Asterdam (1977)

Kernel Networks for Function Approximation

David Coufal[(✉)]

Institute of Computer Science, The Czech Academy of Sciences, Pod Vodárenskou věží 2, 182 07 Prague, Czech Republic
david.coufal@cs.cas.cz

Abstract. Capabilities of radial convolution kernel networks to approximate multivariate functions are investigated. A necessary condition for universal approximation property of convolution kernel networks is given. Kernels that satisfy the condition in arbitrary dimension are investigated in terms of their Hankel and Fourier transforms. A computational example is presented to assess approximation capabilities of different convolution kernel networks.

Keywords: Kernel networks · Convolution · Universal approximation

1 Introduction

It is the classical result that RBF neural networks possess the universal approximation property [5,9]. Roughly speaking, it means that any reasonable function can be arbitrarily well approximated by a feed-forward three-layered RBF neural network, provided that computational units satisfy certain mild conditions.

Mathematically, the result draws on the fact that convolving a function with the translated Dirac function yields pointwise the original function. Computational units forming the RBF network are then selected to approximate the translated Dirac functions by locating their centers at the points of translation (they usually correspond to training data) and setting their widths near to zero. In RBF networks, the possibility of varying the widths of computational units is crucial for obtaining the approximation result.

It was recently proved that the universal approximation property in $\mathcal{L}^2(\mathbb{R}^d)$ space holds under certain conditions also for *kernel networks* [6]. Radial convolution kernel networks form a subset of RBF networks obtained by fixing widths of computational units. Clearly, different proof techniques had to be used in [6] than approximating the Dirac delta function. Namely, the proof draws on techniques from functional and Fourier analysis. The result has been already known for Gaussians [8], but Kůrková has shown that it holds for all kernels in $\mathcal{L}^1(\mathbb{R}^d) \cap \mathcal{L}^2(\mathbb{R}^d)$ such that zeros of their Fourier transforms form a set of Lebesgue measure zero [6]. These kernels are then suitable for use in practical applications.

In this paper, we show that the above condition is not only sufficient, but also necessary for the universal approximation property. This conveniently complements the original result in [6].

C. Jayne and L. Iliadis (Eds.): EANN 2016, CCIS 629, pp. 295–306, 2016.
DOI: 10.1007/978-3-319-44188-7_22

Further, we focus on concrete examples of kernels that satisfy the universal approximation condition. Special interest is put on kernels for which the condition holds in an arbitrary dimension and so they are suitable for processing high-dimensional data. We investigate their Fourier transforms in terms of their Hankel transforms [2].

The paper is organized as follows. The next section briefly reviews the relevant mathematical background together with the statement of the sufficiency result. In the third section, we prove the necessity of the condition for the universal approximation property. The fourth section discusses radial convolution kernels that satisfy the condition in any dimension. Several examples of kernels are presented. The fifth section contains a computational example and the sixth one concludes the paper.

2 Convolution Kernel Networks

A feedforward three-layered neural network with one hidden layer computes functions from the set

$$\operatorname{span} G = \left\{ \sum_{i=1}^{n} w_j g_j \,|\, w_i \in \mathbb{R},\; g_j \in G, n \in \mathbb{N}_+ \right\},$$

where the set of functions G is called a dictionary.

Dictionaries in kernel networks correspond to parametrized families of functions specified by kernel K,

$$G_K = \{K(\,\cdot\,, y) \,:\, \mathbb{R}^d \to \mathbb{R}, y \in \mathbb{R}^d\}.$$

The kernel networks are based on computational units that have fixed width. It means that no varying scaling parameter is allowed here in contrast to more flexible RBF networks. The only varying parameter in a kernel computational unit is the parameter of location $y \in \mathbb{R}^d$, $d \in \mathbb{N}_+$.

A class of the *convolution kernel networks*[1] These are the kernel networks based on kernels of the form $K(x, y) = k(x - y)$, where k is a suitable function from $\mathbb{R}^d \to \mathbb{R}$. These kernels are called the *convolution kernels* [6]. The dictionary for the convolution kernel network has the form

$$G_K = \{k(\,\cdot\, - \,y) \colon \mathbb{R}^d \to \mathbb{R}, y \in \mathbb{R}^d\}.$$

In spite of the fact that the convolution kernel networks are less flexible than more general RBF networks, they still exhibit the universal approximation property. The following theorem proved in [6] specifies this fact formally. The theorem draws on properties of the Fourier transform of the convolution kernel. Recall that the Fourier transform $\hat{f}$ is specified for functions f in $\mathcal{L}^1(\mathbb{R}^d)$ space with standard extension to $\mathcal{L}^2(\mathbb{R}^d)$ space [10].

[1] Note that the convolution kernel networks represent a different concept from the nowadays very popular concept of the *convolutional neural networks* [7]. The first has a shallow architecture in contrast to the deep one of the latter case.

Definition 1. *Let function* $f:\mathbb{R}^d \to \mathbb{R}$ *be in* $\mathcal{L}^1(\mathbb{R}^d)$. *Its multivariate Fourier transform is specified as*

$$\hat{f}(s) = \frac{1}{(2\pi)^{d/2}} \int_{\mathbb{R}^d} e^{\,\mathrm{i}\langle x,s\rangle} f(s)\, dx.$$

Theorem 1 (Kůrková [6]). *Let* d *be a positive integer,* λ^d *the Lebesgue measure on* $\mathbb{R}^d$, $k \in \mathcal{L}^1(\mathbb{R}^d) \cap \mathcal{L}^2(\mathbb{R}^d)$ *be such that the set* $\lambda^d(\{s \in \mathbb{R}^d \mid \hat{k}(s) = 0\}) = 0$, *and* $K:\mathbb{R}^d \times \mathbb{R}^d \to \mathbb{R}$ *be defined as* $K(x,y) = k(x-y)$. *Then* span G_K *is dense in* $(\mathcal{L}^2(\mathbb{R}^d), ||\cdot||_{\mathcal{L}^2(\mathbb{R}^d)})$.

The theorem provides guidance for selecting kernels to design the convolution kernel networks suitable for approximation. We should use kernels that have fully supported Fourier transforms except sets of Lebesgue measure zero. In that case, we are theoretically assured that we may achieve an arbitrary accuracy of approximation in $\mathcal{L}^2$ sense.

In what follows we are going to deal with two issues. First, we show that the condition on zeros of the Fourier transform is also necessary. Hence, the condition is the best possible because it is the least restrictive. Secondly, we discuss examples of suitable kernels, especially in terms of radial functions that are suitable for approximation in arbitrary dimension $d \in \mathbb{N}_+$.

3 Necessity

Lemma 1. *Let* d *be a positive integer,* $k \in \mathcal{L}^1(\mathbb{R}^d) \cap \mathcal{L}^2(\mathbb{R}^d)$, $K:\mathbb{R}^d \times \mathbb{R}^d \to \mathbb{R}$ *be defined as* $K(x,y) = k(x-y)$ *and* span G_K *be dense in* $(\mathcal{L}_2(\mathbb{R}^d), ||\cdot||_{\mathcal{L}_2(\mathbb{R}^d)})$. *Then* $\lambda^d(\{s \in \mathbb{R}^d | \hat{k}(s) = 0\}) = 0$.

Proof. We prove the reverse implication, i.e., if for the kernel k is $\lambda^d(\{s \in \mathbb{R}^d \mid \hat{k}(s) = 0\}) > 0$, then G_K cannot be dense in $(\mathcal{L}_2(\mathbb{R}^d), ||\cdot||_{\mathcal{L}_2(\mathbb{R}^d)})$.

Let $\lambda^d(\{s \in \mathbb{R}^d \mid \hat{k}(s) = 0\}) > 0$. Standard Euclidean space $\mathbb{R}^d$ is σ-compact so there exists a hypercube $H_{n^*} = [-n^*, n^*]^d \subseteq \mathbb{R}^d$, $n^* \in \mathbb{N}_+$ such that $\lambda^d(\{s \in \mathbb{R}^d \mid \hat{k}(s) = 0\} \cap H_{n^*}) > 0$. For such the hypercube H_{n^*} we denote $H^0_{n^*} = \{s \in \mathbb{R}^d \mid \hat{k}(s) = 0\} \cap H_{n^*}$; and therefore $\lambda^d(H^0_{n^*}) > 0$.

Now, let us consider the function $f_{n^*}(x) = \prod_{i=1}^d n^* \sqrt{2/\pi}\,\mathrm{sinc}(n^* x_i)$, where $\mathrm{sinc}(z) = \sin(z)/z$ for $z \neq 0$, $z \in \mathbb{R}$, $\mathrm{sinc}(0) = 1$ and $x = (x_1, \ldots, x_d)$. This function is in $\mathcal{L}^2(\mathbb{R}^d)$ because $||\prod_{i=1}^d n^* \sqrt{2/\pi}\,\mathrm{sinc}(n^* x_i)||^2_{\mathcal{L}_2(\mathbb{R}^d)} = (2n^*)^d$. The Fourier transform of f_{n^*} writes $\hat{f}_{n^*}(s) = \prod_{i=1}^d \mathrm{rect}_{[-n^*,n^*]}(x_i)$, where $\mathrm{rect}_{[-n^*,n^*]}$ is the unit pulse on interval $[-n^*, n^*]$. Hence $\hat{f}_{n^*}(s)$ corresponds to the characteristic function of the hypercube H_{n^*} and $||\hat{f}_{n^*}||^2_{\mathcal{L}_2(\mathbb{R}^d)} = (2n^*)^d$.

Suppose that there exists an approximant $F = \sum_{j=1}^N w_j k(x - y_j)$ of f_{n^*} such that $||f_{n^*} - F||^2_{\mathcal{L}^2(\mathbb{R}^d)} < \lambda^d(H^0_{n^*})$. The Fourier transform of F writes $\hat{F}(s) = \sum_{j=1}^N w_j e^{\,\mathrm{i}\langle y_j, s\rangle} \hat{k}(s)$. Using Parseval's theorem gives

$$\begin{aligned}||f_{n^*} - F||^2_{\mathcal{L}^2(\mathbb{R}^d)} &= ||\hat{f}_{n^*} - \hat{F}||^2_{\mathcal{L}^2(\mathbb{R}^d)} \\ &= \int_{\mathbb{R}^d} |\hat{f}_{n^*}(s) - \sum_{j=1}^N w_j e^{\mathrm{i}\langle y_j, s\rangle} \hat{k}(s)|^2 \, ds \\ &\geq \int_{H^0_{n^*}} |\hat{f}_{n^*}(s) - \sum_{j=1}^N w_j e^{\mathrm{i}\langle y_j, s\rangle} \hat{k}(s)|^2 \, ds.\end{aligned}$$

But for $x \in H^0_{n^*}$, one has $\sum_{j=1}^N w_j e^{\mathrm{i}\langle y_j, s\rangle} \hat{k}(s) = 0$ and also $\hat{f}_{n^*}(s) = 1$. Thus, $||f_{n^*} - F||^2_{\mathcal{L}_2(\mathbb{R}^d)} \geq \lambda^d(H^0_{n^*}) > 0$ and F cannot be the approximant of f_{n^*}. □

The lemma establishes necessity of condition $\lambda^d(\{s \in \mathbb{R}^d | \hat{k}(s) = 0\}) = 0$ for dense approximation in $\mathcal{L}^2(\mathbb{R}^d)$ space. Considering the concrete example of $\mathcal{L}^1(\mathbb{R}^d) \cap \mathcal{L}^2(\mathbb{R}^d)$ kernel with the compactly supported Fourier transform, the sinc^2 function is a good example in one dimension. For the d-dimensional space we can consider product $\prod_{i=1}^d \mathrm{sinc}^2(x_i)$. It is well know that this multidimensional kernel has the Fourier transform $\prod_{i=1}^d \mathrm{tri}(x_i)$, where $\mathrm{tri}(x_i)$ is the properly scaled triangular pulse on unit interval $[-1, 1]$. Simply put, we may conclude that a linear combination of translated sinc^2 kernels cannot approximate the rectangular pulse to arbitrary precision in $|| \cdot ||_{\mathcal{L}_2(\mathbb{R}^d)}$ norm.

4 Radial Convolution Kernels

Radial functions are functions that are invariant with respect to rotation. The formal definition reads as follows:

Definition 2. *A function $\Phi : \mathbb{R}^d \to \mathbb{R}$ is called radial if there exists a function $\varphi : \mathbb{R} \to \mathbb{R}$ such that $\Phi(x) = \varphi(||x||_2)$, where $|| \cdot ||_2$ is the Euclidean norm on $\mathbb{R}^d$.*

The Fourier transform of a radial function is also a radial function. It can be expressed in terms of the Hankel transform [2,12].

Definition 3. *The Hankel transform of order ν of a function $\varphi : [0, \infty) \to \mathbb{R}$ is defined as*

$$\mathscr{H}_\nu\{\varphi(r)\}(s) = \int_0^\infty \varphi(r) J_\nu(sr) r \, dr.$$

where J_ν is the Bessel function of the first kind of order $\nu > -\frac{1}{2}$;

The proof of the following theorem can be found in [11] (Theorem 3.3).

Theorem 2. *Let $\Phi \in \mathcal{L}^1(\mathbb{R}^d)$ be continuous and radial, i.e., $\Phi(x) = \varphi(||x||_2)$. Then its Fourier transform $\widehat{\Phi}(s)$ is also radial $\widehat{\Phi}(s) = \varphi^{\mathscr{H}}(||s||_2)$ where*

$$\varphi^{\mathscr{H}}(s) = \frac{1}{\sqrt{s^{d-2}}} \int_0^\infty \varphi(r) r^{\frac{d}{2}} J_{(d-2)/2}(sr) \, dr = s^{-\nu} \mathscr{H}_\nu\{\varphi(r) \cdot r^\nu\}(s)$$

for $\nu = (d-2)/2$, i.e., $\nu = -\frac{1}{2}, 0, \frac{1}{2}, 1, \ldots$ for $d = 1, 2, 3, 4 \ldots$.

Table 1. The Hankel transforms of univariate functions $\varphi(r)\, r^{\nu}$.

Function name	Univariate function $\varphi(r)$	Hankel transform $\mathscr{H}_{\nu}\{\varphi(r)\, r^{\nu}\}$
Gaussian	$\varphi(r) = \exp(-a^2r^2)$	$H_\nu(s) = \frac{s^\nu}{(2a^2)^{\nu+1}} \exp(-\frac{s^2}{4a^2})$
Inv. multiquadric	$\varphi(r) = (a^2 + r^2)^{-(\mu+1)}$	$H_\nu(s) = \frac{a^{\nu-\mu} s^\mu}{2^\mu \Gamma(\mu+1)} K_{\nu-\mu}(as)$
Cut power	$\varphi(r) = (a^2 - r^2)_+^\mu$	$H_\nu(s) = \frac{2^\mu \Gamma(\mu+1) a^{\nu+\mu+1}}{s^{\mu+1}} J_{\nu+\mu+1}(as)$
Rect. pulse $1_{[0,1]}$	$\varphi(r) = \begin{cases} 1 & 0 \le r \le 1 \\ 0 & \text{otherwise} \end{cases}$	$H_\nu(s) = s^{-1} J_{\nu+1}(s)$

Theorem 2 gives us a convenient tool for investigating the Fourier transforms of multidimensional radial functions by employing the Hankel transforms of suitable univariate functions multiplied by the term r^ν.

In Table 1, there are presented several univariate functions with their Hankel transforms. They contain special functions, namely, Bessel functions of the first kind J_α and modified Bessel functions of the second kind K_α. The Hankel transforms were obtained from various sources, mainly from [2,12]. The comprehensive source is [1]. Note that in [1], the Hankel transform is defined as $h_\nu\{f(x)\}(y) = \int_0^\infty f(x) J_\nu(xy)(xy)^{1/2}\, dx$. The relation between this and our version writes $\mathscr{H}_\nu\{f(x)\}(s) = s^{-1/2} h_\nu\{x^{\nu+1/2} f(x)\}(s)$.

For the purpose of investigating the Fourier transforms of radial convolution kernels, see Definition 4 below, we scale the univariate functions in Table 1 by the parameter $b > 0$. To do this, note that $\mathscr{H}_\nu\{f(r/b) r^\nu\}(s) = b^{\nu+2} \mathscr{H}_\nu\{f(r) r^\nu\}(bs)$ which can be directly derived from the definition formula of the Hankel transform.

Table 2. The Hankel transforms of scaled univariate functions $\varphi_b(r)\, r^{\nu}$.

Function name	Scaled function $\varphi_b(r) = \varphi(r/b)$	Hankel transform $\mathscr{H}_{\nu}\{\varphi_b(r)\, r^{\nu}\}$
Gaussian	$\varphi_b(r) = \exp(-(r/b)^2)$	$H_\nu(s) = \frac{b^{2(\nu+1)} s^\nu}{2^{\nu+1}} \exp(-\frac{1}{4}(bs)^2)$
Inv. multiquadric	$\varphi_b(r) = (a^2 + (r/b)^2)^{-(\mu+1)}$	$H_\nu(s) = \frac{a^{\nu-\mu} b^{\nu+\mu+2} s^\mu}{2^\mu \Gamma(\mu+1)} \cdot K_{\nu-\mu}(abs)$
Cut power	$\varphi_b(r) = (a^2 - (r/b)^2)_+^\mu$	$H_\nu(s) = \frac{2^\mu \Gamma(\mu+1) a^{\nu+\mu+1} b^{\nu-\mu+1}}{s^{\mu+1}} \cdot J_{\nu+\mu+1}(abs)$
Rect. pulse $1_{[0,1]}$	$\varphi_b(r) = \begin{cases} 1 & 0 \le r/b \le 1 \\ 0 & \text{otherwise} \end{cases}$	$H_\nu(s) = b^{\nu+1} s^{-1} J_{\nu+1}(bs)$

In the Hankel transform, the source function φ is regarded as a function from $[0,\infty)$ to $\mathbb{R}$. An image under the Hankel transform is again a function from $[0,\infty)$ to $\mathbb{R} \cup \infty$ (the image might be unbounded at 0); and its Hankel transform is the original function. That is, the Hankel transform is self-inverse [2,12] and therefore images under the Hankel transform can be taken as other candidates to define the radial convolution kernels.

Definition 4. *A kernel $K_b : \mathbb{R}^d \to \mathbb{R}^d$ is called the radial convolution kernel if there exists a univariate function φ such that $K_b(x, y) = \varphi_b(||x - y||_2) = \varphi(||(x - y)/b||_2)$ for $b > 0$.*

Clearly, the radial convolution kernels are induced by translations of the scaled radial functions, i.e., $K_b(x, y) = \Phi((x - y)/b)$.

In Table 3, we have computed the Fourier transforms of the scaled multivariate radial functions $\Phi(x) = \varphi(||x||_2/b)$ for the univariate functions φ_b in Table 2. The Fourier transforms were computed according to Theorem 2 with the order of the Hankel transform set to $\nu = (d - 2)/2 = d/2 - 1$, where $d \in \mathbb{N}_+$ is the dimension.

Table 3. Scaled multivariate radial functions and their Fourier transforms.

Function name	Multivariate expression	Fourier transform
Gaussian	$\varphi_b(\lvert\lvert x\rvert\rvert_2) = \exp(-\lvert\lvert x\rvert\rvert_2^2/b^2)$	$F(s) = (b^2/2)^{d/2} \exp(-\frac{b^2}{4}\lvert\lvert s\rvert\rvert_2^2)$
Inverse multiquadric	$\varphi_b(\lvert\lvert x\rvert\rvert_2) = (a^2 + \lvert\lvert x\rvert\rvert_2^2/b^2)^{-(\mu+1)}$	$\mathcal{F}(s) = \frac{a^{d/2-(\mu+1)}\, b^{d/2+(\mu+1)}}{2^\mu \Gamma(\mu+1)\, \lvert\lvert s\rvert\rvert_2^{d/2-(\mu+1)}} \cdot K_{d/2-(\mu+1)}(ab\lvert\lvert s\rvert\rvert_2)$
Cut power	$\varphi_b(\lvert\lvert x\rvert\rvert_2) = (a^2 - \lvert\lvert x\rvert\rvert_2^2/b^2)_+^\mu$	$\mathcal{F}(s) = \frac{2^\mu \Gamma(\mu+1)\, a^{d/2+\mu}\, b^{d/2-\mu}}{\lvert\lvert s\rvert\rvert_2^{d/2+\mu}} \cdot J_{d/2+\mu}(ab\lvert\lvert s\rvert\rvert_2)$
Circ function	$\varphi_b(\lvert\lvert x\rvert\rvert_2) = \begin{cases} 1 & 0 \leq \lvert\lvert x\rvert\rvert_2/b \leq 1 \\ 0 & \text{otherwise} \end{cases}$	$\mathcal{F}(s) = (b/\lvert\lvert s\rvert\rvert_2)^{d/2} J_{d/2}(b\lvert\lvert s\rvert\rvert_2)$

The purpose of this computation is to investigate behavior of the Fourier transforms to check if the condition of Theorem 1 is satisfied or not for the corresponding radial convolution kernels. In Fig. 1, there are presented graphically the computed Fourier transforms for $a = 0, b = 1$ and $d = 2$. For the inverse multiquadric we set $\mu = 0$ and for the cut power $\mu = 1$.

Gaussian. This central function of real and complex analysis cannot be omitted in our list of radial kernels. The Gaussian is an eigenvalue of the Fourier transform, which is confirmed by its computation using the Hankel transform. Since the Gaussian is a positive function and the Fourier transform of itself we have the condition of Theorem 1 satisfied in any dimension. Formally, one has $\{\hat{k}(s) = 0\} = \emptyset$ in any dimension and therefore $\lambda^d(\{\hat{k}(s) = 0\}) = 0$ for any $d \in \mathbb{N}_+$.

The Fourier transform of a d-variate Gaussian can be computed by employing the characteristic function of the d-variate normal distribution $\mathcal{N}_d(0, \Sigma)$. The density of $\mathcal{N}_d(0, \Sigma)$ writes $f(x) = ((2\pi)^d|\Sigma|)^{-1/2} \exp(-\frac{1}{2}x'\Sigma^{-1}x)$ for a positive definite symmetric matrix Σ. The equality $\exp(-\frac{1}{2}x'\Sigma^{-1}x) = \exp(-||x||_2^2/b^2)$ is reached for $\Sigma^{-1} = diag_d(2/b^2, \ldots, 2/b^2)$, where $diag_d(\cdot)$ codes a diagonal

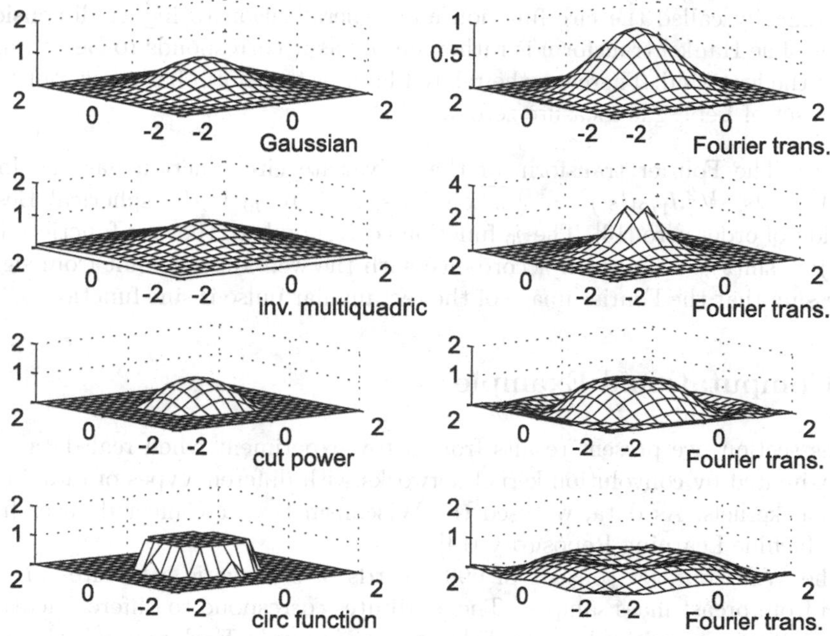

Fig. 1. Examples of radial kernels and their Fourier transforms for $d = 2$.

matrix of order d; and therefore $\Sigma = diag_d(b^2/2, \ldots, b^2/2)$. Further, the following relation holds between the characteristic function χ_f of density f and Fourier transform (1) of $\exp(-||x||_2^2/b^2)$: $\chi_f(s) = |\Sigma|^{-1/2}\mathcal{F}\{f\}$. It is well known that $\chi_f(s) = \exp(-\frac{1}{2}s'\Sigma s)$ and the determinant of diagonal matrix Σ reads as $|\Sigma| = (b^2/2)^d$. Putting all these facts together one gets that $\mathcal{F}\{\exp(-||x||_2^2/b^2)\} = |\Sigma|^{1/2} \cdot \exp(-\frac{1}{2}s'\Sigma s)$ which in turn gives the formula in Table 2.

Inverse Multiquadric. The Hankel transform of an inverse multiquadric involves modified Bessel function of the second kind $K_\alpha : \mathbb{R} \to \mathbb{R}, \alpha \in \mathbb{R}$. This function is positive (and unbounded at the origin) hence the set of zeros of the related Fourier transform is again empty and the inverse multiquadrics are suitable for constructing the convolution kernel networks that possess the $\mathcal{L}_2(\mathbb{R}^d)$ universal approximation property in sense of Theorem 1.

Cut Power. In contrast to the above two cases, the cut power has compact support. That implies that the Fourier transform cannot have a compact support, neither. This is confirmed by the Hankel transform in Table 1. The Hankel transform involves Bessel function of the first kind $J_\alpha : \mathbb{R} \to \mathbb{R}$, $\alpha \in \mathbb{R}$. This function is no more positive, but set of its zeros is countable. Hence the cut powers are also suitable for approximation in $\mathcal{L}^2$ sense.

Circ Function. The circ function is, in fact, the indicator function of the unit ball in the corresponding d-dimensional space. That is, in the univariate case it corresponds to the rectangular pulse $1_{[-1,1]}$, in two dimensional case to what is canonically called the circ function and generalization to higher dimensions applies. The Hankel transform is rather simply as it corresponds to Bessel function of the first kind. Therefore the related Fourier transform has zeros, but they form a set of Lebesgue measure zero.

Remark. The Fourier transform of the univariate circ function has the form $\mathcal{F}(|s|) = |s|^{-1/2}J_{1/2}(|s|) = \sqrt{2/\pi} \cdot j_0(|s|)$, where j_0 is the spherical Bessel function of order zero [13]. The j_0 function corresponds to the sinc function, i.e., $j_0(|s|) = \sin(s)/s$. This is in accordance with the well-known result from signal processing that the Fourier image of the rectangular pulse is sinc function.

5 Computational Example

In this section, we present results from a toy experiment when real data were approximated by convolution kernel networks with different types of radial convolution kernels. As data, we used the Wisconsin Breast Cancer dataset from UCI Machine Learning Repository [14].

The WBC dataset consists of 699 records of 9 numeric attributes of cells nuclei from breast mass samples. The attributes correspond to different features computed from a digitized image of the analyzed sample. Each record is classified as benign or malignant. 16 records contain missing information. We had excluded them from the dataset and worked only with the remaining 683 ones.

We applied the following further steps to preprocess the dataset. We split 683 records into two halves. The first consisted of the records located at odd rows, i.e., having odd row numbers $1, 3, \ldots, 681, 683$. The second half consisted of the records with even row numbers $2, 4, \ldots, 680, 682$. Thus the first group, which we used as the training data, contained 342 records, and the second one 341 records. We used this split to retain its reproducibility. The split has 65 % proportion of the benign class in both groups, which corresponds to the proportion of the bening class in the entire dataset (444/683). Further, the data in both groups were normalized by subtracting means and dividing by the standard deviations of the data in the first group (separately for each column). Finally, the benign cases were recoded as 1 and malignant as -1.

The normalized training data were used to learn kernel convolution networks in the MATLAB computational environment. An individual network was characterized by (1) the type of the kernel used - we used the kernels presented in Table 3; (2) the number of computational units - $N_c \in \mathbb{N}_+$; (3) centers of computational units - $a_j \in \mathbb{R}^9$, $j = 1, \ldots, N_c$; (4) specification of the width parameter $b \in \mathbb{R}$ and (5) by the vector of weights $w^b \in \mathbb{R}^{N_c}$. The j-th computational unit of the network computes the function $k_j : \mathbb{R}^9 \to \mathbb{R}$ of the form $k_j(x) = \varphi_b(||x - a_j||_2)$. Computation of the network then corresponds to the formula $NN_{\varphi,b}(x) = \sum_{j=1}^{N_c} w_j^b k_j(x) = \sum_{j=1}^{N_c} w_j^b \varphi_b(||x - a_j||_2)$.

Using the training data, we specified the parameters as follows: (2 & 3) the number of computational units was set to $N_c = 3$ and the centers were identified by the fuzzy c-means clustering algorithm [4]. (4 & 5) For b fixed and the selected kernel, following the theory presented in Sect. 5.7 of [3] on generalized radial basis neural networks, we formed 342×9 Green's matrix G of entries $G_{ij} = k_j(x_i) = \varphi_b(||x_i - a_j||_2)$ for $i = 1, \ldots, 342$ and $j = 1, \ldots, 9$. The column vector of weights was then given as $w^b = G^+ y$, where G^+ is the pseudo-inverse of G and y is the column vector of 1 and -1 containing the classification of records in the training data. In [3], there is shown that this learning of weights is optimal in terms of minimization the error of a regularized network. The structure of the network is presented in Fig. 2.

To make the above procedure clearer let us work with a concrete piece of data. The data entry on the first row in WBC dataset reads as (x_1, y_1) with $x_1 = (5, 1, 1, 1, 2, 1, 3, 1, 1)$ for attributes and $y_1 = 1$ marking the first entry classified as a bening case. After normalization, we get $x_1^n = (0.18, -0.72, -0.75, -0.64, -0.55, -0.70, -0.22, -0.62, -0.36)$. The centers a_j identified by FCM clustering algorithm applied on the normalized data have (after rounding) the coordinates $a_1 = (-0.52, -0.63, -0.63, -0.54, -0.53, -0.61, -0.58, -0.56, -0.33)$, $a_2 = (0.91, 1.14, 1.15, 0.99, 0.94, 1.10, 1.05, 1.00, 0.53)$ and $a_3 = (0.92, 1.19, 1.19, 1.04, 1.01, 1.09, 1.10, 1.07, 0.59)$. Hence the vector of the Euclidean distances of x_1^n from centers a_j, $j = 1, \ldots, 3$ reads as $(0.82, 4.56, 4.68)$. Selecting, for example, the inverse multiquadric with the parameters $\mu = 0$, $a = 1$ and the width $b = 1$, i.e., $\varphi_{b=1}(||x||_2) = 1/(1 + (||x||_2^2)$, one gets the first row of the Green's matrix as $G_{1j} = [0.60, 0.05, 0.04]$. The other rows were obtained by the same computations. The vector of weights was then obtained using the standard MATLAB command for computing the pseudo-inverse as `w=pinv(G)*y`. The values of weights then reads as $w_1 = 1.78$, $w_2 = 9.42$ and $w_3 = -15.43$.

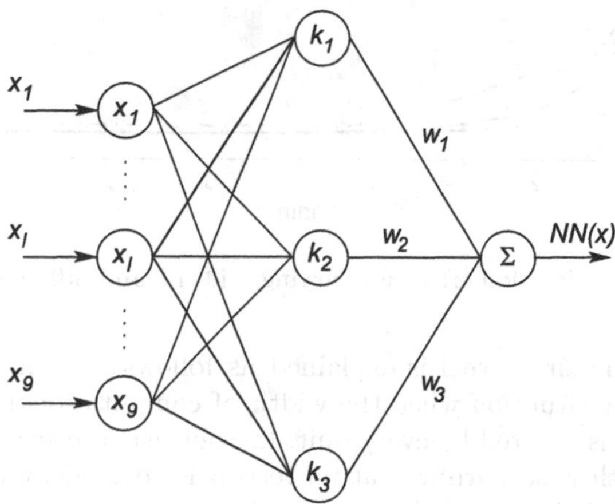

Fig. 2. The convolution kernel network for WBC dataset.

We learned the network weights w for different values of the width parameter b. Namely, we varied b from $b = 0.05$ to $b = 10$ with the step $\Delta b = 0.05$. For each setting of b we computed the error of approximation as

$err_\varphi(b) = 1/N \cdot \sum_{i=1}^{N}(NN_{\varphi,b}(x_i)) - y_i)^2$. The error was computed for both learning and testing phases with $N = 342$ and $N = 341$, respectively. In the testing phase for each b, the network developed on the training data ($N = 341$) was applied to the testing data ($N = 342$). The development of both errors is presented in Fig. 3 for different types of the radial convolution kernels.

Inspecting the graphs, we see that development of approximation errors is similar in both phases. The error is higher in the testing phase which is expectable. The errors differ when it comes to different kernels. We see that the best performing is the inverse multiquadric, however, performance of the Gaussian and cut power kernels are comparable.

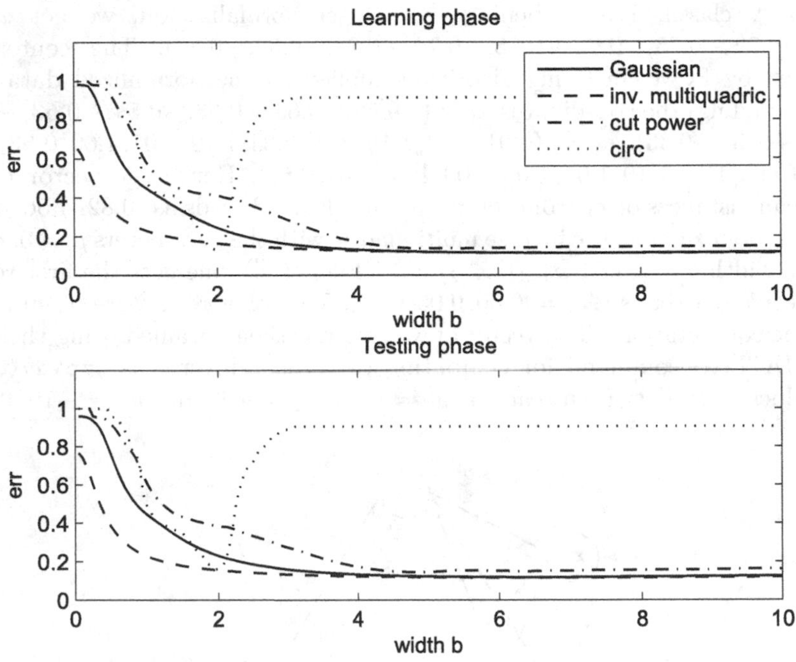

Fig. 3. Approximation errors for varying width b and different kernels.

Behavior of the circ kernel is explained as follows. The first constant part corresponds to the situation when the width of computational units is so wide that every record is covered by every unit. In contrast, the second part is opposite, i.e., the width is so narrow that no record is covered by no unit. Clearly, the optimal width is located in between both extrema. We see that for all kernels, the optimal width from the learning phase more or less corresponds to the minimum of errors in the testing phase.

In the triplet consisting of the Gaussian, inverse multiquadric and cut power the last is performing the worst, however, because the cut powers have compact support for each $b > 0$ they are connected with less computational effort, which might be an advantage in certain situations.

6 Conclusions

In the paper, we have discussed the sufficient and necessary condition for the convolution kernel network to possess the universal approximation property in $\mathcal{L}^2(\mathbb{R}^d)$ space. The condition is stated in terms of the Fourier transform of employed kernel. The condition says that the null set of the Fourier transform must be of Lebesgue measure zero. We investigated the condition from the practical point of view. Namely, we were interested in what kernels satisfy this condition in an arbitrary dimension. In order to be able to compute the respective Fourier transforms we investigated the radial convolution kernels. For these kernels, the Fourier transform can be computed via the Hankel transform.

Following this fact, we computed explicitly the Fourier transforms for four types of the radial convolution kernels - the Gaussian, the inverse multiquadric, the cut power and the circ kernel. We found that the first two types have their Fourier transforms strictly positive in any dimension. The Fourier transforms of other two types might be negative, however, in both cases the sets of zeros are countable and therefore of Lebesgue measure zero. These findings show that all four types can be used for building reasonable convolution kernel networks.

The paper includes an experimental part where we checked the theoretical results in the toy experiment. We found that the best performing kernel was the inverse multiquadric followed by the Gaussian, the cut power and the circ kernel with the worst performance. However, more extensive experiments should be performed to assess if this ordering also holds in a more general context.

Acknowledgments. This work was supported by the COST Grant LD13002 provided by the Ministry of Education, Youth and Sports of the Czech Republic and institutional support of the Institute of Computer Science RVO 67985807.

References

1. Bateman, H.: Tables of Integral Transforms [vol. I & II]. McGraw-Hill Book Company, New York (1954). http://authors.library.caltech.edu/43489
2. Debnath, L., Bhatta, D.: Integral Transforms and their Applications. Chapman & Hall/CRC, Boca Raton (2007)
3. Haykin, S.S.: Neural Networks and Learning Machines, 3rd edn. Prentice Hall, Upper Saddle River (2009)
4. Höppner, F., et al.: Fuzzy Cluster Analysis: Methods for Classification, Data Analysis and Image Recognition. Wiley, Chichester (1997)
5. Park, J., Sandberg, I.: Approximation, radial basis function networks. Neural Comput. **5**, 305–316 (1993)
6. Kůrková, V.: Capabilities of radial and kernel networks. In: Matoušek, R. (ed.) 19th International Conference on Soft Computing MENDEL (2013)
7. LeCun, Y., Bengio, Y., Hinton, G.: Deep learning. Nature **521**(7553), 436–444 (2015)
8. Mhaskar, N.H.: Versatile Gaussian networks. In: Proceedings of IEEE Workshop of Nonlinear Image Processing, pp. 70–73 (1995)

9. Park, J., Sandberg, I.: Universal approximation using radial-basis-function networks. Neural Comput. **3**, 246–257 (1991)
10. Rudin, W.: Functional Analysis, 2nd edn. McGraw-Hill, New York (1991)
11. Stein, E.M., Weiss, G.: Introduction to Fourier Analysis on Euclidean Spaces. Princeton University Press, Princeton (1971)
12. Poularikas, A.D. (ed.): The Transforms and Applications Handbook, 2nd edn. CRC Press LLC, Boca Raton (2009)
13. Weisstein, E.W.: Spherical Bessel Function of the First Kind. http://mathworld.wolfram.com/SphericalBesselFunctionoftheFirstKind.html
14. Wolberg, W.H., Street, W.N., Mangasarian, O.L.: Breast Cancer Wisconsin (Diagnostic) Data Set (1995). http://archive.ics.uci.edu/ml/datasets/Breast+Cancer+Wisconsin+%28Diagnostic%29

Short Papers

Simple and Stable Internal Representation by Potential Mutual Information Maximization

Ryotaro Kamimura(✉)

IT Education Center and Graduate School of Science and Technology, Tokai Univerisity, 4-1-1 Kitakaname, Hiratsuka, Kanagawa 259-1292, Japan
ryo@keyaki.cc.u-tokai.ac.jp

Abstract. The present paper aims to interpret final representations obtained by neural networks by maximizing the mutual information between neurons and data sets. Because complex procedures are needed to maximize information, the computational procedures are simplified as much as possible using the present method. The simplification lies in realizing mutual information maximization indirectly by focusing on the potentiality of neurons. The method was applied to restaurant data for which the ordinary regression analysis could not show good performance. For this problem, we tried to interpret final representations and obtain improved generalization performance. The results revealed a simple configuration where just a single important feature was extracted to explicitly explain the motivation to visit the restaurant.

1 Introduction

Information-theoretic methods have been developed to control the amount of information contained in neural networks since Linsker [1–4] stated the maximum information preservation principle for perceptual systems. Though many attempts [5,6] have been made to apply them to neural networks, they have not necessarily been employed to their fullest extent.

In addition, there have been few attempts to interpret final representations in neural networks due to three main reasons, namely, complex procedures, excessive information and stability. First, information-theoretic methods require complex procedures for computing entropy or information functions. Though many methods have been developed for simplifying these procedures [3–8], they have not necessarily been simplified enough to be applied to practical problems. Second, there is the problem of excessive information acquisition that comes with information maximization [9]. Information maximization can be used to compress information content into a small number of neurons; even unnecessary information can be compressed, which can decrease generalization performance. Thus, these information-theoretic methods have tried to decrease information content for the sake of improving generalization [9]. For interpreting neural networks, it is necessary to increase the amount of information and in particular to condense it into a smaller number of neurons [10] without acquiring unnecessary information. Third, there is a problem with stability of the results. Even if we

C. Jayne and L. Iliadis (Eds.): EANN 2016, CCIS 629, pp. 309–316, 2016.
DOI: 10.1007/978-3-319-44188-7_23

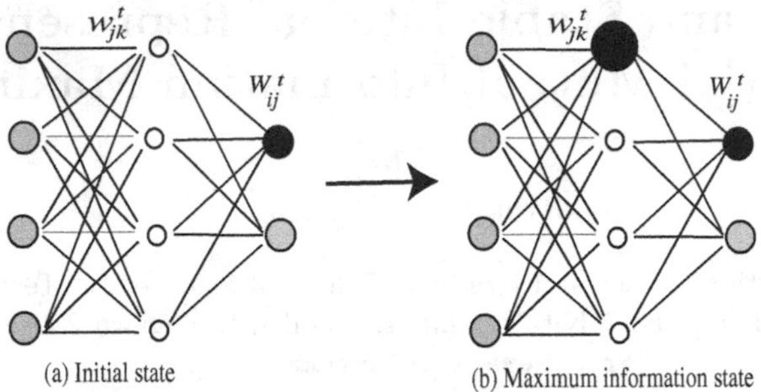

Fig. 1. Mutual information maximization for hidden neurons.

succeed in simplifying internal representations for interpretation, they are of no use if completely different internal representations are produced, depending on data sets and initial conditions.

The present method tries to solve these problems in the following ways. First, it aims to realize mutual information not directly, but indirectly. More exactly, we do not try to directly achieve maximum mutual information states. To realize this, we have proposed potential learning [11–15]. The potentiality of neurons represents their ability to deal with as many different situations as possible. The potentiality is approximated by the variance of neurons. By using this potentiality, mutual information is easily increased by increasing the parameter to produce a smaller number of neurons with higher potentiality. Second, better generalization can be expected by the present method. Since the method tries to increase mutual information as much as possible, connection weights will be forced to be smaller. Thus, maximum information states are realized over sparse connection weights and neurons, which can improve generalization performance. Finally, stability can be improved as well. In the process of information maximization, neural networks are forced to use fewer connection weights and neurons, which severely limits the production of different internal representations, leading to stability.

2 Theory and Computational Methods

We here explain how to formulate potential mutual information only for hidden neurons; mutual information for input neurons can be computed in the same way. Figure 1 shows the process of information maximization for a given data set. In the maximum information state, only one neuron fires, while all the others cease to do so. Potentiality has been used to determine which neurons should be fired, and represents the relative importance of neurons [11–15]. For the first approximation, the potentiality of neurons is defined by using their variance.

Let us concretely define the hidden neurons' potentiality. As shown in Fig. 1, w_{jk}^t denotes connection weights from the kth input neuron to the jth hidden neuron for the tth data set. Then, the potentiality is defined by using the variance

$$v_j^t = \frac{1}{L-1}\sum_{k=1}^{L}(w_{jk}^t - w_j^t)^2, \tag{1}$$

where L is the number of input neurons and the average weight is computed by

$$w_j^t = \frac{1}{L}\sum_{k=1}^{L} w_{jk}^t. \tag{2}$$

Then, the potentiality is normalized as

$$p(j|t) = \frac{v_j^t}{\sum_{m=1}^{M} v_m^t}, \tag{3}$$

where M is the number of hidden neurons. Finally, we have potential mutual information for hidden neurons

$$PI_{hid} = -\sum_{j=1}^{M} p(j)\log p(j) + \sum_{t=1}^{T} p(t)\sum_{j=1}^{M} p(j|t)\log p(j|t). \tag{4}$$

When this potential mutual information increases, hidden neurons fire uniformly on average, while each hidden neuron tries to specialize on the specific sets of input patterns.

In addition, the potentiality in terms of variance is further simplified to facilitate the computation, namely, pseudo-potentiality. The hidden pseudo-potentiality is defined by

$$\phi_j^{t,r} = \left(\frac{v_j^t}{v_{max}^t}\right)^r, \tag{5}$$

where v_{max} is the maximum potentiality and r is the potentiality parameter and $r \geq 0$. By normalizing this pseudo-potentiality, we have the pseudo-firing probability

$$p(j|t;r) = \frac{\phi_j^{t,r}}{\sum_{m=1}^{M} \phi_m^{t,r}} \tag{6}$$

Then, we have pseudo-potential information

$$PPI_{hid}^r = \log M + \frac{1}{T}\sum_{t=1}^{T}\sum_{j=1}^{M} p(j|t;r)\log p(j|t;r), \tag{7}$$

where for simplicity, the average firing probability $p(j)$ is supposed to be uniformly distributed. As expected, the pseudo-information can be increased just by increasing the parameter r. This is because when the parameter r increases, the number of strongly fired neurons decreases, corresponding to an increase in mutual information.

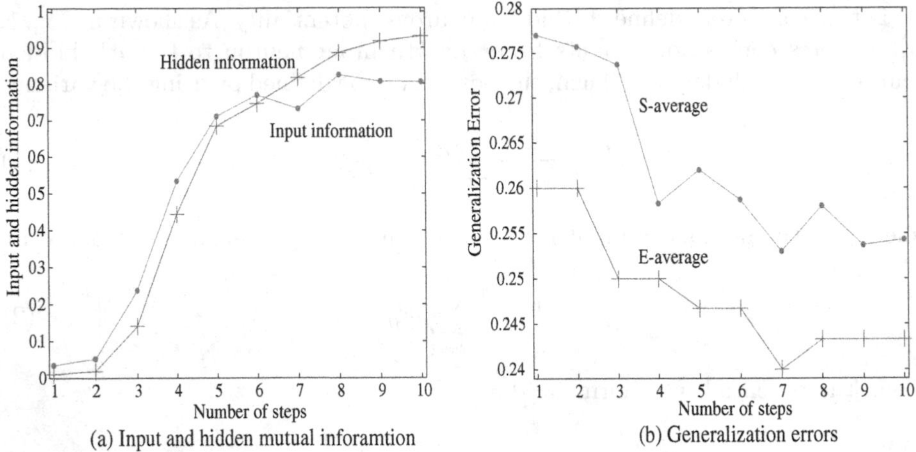

Fig. 2. Input and hidden information and generalization errors with 10 hidden neurons for the restaurant data set.

3 Results and Discussion

3.1 Experimental Outline

The restaurant data was extracted from the customer data of a restaurant [16], where there were 1,000 customers with 12 variables [16]. The problem with this dataset was to determine what makes people come to the restaurant, namely, what are the critical factors which attract people and make them visit. The conventional regression analysis had difficulty in dealing with this problem [16], even for training data. We computed two types of generalization errors, namely, S-average error and E-average error. The S-average is simply the average of generalization errors over ten different sets of input patterns. On the other hand, E-average is the average obtained by computing the differences between the targets and the average outputs over ten different data sets.

3.2 Mutual Information and Generalization

The results showed that generalization performance was in direct proportion to increases in mutual information. Figure 2(a) shows input and hidden information as a function of the number of steps. When the number of steps increased, the potentiality parameter r increased from 0.1 to 1. As can be seen in the figure, input and hidden mutual information increased gradually, and hidden information reached its maximum value of approximately 1, but the input information was slightly below the hidden information in the end. This is because input neurons are much more related to input patterns than hidden neurons. However, because we could not detect much difference between them, we could predict that the number of strongly activated hidden and input neurons became smaller along with the number of strong connection weights.

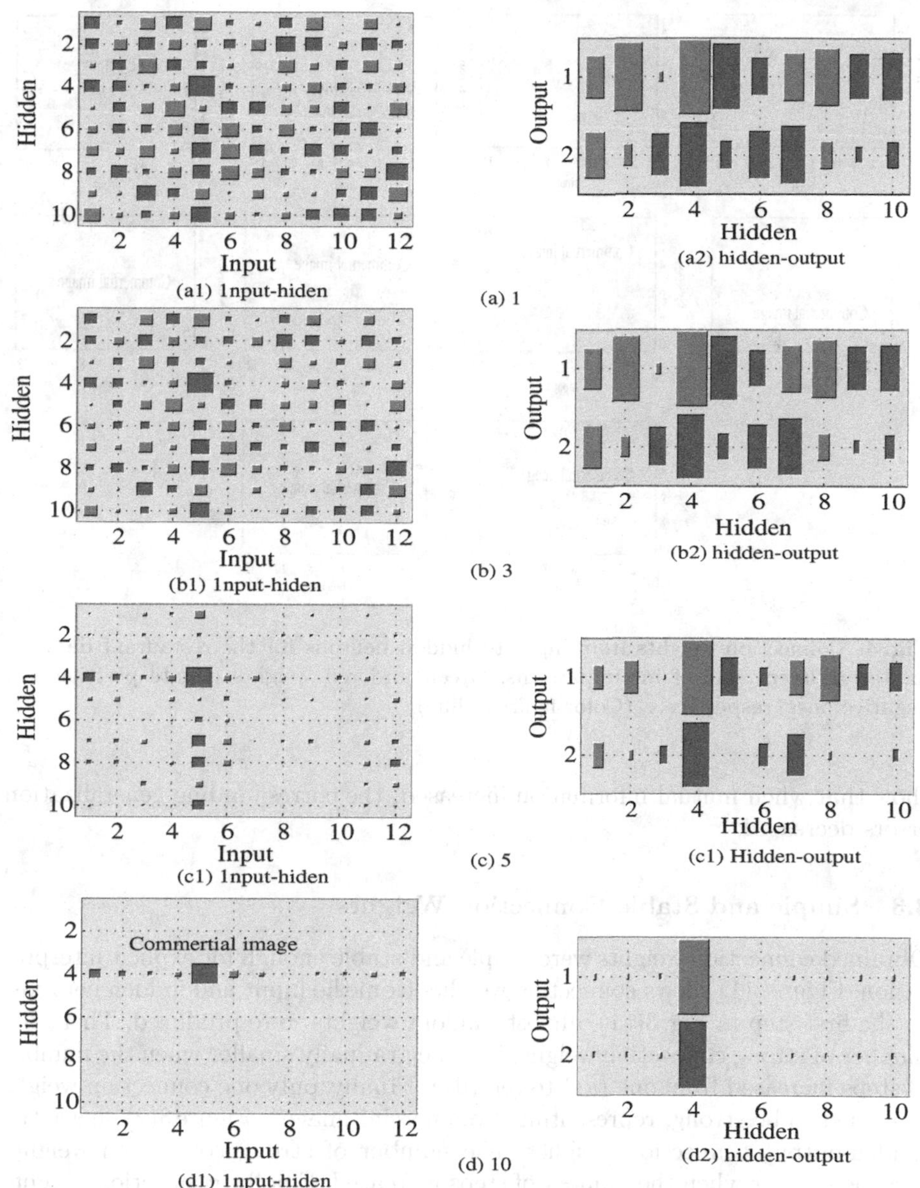

Fig. 3. Connection weights from input to hidden neurons with 10 hidden neurons for the restaurant data set. Green and red weights represent positive and negative ones. (Color figure online)

Figure 2(b) shows generalization errors as a function of the number of steps. The errors decreased gradually when the number of steps increased. It could be seen that the E-average was better than S-average over all steps. The results

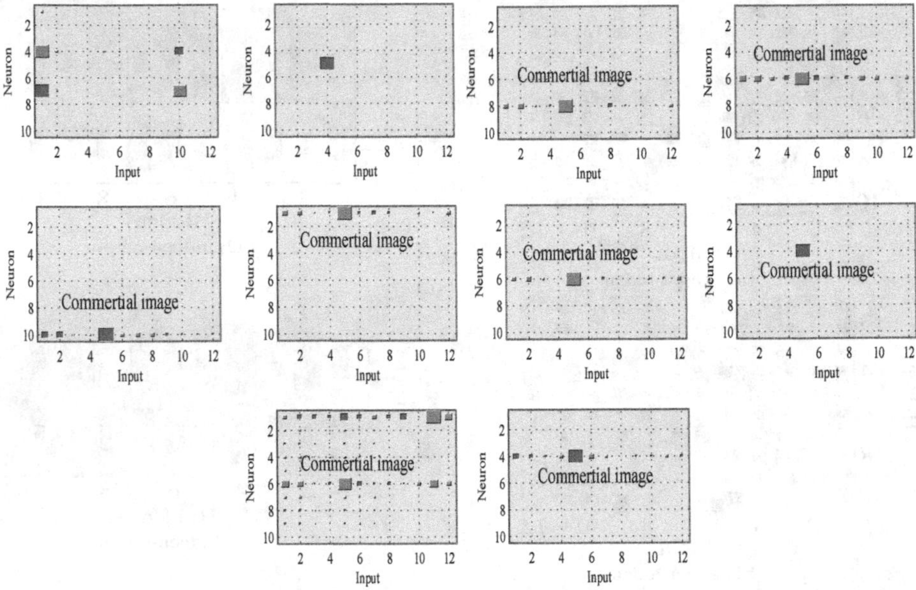

Fig. 4. Connection weights from input to hidden neurons for the restaurant data set for ten different sets of input patterns. Green and red weights denote positive and negative ones, respectively. (Color figure online)

show that when mutual information increased, the corresponding generalization errors decreased.

3.3 Simple and Stable Connection Weights

Obtained connection weights were simple and stable enough for explicit interpretation. Figure 3(1) shows connection weights from the input and hidden neurons. In the first step in Fig. 3(a1), almost random weights were produced. Then, the number of strong connection weights became gradually smaller when the number of steps increased from one (a1) to ten (d1). Finally, only one connection weight remained to be strong, representing "commercial image". Figure 3(2) shows the hidden-output connection weights. The number of strong connection weights became smaller when the number of steps increased. Finally, connection weights from the fourth hidden neuron became the strongest. This means that as the connection weights from the hidden neuron become more negative, shown in red, people are likely to visit the restaurant more frequently, because the corresponding strongest input-hidden weight to the fourth hidden neuron is strongly negative. Because the connection weight from the fifth input neuron represents the commercial images of the restaurant, this can be considered as one of the main features attracting customers. Finally, we checked the stability of final representations. Figure 4 shows the input-hidden connection weights obtained with ten different data sets. Eight out of the ten different representations showed

Table 1. Summary of experimental results in terms of generalization performance for the restaurant data set. The bold face numbers show the best values.

Method	Step	S-average	E-average	Std dev	Min	Max	Input inf	Output inf
MI	10	0.2530	**0.2400**	**0.0264**	**0.2167**	**0.3000**	0.7322	0.8176
BP(ES)		0.2770	0.2600	0.0284	0.2333	0.3233	0.0344	0.0107
SVM		0.3340		0.0366	0.2733	0.3867		

that connection weights from the fifth input neuron, the commercial image, were strongest.

3.4 Generalization Performance Comparison

The present method produced the best generalization performance compared to other conventional methods. Table 1 shows the generalization comparison between four methods. The best generalization errors in terms of S- (0.2530) and E-average (0.2400), standard deviation (0.0264), minimum (0.2167) and maximum value (0.3) were obtained by the present method with ten steps. The second best of 0.26 in terms of E-average was obtained by the BP with the early stopping. Though an extensive parameter search was conducted to obtain the best possible results, the worst case of 0.3340 was obtained by the support vector machine. When the best error was obtained by the present method, hidden mutual information was 0.8176, and input mutual information was 0.7322. This means that higher mutual information equates with lower generalization errors.

4 Conclusion

The present paper proposed a new type of information-theoretic method for interpretation which works by maximizing mutual information for hidden and input neurons. Though information-theoretic methods have been applied to many problems of neural networks, the complexity of the methods has prevented them from being applied practically. The present method used a simplified mutual information method with potentiality. When the potentiality parameter increased, mutual information increased correspondingly, and the number of strongly fired neurons decreased to produce simplified internal representations. The method was applied to find the main reason why visitors chose a certain restaurant. The results showed that mutual information could be increased, and improved generalization could be obtained by the present method. In addition, the commercial image of the restaurant was identified as playing the most important role in attracting visitors. The present paper shows the possibility of simplifying information-theoretic methods and using them to improve generalization as well as interpretation for large scale data sets.

References

1. Linsker, R.: Self-organization in a perceptual network. Computer **21**(3), 105–117 (1988)
2. Linsker, R.: How to generate ordered maps by maximizing the mutual information between input and output signals. Neural Comput. **1**(3), 402–411 (1989)
3. Linsker, R.: Local synaptic learning rules suffice to maximize mutual information in a linear network. Neural Comput. **4**(5), 691–702 (1992)
4. Linsker, R.: Improved local learning rule for information maximization and related applications. Neural Netw. **18**(3), 261–265 (2005)
5. Principe, J.C., Xu, D., Fisher, J.: Information theoretic learning. Unsupervised Adapt. Filter. **1**, 265–319 (2000)
6. Nenadic, Z.: Information discriminant analysis: feature extraction with an information-theoretic objective. IEEE Trans. Pattern Anal. Mach. Intell. **29**(8), 1394–1407 (2007)
7. Principe, J.C.: Information Theoretic Learning: Renyi's Entropy and Kernel Perspectives. Springer, New York (2010)
8. Torkkola, K.: Nonlinear feature transforms using maximum mutual information. In: Proceedings of International Joint Conference on Neural Networks, IJCNN 2001, vol. 4, pp. 2756–2761, IEEE (2001)
9. Deco, G., Finnoff, W., Zimmermann, H.: Unsupervised mutual information criterion for elimination of overtraining in supervised multilayer networks. Neural Comput. **7**(1), 86–107 (1995)
10. Kamimura, R., Nakanishi, S.: Hidden information maximization for feature detection and rule discovery. Netw. Comput. Neural Syst. **6**(4), 577–602 (1995)
11. Kamimura, R.: Self-organizing selective potentiality learning to detect important input neurons. In: 2015 IEEE International Conference on Systems, Man, and Cybernetics (SMC), pp. 1619–1626, IEEE (2015)
12. Kamimura, R., Kitajima, R.: Selective potentiality maximization for input neuron selection in self-organizing maps. In: 2015 International Joint Conference on Neural Networks (IJCNN), pp. 1–8, IEEE (2015)
13. Kamimura, R.: Supervised potentiality actualization learning for improving generalization performance. In: Proceedings on the International Conference on Artificial Intelligence (ICAI), p. 616. The Steering Committee of The World Congress in Computer Science, Computer Engineering and Applied Computing (WorldComp) (2015)
14. Kitajima, R., Kamimura, R.: Simplifying potential learning by supposing maximum and minimum information for improved generalization and interpretation. In: 2015 International Conference on Modelling, Identification and Control, IASTED (2015)
15. Kamimura, R.: Self-organized potential learning: enhancing SOM knowledge to train supervised neural networks with improved interpretation and generalization performance (under submission). J. Comput. Eng. Inf. Technol. (2016)
16. Nishiuchi, H.: Statistical Analysis for Billion People (in Japanese). Nikkei BP Marketing, Tokyo (2014)

Urdu Speech Corpus and Preliminary Results on Speech Recognition

Hazrat Ali[1(✉)], Nasir Ahmad[2], and Abdul Hafeez[2]

[1] Department of Electrical Engineering, COMSATS Institute of Information Technology, Abbottabad, Pakistan
hazratali@ciit.net.pk

[2] Department of Computer Systems Engineering, University of Engineering and Technology, Peshawar, Pakistan
{n.ahmad,abdul.hafeez}@uetpeshawar.edu.pk

Abstract. Language resources for Urdu language are not well developed. In this work, we summarize our work on the development of Urdu speech corpus for isolated words. The Corpus comprises of 250 isolated words of Urdu recorded by ten individuals. The speakers include both native and non-native, male and female individuals. The corpus can be used for both speech and speaker recognition tasks. We also report our results on automatic speech recognition task for the said corpus. The framework extracts Mel Frequency Cepstral Coefficients along with the velocity and acceleration coefficients, which are then fed to different classifiers to perform recognition task. The classifiers used are Support Vector Machines, Random Forest and Linear Discriminant Analysis. Experimental results show that the best results are provided by the Support Vector Machines with a test set accuracy of 73 %. The results reported in this work may provide a useful baseline for future research on automatic speech recognition of Urdu.

1 Introduction

Urdu is the national language of Pakistan understood by approximately 75 % population of the country. Globally, Urdu speakers accumulate to around 70 million speakers [1]. Urdu language shares its vocabulary with many other Asian languages including Arabic, Farsi, and Turkish. A framework for automatic speech recognition of Urdu can be helpful to contribute towards speech recognition of other similar languages. Unfortunately, for Urdu, lack of standard corpora and baseline approaches have been the bottleneck to make advancements on speech recognition research of Urdu.

Recently, there has been some work reported on the automatic speech recognition of Urdu. While these works have their own significance, either the corpus used in the work has not been specified or it is too limited to be generalized for diverse set of speakers. For example, Sarfraz et al. [2] has presented an Urdu corpus covering speakers only from a single city. Similarly, another speech corpus for Urdu has been presented in [3] however, it is not clear if the corpus is

C. Jayne and L. Iliadis (Eds.): EANN 2016, CCIS 629, pp. 317–325, 2016.
DOI: 10.1007/978-3-319-44188-7_24

available for public use. Akram et al. [4] have presented a continuous speech recognition system for Urdu however, the corpus used in the work is not identified. Information on training and test sets size is also missing. Besides, the accuracy reported by [4] does not exceed 54 %. For Urdu digits recognition, a multilayer perceptron has been used by Ahad et al. [5], presenting a framework for speech recognition of digits from 0 to 9. However, the work in [5] is based on speech data from a single speaker and thus, cannot be generalized for a diverse set of speakers. Another work reported for Urdu digits recognition is by Hasnain et al. [6] with higher accuracy performance. It is not clear if the accuracy measures in [6] are reported for training set only or for unknown test set. The use of hidden markov models for Urdu speech recognition has been reported in [7]. The model used in [7] treats every single word as a single phoneme. This may work for words of shorter duration but may undergo degradation if the words have longer duration.

For the Urdu dataset presented in this work, previous work has used features from discrete wavelet transform with linear discriminant analysis (LDA) [8], MFCC features with LDA [9,10]. In this work, we describe the Urdu corpus for the general understanding of the reader, and make it freely available for academic research use. Further, we report results on speech recognition task for this corpus with three different classifiers namely; Support Vector Machine (SVM), Linear Discriminant Analysis (LDA) and Random Forests (RF). The rest of the paper is organized as follows: In Sect. 2, we describe the development of the corpus and the way the audio files are organized. In Sect. 3, we discuss the extraction of MFCC features as well as the three classifiers used on the features. The results obtained are provided in Sect. 4. Finally, the paper is concluded in Sect. 5.

2 The Corpus

2.1 Corpus Development

The words recorded for this corpus are selected from the most frequently used words in Urdu literature, as summarized by the center of language engineering (CLE) [11]. These words include those which are used in everyday life, and digits from 0 to 9. Wherever possible, an attempt has been made to include antonyms or synonyms of various words. These words were then recorded by ten speakers with Sony Linear PCM Recorder. Any mistake in recording process was compensated by re-recording. The recording was accomplished in multiple sessions. Speakers coming for recording vary in age, origin and first language, ensuring that a diversity is achieved in the corpus. The recorded files are stored with sampling rate of 16000 Hz in *.wav* format. Average duration for each recording is half a second.

2.2 Corpus Organization

The master directory in this corpus contains ten sub-directories and each sub-directory corresponds to the individual speaker. Each sub-directory contains

250 audio files in *.wav* format. The information about each individual speaker is available in the sub-directory name. For example, the sub-directory named AKMNG2 corresponds to speaker AK (speakers are represented by combination of two letters, thus ranging from AA to AK and can be extended as well). The speaker gender information is contained in the third letter M (M corresponds to male and F corresponds to female). The fourth letter N in the sub-directory name denotes that the speaker is a non-native speaker (N represents that the speaker is non-native while Y represents that the speaker is a native speaker). The last two letters comprising of a character and a number correspond to the age of the speaker. Age ranges are from G1 (20–25 years) through G2 (26–30 years). Each file name provides information on speaker as well as the word number. The words are numbered from 001 to 250, appended to the sub-directory name to form the file name. An overview of the corpus organization is shown in Fig. 1. Access to the corpus can be requested by writing email to the first author.

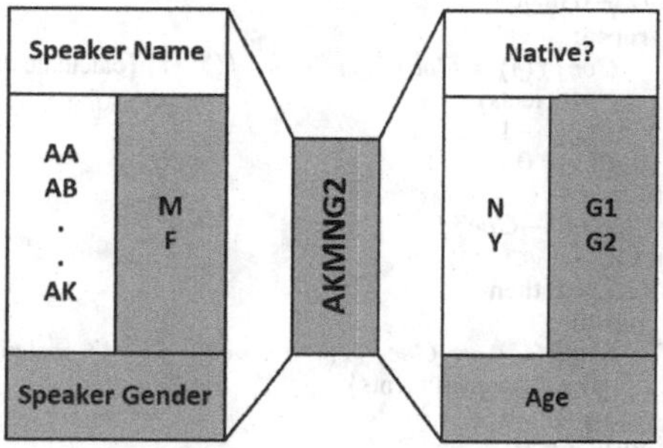

Fig. 1. Speakers are named from AA to AK. Speaker gender is defined by M for male and F for female. In the native field, N represents that speaker is non-native speaker and Y represents that speaker is native. Speakers belong to age group G1 or G2.

3 Experimental Setup

3.1 Features Extraction

For the dataset, we randomly divide the audio files into training and test sets with a ratio of 7:3. We then calculate the mel frequency cepstral coefficients (MFCC) for each audio file. The mel frequency cepstral coefficients have been in wide use by the speech processing community both for speech and speaker recognition applications [10, 12–14]. The MFCCs are based on mel-scale, a non-linear scale with logarithmic behavior [12]. Frequency mapping on a mel scale is given by equation:

$$f_{mel} = 2595 \times \log\left(1 + \frac{f}{700\,Hz}\right) \quad (1)$$

where, f_{mel} is the mel-scale frequency and f is the linear frequency in Hz. Different methods for calculation of the MFCCs can be seen in [12–14]. For MFCC calculation in this work, the Malcom's implementation has been used, as also used in [10,21]. The steps involved in MFCC features extraction are demonstrated through algorithm shown in Fig. 2. For each audio file, 12 coefficients are computed followed by concatenation of delta and delta-delta coefficients. Thus, each file is represented by 36 features set.

Algorithm 1 Algorithm for MFCC calculation

```
for i = 0 to No. of Frames do
    Calculate Power Spectrum
end for
for i = 0 to No. of Filter Coefficients do
    Mel Filter Bank Calculation
    Apply the filter bank to the spectrum
    sumE ← ∑ the energy in each filter
    logE ← log(sumE)
end for
Discrete Cosine Transformation for the logE
Retain N coefficients
if D ≠ 0 then
    repeat
        Coeff(j) = Coeff(j)−Coeff(j−1) {calculate delta coefficients}
        j ← j − 1
    until j = 0
else
    Coeff ← Coeff
end if
if DD ≠ 0 then
    repeat
        Coeff(j) = Coeff(j) − Coeff(j − 1) {calculate delta-delta coefficients}
        j ← j − 1
    until j = 0
else
    Coeff ← Coeff
end if
```

Fig. 2. MFCC Calculation (as in [10,21])

3.2 Support Vector Machines

Support Vector Machine (SVM) is a kernel based algorithm. SVMs are popularly used for discriminative classification. SVMs can be traced back to the work Boser et al. [15]. They were used for automatic recognition of handwritten characters [16] and thus, became popular. In SVMs, the data of different classes is separated by hyper planes such that the distance for data of each class is maximized (for binary classification, the distance of samples of both the classes from the hyper plane will be maximized). Thus, SVMs are classifiers with large-margin boundary. For SVMs, the important feature is the kernel function used. The kernel function might be linear, polynomial or Gaussian. The strength of SVMs lie in the fact that they do not suffer the problem of local optima. However, attention is required to select the suitable kernel function. For SVMs, the function is given by sums of the kernel function $K(x_m, x_n)$:

$$f(x) = \sum_{m=1}^{N} \alpha_m t_m K(x_m, x_n) + d \tag{2}$$

where t_m denotes the ideal outputs, $\sum_{m=1}^{N} \alpha t_m = 0$ and α_m is greater than zero. Ideally, the outputs are +1 or −1 representing the corresponding class to which the data sample belongs. The output class for any data sample is decided by comparison of value of $f(x)$ with a threshold value. Generally, the one-vs-all approach is used if we have more than two classes of data (i.e., a multi-class problem). In our work on the use of SVM, we utilize the libSVM library [17]. We use the Gaussian RBF kernel, which for two data points, can be defined as below:

$$K(x_m, x_n) = exp(\gamma(\|x_m - x_n\|)^2) \tag{3}$$

We run a grid search and choose the γ and regularization constant C (hyper-parameters) after running the experiment over multiple iterations.

4 Random Forest

In computer vision, decision trees have been remarkable and successful for classification as well as regression tasks. Decision trees have previously been used as stand-alone approach. When an ensemble of multiple decision trees is used for decision making, they form a random forest classifier (or random decision forest classifier). RF has been successfuly used on hand-written digits recognition task as reported in [18], Other work on the use of RF classification is reported in [19]. For classification through RF classifier, the process involves training of the trees with features selected randomly. In order to make a final prediction, average is then calculated for the posteriors of each class output. To perform speech recognition using a RF classifier, we feed the MFCCs to train the classifier comprising of 300 trees.

4.1 Linear Discriminant Analysis

Linear Discriminant Analysis (LDA) [20] is popular for dimensionality reduction as well as for classification tasks. When LDA is applied to a data, it transforms the data into a matrix Θ. *"LDA tends to maximize the ratio between the inter-class variance and intra-class variance"* [10]. Classification is achieved such that for each test example, calculation of Euclidean distance is performed. So, for a particular problem, if we have n distinct classes, there will be n number of Euclidean distances to be calculated over each test example. The class is predicted for the prediction for which the corresponding distance is the smallest. LDA transformation can be represented by $S(\Theta)$;

$$S(\Theta) = \frac{|\Theta^T \Psi \Theta|}{|\Theta^T W \Theta|} \tag{4}$$

where, the within-class variance is given by W and variance matrix is given by Ψ, $|.|$ is the value of the determinant. For the speech recognition task, we use LDA with the MFCC features and compare the results with those obtained for RF and SVM classifiers.

5 Experimental Results

Once the recognition is performed, the prediction results are put into a confusion matrix for the test data. For N number of words, the size of the confusion matrix is $N \times N$ matrix. $ConfM$ provides a general representation of the confusion matrix.

$$ConfM = \begin{matrix} c_{11} & c_{12} & c_{13} \cdots & c_{1N} \\ c_{21} & c_{22} & c_{23} \cdots & c_{2N} \\ c_{31} & c_{32} & c_{33} \cdots & c_{3N} \\ . & . & \cdots & . \\ . & . & \cdots & . \\ c_{N1} & c_{N2} & c_{N3} \cdots & c_{NN} \end{matrix} \tag{5}$$

Table 1. Recognition accuracy in percentage

S. No	Word number	Recognition rate (SVM classifier)	Recognition rate for RF	Recognition rate for LDA
1	001	100 %	66.67 %	100 %
2	002	66.67 %	33.33 %	33.33 %
3	003	66.67 %	100 %	100 %
4	004	100 %	66.67 %	66.67 %
5	005	66.67 %	66.67 %	66.67 %
6	006	33.33 %	100 %	66.67 %
7	007	66.67 %	66.67 %	66.67 %
8	008	100 %	66.67 %	0 %
9	009	66.67 %	33.33 %	33.33 %
10	010	66.67 %	33.33 %	100 %

In the above confusion matrix, $ConfM$, correct word recognition is shown by the values in the diagonal entries i.e., c_{ij} for $i = j$. Conversely, the number of false predictions for a test word is provided by the enteries in the non-diagonal position of the matrix, i.e., c_{ij} for $i \neq j$. The SVM classifier has resulted in an overall test accuracy of 73 %. Compared to this, the overall accuracy obtained by the random forest classifier as well as the LDA classifier is 63 %. Figures 3, 4 and 5 show the confusion matrix plots for the three classification methods namely, SVM classification, Random Forest classification and LDA classification respectively. For each digit, the corresponding recognition rates for SVM classifier,

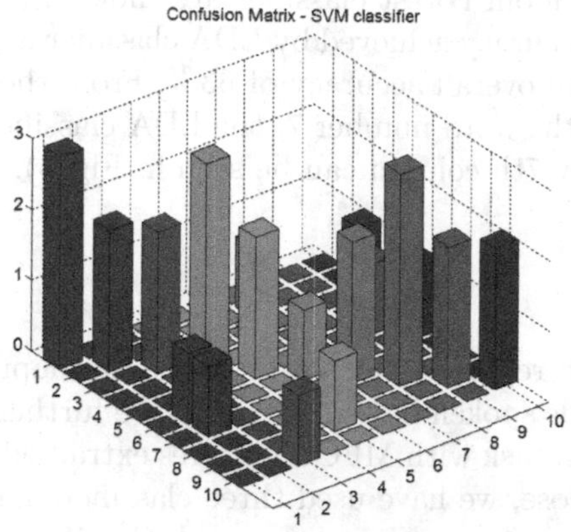

Fig. 3. Confusion matrix plot (For SVM classifier)

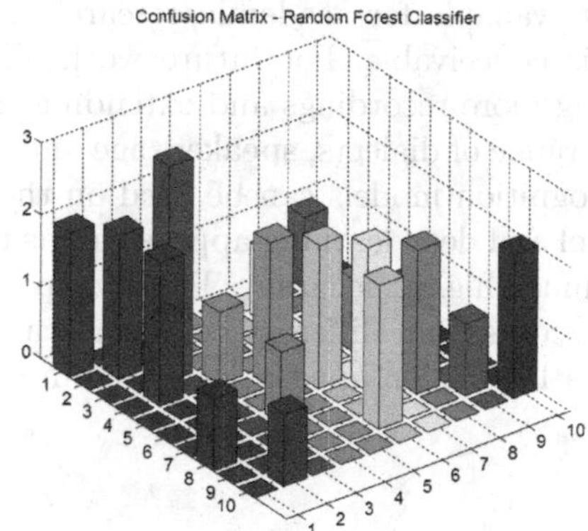

Fig. 4. Confusion matrix plot (for Random Forest classifier)

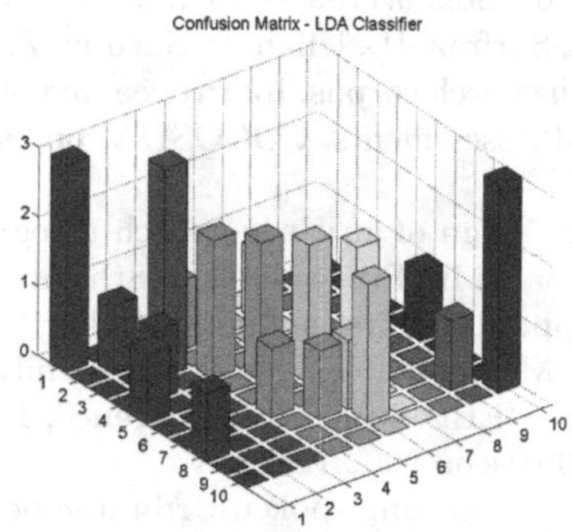

Fig. 5. Confusion matrix plot (for LDA classification)

LDA classifier and Random Forest classifier are shown in Table 1. It is obvious from the results that accuracy achieved by LDA classifier is same as the accuracy for RF classifier, i.e., an overall accuracy of 63 %. From the confusion matrix, it can be noted that for the word number 7, the LDA classifier has resulted in 0 % accuracy (as the empty 7th column can be seen in Fig. 5).

6 Conclusion

In this paper, we have reported our work on the development of Urdu corpus comprising of 250 words spoken by ten speakers. We further reported our results for a speech recognition task with MFCC features extracted from the audio data. For classification purpose, we have used three classifiers namely; SVM, RF and LDA and reported percentage accuracy for each classifier. Experimental results have shown that SVM has performed well on this particular dataset with a 73 % recognition accuracy compared with the 63 % accuracy for RF and LDA. These results can serve as a reference baseline for further advancement on the Urdu dataset. The dataset is available for academic/research use and thus, a direct comparison of results is conceivable. For future work, firstly, the corpus can be extended by including more recordings and extending the list of words thus, covering a more diverse range of dialects, speakers age and vocabulary. Secondly, more robust speech recognition models can be used on the Urdu data set, such as Hidden Markov Model and deep learning approaches as these can arguably be more robust providing much higher accuracy. Thirdly, an ensemble model which combines classification scores from different classifiers can also be explored for this data, for example, a late fusion approach as used in [22].

References

1. Ethnologue. http://www.ethnologue.com/show_country.asp?name=PK
2. Sarfraz, H., et al.: Speech corpus development for a speaker independent spontaneous Urdu speech recognition system. In: Proceedings of the O-COCOSDA, Kathmandu, Nepal (2010). doi:10.1109/ivtta.1994.341535
3. Raza, A.A., Hussain, S., Sarfraz, H., Ullah, I., Sarfraz, Z.: Design and development of phonetically rich Urdu speech corpus. In: Proceeding of International Conference on Speech Database and Assessments, COCOSDA, pp. 38–43 (2009). doi:10.1109/icsda.2009.5278380
4. Akram, M.U., Arif, M.: Design of an Urdu speech recognizer based upon acoustic phonetic modeling approach. In: Proceedings of 8th International Multitopic Conference (INMIC 2004), pp. 91–96, December 2004. doi:10.1109/inmic.2004.1492852
5. Ahad, A., Fayyaz, A., Mehmood, T.: Speech recognition using multilayer perceptron. In: Proceedings. IEEE Students Conference, ISCON 2002, pp. 103–109, August 2002. doi:10.1109/iscon.2002.1215948
6. Hasnain, S., Awan, M.: Recognizing spoken Urdu numbers using fourier descriptor and neural networks with matlab. In: Second International Conference on Electrical Engineering (ICEE 2008), pp. 1–6, March 2008. doi:10.1109/icee.2008.4553937

7. Ashraf, J., Iqbal, N., Sarfraz Khattak, N., Mohsin Zaidi, A.: Speaker independent Urdu speech recognition using HMM. In: The 7th International Conference on Informatics and Systems (INFOS 2010), pp. 1–5, March 2010. doi:10.1007/978-3-642-13881-2_14
8. Ali, H., Ahmad, N., Zhou, X., Iqbal, K., Ali, S.M.: DWT features performance analysis for automatic speech recognition of Urdu. SpringerPlus **3**(1), 204 (2014). doi:10.1186/2193-1801-3-204
9. Ali, H., Ahmad, N., Zhou, X.: Automatic speech recognition of Urdu words using linear discriminant analysis. J. Intell. Fuzzy Syst. **28**(5), 2369–2375 (2015). doi:10.3233/ifs-151554
10. Ali, H., Jianwei, A., Iqbal, K.: Automatic speech recognition of Urdu digits with optimal classification approach. Int. J. Comput. Appl. **118**(9), 1–5 (2015). doi:10.5120/20770-3275
11. Center for Language Engineering. www.cle.org.pk
12. Molau, S., Pitz, M., Schluter, R., Ney, H.: Computing Mel-frequency cepstral coefficients on the power spectrum. In: IEEE International Conference on Acoustics, Speech, and Signal Processing (ICASSP 2001), pp. 73–76 (2001). doi:10.1109/icassp.2001.940770
13. Han, W., Chan, C.F., Choy, C.S., Pun, K.P.: An efficient MFCC extraction method in speech recognition. In: Proceedings. IEEE International Symposium on Circuits and Systems, ISCAS 2006, May 2006. doi:10.1109/iscas.2006.1692543
14. Kotnik, B., Vlaj, D., Horvat, B.: Efficient noise robust feature extraction algorithms for distributed speech recognition (DSR) systems. Int. J. Speech Technol. **6**(3), 205–219 (2003)
15. Boser, B.E., Guyon, I.M., Vapnik, V.N.: A training algorithm for optimal margin classifiers. In: Proceedings of the Fifth Annual Workshop on Computational Learning Theory, COLT 1992, pp. 144–152 (1992). doi:10.1145/130385.130401
16. Bottou, L., Cortes, C., Denker, J., Drucker, H., Guyon, I., Jackel, L., LeCun, Y., Muller, U., Sackinger, E., Simard, P., Vapnik, V.: Comparison of classifier methods: a case study in handwritten digit recognition. In: Proceedings of the 12th IAPR International Conference on Pattern Recognition, pp. 77-82, October 1994. doi:10.1109/icpr.1994.576879
17. Chang, C.C., Lin, C.J.: LIBSVM: a library for support vector machines. ACM Trans. Intell. Syst. Technol. **2**, 27:1–27:27 (2011). doi:10.1145/1961189.1961199. Software available at http://www.csie.ntu.edu.tw/~cjlin/libsvm
18. Ho, T.K.: Random decision forests. In: Proceedings of the Third International Conference on Document Analysis and Recognition, vol 1. pp. 278–282, August 1995. doi:10.1109/icdar.1995.598994
19. Caruana, R., Karampatziakis, N., Yessenalina, A.: An empirical evaluation of supervised learning in high dimensions. In: Proceedings of the 25th International Conference on Machine Learning, ICML 2008, pp. 96–103 (2008). doi:10.1145/1390156.1390169
20. Balakrishnama, S., Ganapathiraju, A.: Linear discriminant analysis: a brief tutorial. http://www.music.mcgill.ca. Accessed 10 Feb 2016
21. Ali, H., Zhou, X., Tie, S.: Comparison of MFCC and DWT features for automatic speech recognition of Urdu. In International Conference on Cyberspace Technology (CCT 2013), Beijing, China, pp. 154–158, November 2013. doi:10.1049/cp.2013.2112
22. Ali, H., d'Avila Garcez, A.S., Tran, S.N., Zhou, X., Iqbal, K.: Unimodal late fusion for NIST i-vector challenge on speaker detection. Electron. Lett. **50**(15), 1098–1100 (2014). doi:10.1049/el.2014.1207

Bio-inspired Audio-Visual Speech Recognition Towards the Zero Instruction Set Computing

Mario Malcangi(✉) and Hao Quan

Department of Computer Science, Università degli Studi di Milano, Milan, Italy
{malcangi,quan}@di.unimi.it

Abstract. The traditional approach to automatic speech recognition continues to push the limits of its implementation. The multimodal approach to audio-visual speech recognition and its neuromorphic computational modeling is a novel data driven paradigm that will lead towards zero instruction set computing and will enable proactive capabilities in audio-visual recognition systems. An engineering-oriented deployment of the audio-visual processing framework is discussed in this paper, proposing a bimodal speech recognition framework to process speech utterances and lip reading data, applying soft computing paradigms according to a bio-inspired and the holistic modeling of speech.

Keywords: Audio-visual information processing · Automatic speech recognition · Bio-inspired computing · Convolutional neural networks · Evolving fuzzy neural networks

1 Premises

Automatic speech recognition (ASR) will be a key technology for the next generation on information systems, when human-to-machine interaction will be similar to the human-to-human interaction. Experiments to understand speech perception began last century. In 1921, Fletcher and Stainberg had found a functional relations between nonsense's phone sequences (e.g. consonant-vowel-consonant) error-recognition rate and words' recognition rate. This relation demonstrated that the context influences the intelligibility. Allen in his work "How do humans process and recognize speech?" [1] discusses extensively the role of the context in human speech recognition (HSR), citing the famous example of the two questions "How do human recognize speech?" and "How do humans wreck a nice beach?" that can be uttered so that only with appropriate context they can be distinguished. Entropy is higher for simple sounds (phones) and lower for complex words, so two important strategies are in HSR.

2 Introduction

Audio-visual information processing (AVIP) is an interdisciplinary research field that joins computer science and signal processing. It concerns the processing of information that is embedded in physical signals generated by the human beings and by the

C. Jayne and L. Iliadis (Eds.): EANN 2016, CCIS 629, pp. 326–334, 2016.
DOI: 10.1007/978-3-319-44188-7_25

surrounding environment. Most of the research efforts have been targeted audio and visual as individual fields, considering these fields as independent at each other. An emblematic example of this is the ASR problem approached mostly as an audio processing special purpose task. Several investigations [2–9] demonstrated that speech understanding in human beings is a multimodal process where the audio and the visual information concur to successfully complete the correct recognition of communication sounds such as phonemes, phones and words.

The AVIP activity in human beings is not perfect but efficient. This is because it is a biological-based processing model, with evolving inference paradigms performing in adaptive and context aware way. The multimodal nature of both audio production and perception and the relationship between audio and visual information has been investigated and experimental results demonstrates that the bio-inspired approach to the issue of AVIP could be the right way to develop robust and effective AVIP-based applications [10, 11].

There are also several bio-inspired processing processes that are under considered in the development of the ASRs, such as localized time-frequency events, temporal and spatial information (binaural), pitch (for source localization and separation). These processes needs to be considered in order to match the right paradigm to be applied for the ASR development.

Two main bio-inspired soft computing paradigms nicely match audio and visual perception in human beings, the convolutional neural network (CNN) and the evolving connectionst systems (ECOS).

The CNN, a bio inspired variant of the multilayer perceptron (MLP), has been successfully applied to face recognition [12]. CNN embeds the convolution paradigm useful to model spatial and temporal correlations. This apply to speech signal to compensate the translational variance and to capture translational invariance with a reduced set of parameters [13].

CNNs exibit invariance to shifts of speech features along the frequency, dealing with speaker and environment variations. The CNN special network structure (alternation of convolutional and pooling layers) is the main advantage over standard neural network as it demonstrates to be compatible and efficient respect to the way the data can be arranged to be processed efficiently. Considering the voice spectrogram as a 2-D image of features distributed along the frequency and time axes, the same approach of the use of CNN for the image recognition can be extended to speech recognition.

ECOS [14], mainly the evolving fuzzy neural networks (EFuNN), is a bio inspired inference paradigm that meets the capability of the HSR to adapt to noise and the signal filtering by evolving. In a multimodal context such as the audio-visual integration at the higher layers of the HSR hearing model, EFuNN is able to fuse the decision from audio and visual stages [15].

3 Bio Inspired Framework for Audio-Visual Speech Recognition

The bio inspired framework (Fig. 1) for audio-visual speech recognition (AVSR) is a three stage system that apply three bio inspired inferencing (soft computing) paradigms, the convolutional, the evolving and the rule-based. The full framework is completed by two mixed-signal processing (hard computing) frontend and backend stages, to interface the framework by sensing and by actuating towards the physical world.

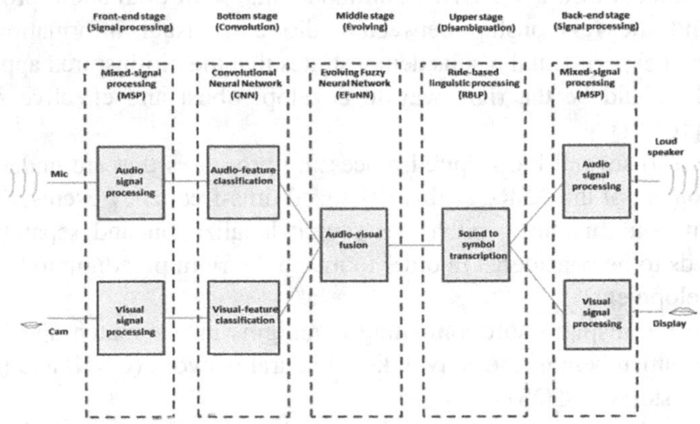

Fig. 1. Framework for the bio-inspired audio-visual speech recognition

3.1 Front End Stage

The front end is a mixed-signal processing (MSP) stage that implements the signal conditioning (linearization, amplification, equalization and filtering) and the extraction of the low level features (time and frequency measurements). It is based on analog and digital signal processing (mixed-signal) models that puts the crisp signal information in a measurement domain suitable to the lower information processing stages. Two distinct MSP front end are available, one for the audio signal, captured by a microphone, and one for the visual signal, captured by a camera.

3.2 Convolutional Stage

The convolutionalstage implements the high-level feature mapping (phoneme and viseme) task by exploiting the temporal and spatial local correlation of the audio and the visual information. Audio and visual low level features from front end stage are inputted to the convolutional layer. This stage consists of two information path, one for audio information and one for visual information. The purpose is to feature the audio and visual information according to the semantic that will be applied at higher stages (e.g. phonemes and visemes featuring in AVSR systems). Each node of the input layer is connected to the inner layer nodes in a spatially contiguous receptive schema (e.g. each node at layer *n* is connected to only 3 adjacent *n-1* layer nodes) (Fig. 2).

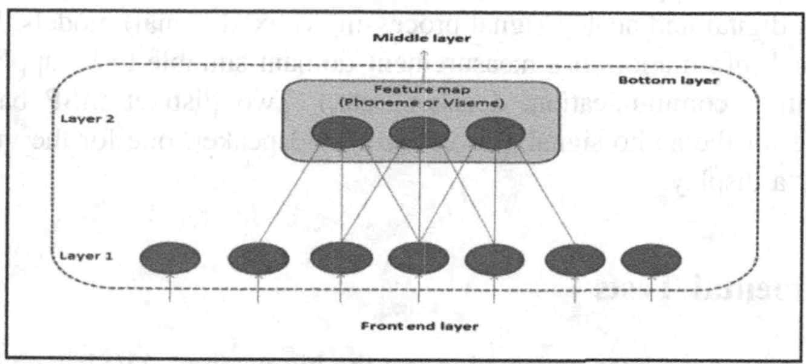

Fig. 2. Feature mapping at convolution stage

3.3 Evolving Stage

The evolving stage implements decision fusion on the audio and visual high-level feature scores by applying the evolving paradigm to enable the adaptation of the AVSR system to the environment variability (e.g. noise) and to the information mismatch (e.g. mismatch of /m/ and /n/ phonemes due to high degree of similarity of time and frequency features).

The evolving stage is implemented by the EFuNN paradigm (Fig. 3).

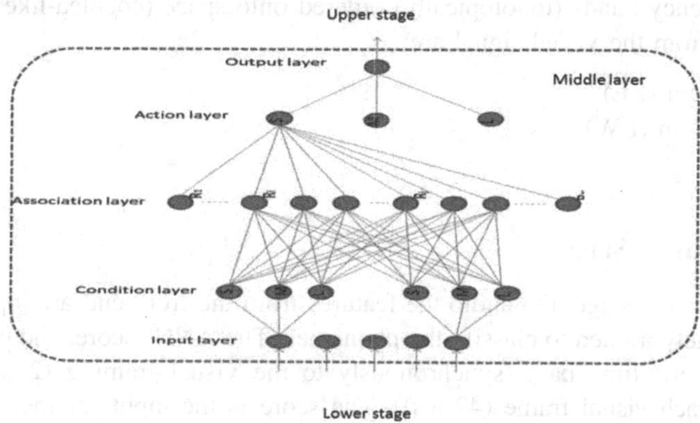

Fig. 3. EFuNN evolving architecture is applied to fuse phoneme-viseme classification and to predict phoneme occurrence

3.4 Linguistic Stage

The linguistic stage implements by rules the process of sound to symbols conversion (e.g. phoneme to grapheme) and of the text disambiguation.

3.5 Back End Stage

The back end is a mixed-signal processing (MSP) stage that implements the signal conditioning (linearization, amplification, equalization and filtering) and physical feature generation to be applied to the physical world (e.g. audio-visual speech synthesis). It is based on digital and analog signal processing (mixed-signal) models that produce the crisp signal information in a measurement domain suitable to be applied to other systems (control, communication, decision, etc.). Two distinct MSP back end are available, one for the audio signal, played by a loudspeaker, one for the visual signal, visualized by a display.

4 Experimental Tests

Some experiments have been executed to test the bio inspired AVSR's ability to adapt to physical context changes (e.g. noise). The test concerned the recovering of the right grapheme from the utterance of a words with two acoustically similar phonemes, /m/ and /n/, and their corresponding visemes.

4.1 Front-End

Two front-end has been programmed to extract the physical features of the captured signal by an audio sensor (microphone) sampled at 16 kHz and 16 bit encoded, and by a visual sensor (camera) 24 fps. A set of digital signal processing (DSP) algorithms has been applied for feature extraction purpose from the audio signal:

- Five frequency bands (tonotopically) ordered onto space (cochlea-like) the feature extracted from the visual signal are:
  - Lips eight (LE)
  - Lips width (LW)

4.2 Convolution Stage

At the convolution stage, the audio the features from the front-end are inputted to the ANNs separately-trained to classify the phonemes. The ANNs scores the phonemes on a frame-by-frame time base, synchronously to the visual framing (2 audio frames (21 ms) for each visual frame (42 ms)). The score is the input for the middle stage (evolving stage). The visual features (visemes) are encoded by LE and LW measurements and passed directly to the middle stage as knowledge related to the current viseme (Fig. 4).

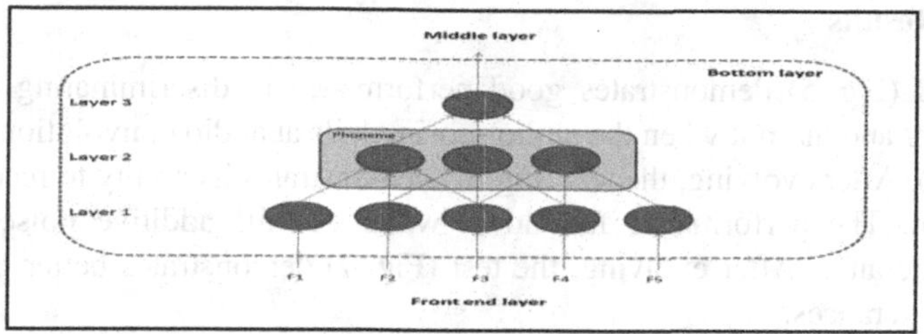

Fig. 4. Phoneme featuring at convolution stage.

4.3 Evolving Stage

At the second stage an evolving fuzzy neural network (EFuNN) has been trainedted to execute the decision fusion of the scoring of the audio and of the visual features. The EFuNN has been trained to predict the phoneme sequence frame-by-frame, streaming the phonetic transcription of the spoken word at output.

The NeuCom [16] environment was used to model and simulate the EFuNN by applying the following setup:

- Sensitivity threshold: 0.95
- Error threshold: 0.05
- Number of membership functions: 5
- Learning rate for W1: 0.1
- Learning rate for W2: 0.1
- Node age: 60

4.4 Linguistic Stage

At the third stage the fuzzy logic engine disambiguates the phonetic transcription executing the phoneme-to-grapheme transcription.

4.5 Test Setup

The test has been executed on 100 utterances of phonemes /m/ and /n/under four acoustic changing conditions (increasing additive white noise):

1. 0 dB
2. +6 dB
3. +12 dB
4. +18 dB

The EFuNN was first trained with 80 % of the utterances at 0 dB noisy condition and tested with the remaining 20 %. Then at next test time, the EFuNN evolved using 80 % previous test utterances and fuses decisions using the new noisy classification from audio and visual featured utterances.

4.6 Test Results

Test at 0 dB (Fig. 5) demonstrates good performance in discriminating two similar phonemes /m/ and /n/, not when the audio scoring fails at audio convolution layer (/m/ / n/ mismatch). After evolving, the test (Fig. 6) demonstrates its ability to recover the /m/ /n/ mismatch. The performance fall down when +18 dB additive noise masks the noise-free utterance. After evolving, the test (Fig. 7) demonstrates better performance with noisy utterances.

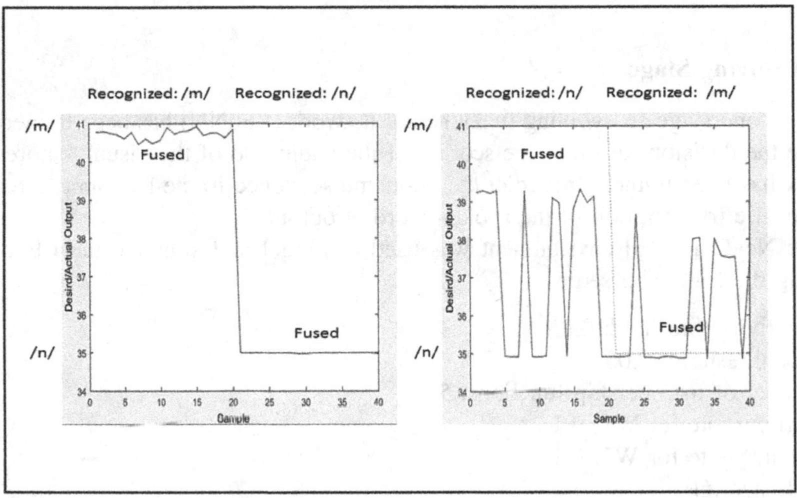

Fig. 5. Decision fusion at evolving stage for /m/ /n/ sequence for correct recognition (left) and for wrong recognition (right).

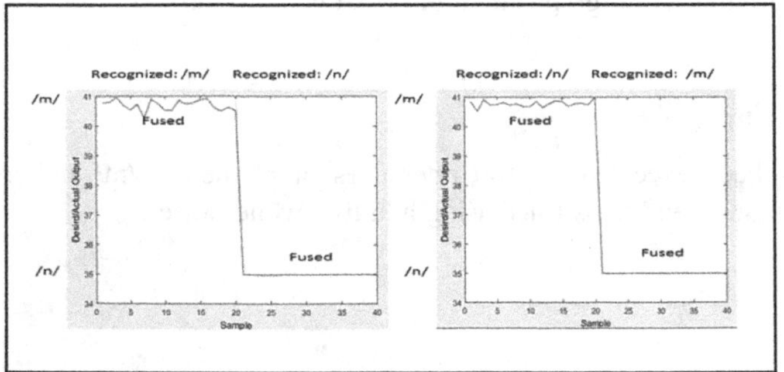

Fig. 6. Decision fusion at evolving stage for /m/ /n/ sequence for correct recognition (left) and for wrong recognition (right) after training on wrong recognition.

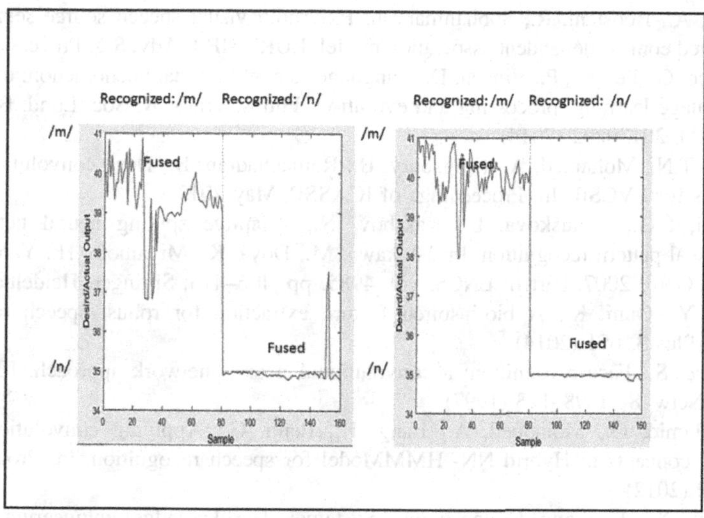

Fig. 7. Decision fusion at evolving stage for /m/ /n/ sequence for noisy input: without adapting (left) and with adapting trought evolving (right).

5 Conclusion and Future Works

The experiment demonstrated that the bio-inspired AVIP framework is effective in harsh conditions keeping low the system complexity. This performance is related the special purpose nature of the subsystems and of their capability to adapt to context changes by data-driven paradigms and intrinsic evolving capabilities.

The purpose of this research is to find which bio-inspired processing and inferencing paradigms could be optimal for the complete computational path from sensing to actuation in audio-visual applications. Bio-inspired signal processing is the next step to extend the paradigm to the front end and the back end of the AVIP framework.

References

1. Allen, J.B.: How do humans process and recognize speech? IEEE Trans. Speech Audio Process. **2**(4), 567–577 (1994)
2. McGurk, H., MacDonald, J.: Hearing lips and seeing voices. Nature **264**, 746–748 (1976)
3. Massaro, D.: Speech perception by ear and eye: A paradigm for psychological enquiry. Erlbaum, London (1987)
4. Norrix, L.W., Green, K.P.: Auditory-visual context effects on the perception of /r/ and /l/ in a stop cluster. J. Acoust. Soc. Am. **99**, 2951 (1996)
5. Bernstain, L.: Visual speech perception. Audio Visual Speech Processing, pp. 21–39 (2012)
6. Cappelletta, L., Harte, N.: Phoneme-to-viseme mapping for visual speech recognition. In: Proceedings of the International Conference on Pattern Recognition Applications and Methods. (IEEE) (2012)

7. Kazemi, A., Boostani, R., Sobhanmanesh, F.: Audio visual speech source separation via improoved context dependent association model. EURASIP J. Adv. Sig. Process., 47 (2014)
8. Vigliocco, G., Perniss, P., Vinson, D.: Language as a multimodal phenomenon: implications for language learning, processing and evolution. Philos. Trans. R. Soc. Lond. B Biol. Sci. **369**(1651), 20130292 (2014)
9. Sainath, T.N., Mohamed, A., Kingsbury, B., Ramabhadran, B.; Deep convolutional neural networks for LVCSR. In: Proceedings of ICASSP, May 2013
10. Wysoski, S.G., Benuskova, L., Kasabov, N.: Adaptive spiking neural networks for audiovisual pattern recognition. In: Ishikawa, M., Doya, K., Miyamoto, H., Yamakawa, T. (eds.) ICONIP 2007, Part II. LNCS, vol. 4985, pp. 406–415. Springer, Heidelberg (2008)
11. Zouhir, Y., Ouni, K.: A bio-inspired feature extraction for robust speech recognition. SpringerPlus **3**, 651 (2014)
12. Lawrence, S.: Face recognition: a convolutional neural- network approach. IEEE Trans. Neural Netw. **8**(1), 98–113 (1997)
13. Abdel-Hamid, O., Mohamed, A., Jiang, H., Penn, G.: Applying convolutional neural network concepts to Hybrid NN- HMMModel for speech recognition. In: Proceedings of ICASSP (2012)
14. Kasabov, N.: Evolving Connectionist Systems: The knoledge engineering approach. Springer, Heidelberg (2007)
15. Malcangi, M., Grew, P.: Evolving fuzzy-neural method for multimodal speech recognition. In: Iliadis, L., et al. (eds.) EANN 2015. CCIS, vol. 517, pp. 216–227. Springer, Heidelberg (2015). doi:10.1007/978-3-319-23983-5_21
16. http://www.kedri.aut.ac.nz/areas-of-expertise/data-mining-and-decision-support-systems/neuco

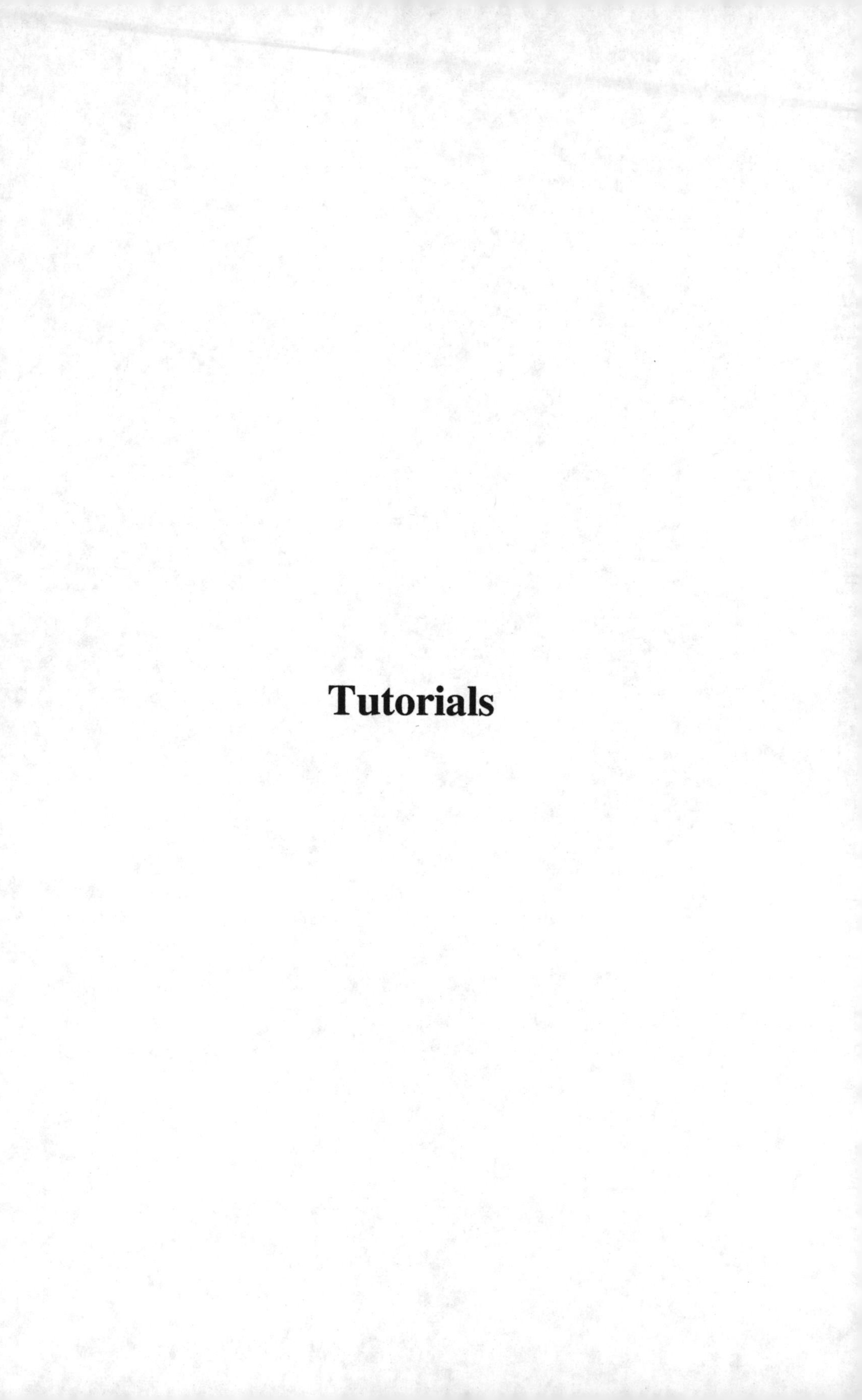

Tutorials

Classification of Unbalanced Datasets and Detection of Rare Events in Industry: Issues and Solutions

Marco Vannucci[(✉)] and Valentina Colla

TeCIP Institute, Scuola Superiore Sant'Anna, via G. Moruzzi, 1, 56124 Pisa, Italy
{mvannucci,colla}@sssup.it

Abstract. Classification of unbalanced datasets is a critical task that is getting interest due to its relevance in many contexts and especially in the industrial one where machine faults, quality deviations belong to the class of rare events whose identification is fundamental. This work introduces and outlines the main themes related to this problem including an analysis of the factors that make the detection of unfrequent events complicated, a list of the metrics used for classifiers assessment and a review of most popular and emerging approaches used for facing class unbalance with a special focus on the detection of rare events.

1 Introduction

In many practical fields classifiers have to cope with unbalanced datasets (also known as *uneven* datasets) within a wide and varied set of contexts and problems. These datasets are characterized by a significant unbalance in the class distribution among the training instances. Normally this kind of datasets are categorized into two groups according to the severity of the class unbalance: *normally unbalanced* datasets are characterized by a lower level of disparity with a the rate of rare samples above 10 % whilst in *highly unbalanced* datasets this rate is (often far) below 10 %. In this latter case the difficulty of the learning process exponentially increases. Within the unbalanced dataset classification nomenclature the majority class is usually also called *negative* and the minority class *positive*. This second class of instances is, in many frameworks, the one whose detection is fundamental since it represents situations whose identification, according to the problem context, is particularly important. There are many and frequent examples in real world applications of this kind of problems in different fields. In the industrial context machine malfunctions detection [24] and product quality assessment [2] are characterized by the rarity of the events (i.e. malfunctions and defects, which are far less frequent than normal situations) that, if correct and systematic, would allow – among the others – money saving and higher quality products delivered to customers (which avoid complaints) respectively. In medicine, rare diseases identification belongs to this category as well as the identification of fraudulent credit card transactions in finance, several speech

C. Jayne and L. Iliadis (Eds.): EANN 2016, CCIS 629, pp. 337–351, 2016.
DOI: 10.1007/978-3-319-44188-7_26

recognition tasks [19] and several bioinformatics related problems [9]. Unfortunately, despite its relevance, the problem of the detection of these patterns is extremely hard due to a multitude of interacting factors that prevent standard classifier to succeed. Nevertheless the strategic importance of this problem pushed researchers to develop smarter approaches of different nature to overcome the criticalities related to class unbalance. In this paper the main issues related to the classification of unbalanced datasets are presented together with a review of most used and state–of–art methods developed for solving this problem.

The paper is organised as follows: firstly the main factors that make this problem so complex are analysed in Sect. 2 and the most common performance measures adopted when facing unbalanced datasets are presented in Sect. 3, subsequently a comprehensive review of methods for unbalanced dataset classification is provided in Sect. 4, finally Sect. 5 is devoted to discussion, conclusions and future perspectives of the described approaches.

2 Difficulties in Unbalanced Data Classification

The detection of unfrequent patterns by means of standard classifiers is a complicate task due to several interacting reasons that limit the efficiency of these methods in such context. The main reason for standard classifiers such as Artificial Neural Networks (ANN), Decision Trees (DT) and Support Vector Machines (SVM) are not successful in the classification of unbalanced datasets is strictly related to the assumption they make on the even distribution of the training samples among classes. The goal of most machine learning algorithms is the achievement of an optimal *overall* performance, that is successful when classes are balanced while, in the case of unbalanced datasets, the decision boundaries defined by the classifiers tend to be biased toward the majority class and, as a consequence, the minority class is misclassified [10]. The unbalance degree directly affects the performance, in facts, as shown in [7], the higher the unbalance ratio the lower the recognition rate of minority class. Although the unbalance ratio and the basic assumption of even distribution of standard classifier is the main critical point, other factors negatively influence classifiers performance.

Lack of data compromises the capabilities of the learners of characterizing minority samples and to find effective boundaries to distinguish them from majority samples [29]. The main effect of lack of minority samples is the reduction of the region of the domain that the classifiers associate to them, in favour of the frequent samples. This condition is quite frequent in real world problems. In [14] it was empirically demonstrated that, even maintaining constant the unbalance rate, as the number of training examples increases the performance of the classifiers enhances due to the informative content brought in by the new samples. This latter result suggests that with a sufficiently large amount of data the detrimental effect of class unbalance could be arbitrarily reduced. Lack of data is a frequent problem in the industrial environment where it is not always possible to pursue durable measuring campaigns within particular plants due, for instance, to harsh conditions. For this reason the few data available for statistical analysis and classifier training contain a minimum amount of rare patterns usable for the class characterization. The complexity of the classification task – which corresponds to the complexity of the boundaries among classes to be calculated by learners – is another key point that affects the classifiers performance. Complexity is associated to the level of separability of the classes. For simple datasets where classes are not overlapping the effect of class unbalance is null, even in presence of strong unbalance rates [14]. On the contrary, in the case of complex problems characterized by highly overlapping classes and complex boundaries, the mission of the learner becomes complicate and standard classifier tend to solve generated conflicts in favour of majority class in order to maximize the overall performance.

Finally, noise in training data, in combination with the other mentioned factors, decreases the ability of classifiers to correctly characterize minority samples. In [29] it was shown that the effect of noise is highly detrimental for rare samples as noise can erroneously lead into modification of the class attribute within the training dataset: this change on one hand has low impact on the majority class but on the other hand is highly detrimental for the minority samples due to a small number of available samples for the extrapolation of class characteristics.

3 Performance Measures

When dealing with the classification of unbalanced datasets the use of classical evaluation metrics such as the overall accuracy for the assessment of classifiers performance is not suitable as it would not reflect the actual level of appreciation of the classifier. In many industrial problems, for instance, the main aim of the classification systems is not the achievement of an optimal overall performance but the recognition of particular rare situations such as machine faults or defective products. In an unbalanced domain there is the need of rewarding the correct detection of minority samples. The role of the performance measure is fundamental when exploited within the learning of the classifier as it is able to drive the training to achieve the desired results, thus many metrics have been developed with the specific aim of evaluating the performance of classifiers on unbalanced domains. Many of these metrics are based on the confusion matrix that reports the classification of samples with respect to their actual belonging classes and which is depicted in Table 1 for a two–classes problem with values for positive (minority) and negative (majority) classes.

From this matrix the following metrics can be derived:

- **Accuracy:** $\frac{TP+TN}{TP+FN+FP+TN}$ overall accuracy
- **False negative rate** (FN_R): $\frac{FN}{TP+FN}$ rate of rare samples misclassified
- **False positive rate** (FP_R): $\frac{FP}{FP+TN}$ rate of frequent samples misclassified (*false alarms*)
- **True negative rate** (TN_R, Specificity): $\frac{TN}{FP+TN}$ rate of rare samples correctly classified
- **True positive rate** (TP_R, Recall, Sensitivity): $\frac{TP}{TP+FN}$ rate of rate samples correctly classified

Table 1. Confusion matrix for a two–classes problem and positive/negative values.

	Positive prediction	Negative prediction
Positive class	True positive (TP)	False negative (FN)
Negative class	False positive (TP)	True negative (TN)

- **Precision:** $\frac{TP}{TP+FP}$ is the rate of correct positive predictions
- **F-measure:** $\frac{2 \cdot TP_R \cdot Precision}{TP_R + Precision}$ that is defined as the harmonic mean of recall and precision and whose high values represent a good performance in terms of both of them

These measures are all in the range [0;1] and have the advantage of being independent of the classes prior probabilities.

More advanced measures based on the ones above have been proposed in literature. Among them it's worth mentioning the Geometric–Mean $G - Mean = \sqrt{Sensityvity \cdot Specificity}$ that jointly takes into account the classification performance on minority and majority samples.

Another very popular method for assessing the performance of a classifier when coping with unbalance is based on the so–called Receiver Operating Characteristics (ROC) curve [29]. The ROC curve is a set of couples of true and false positive rates (TP_{Rt},FP_{Rt}) collected for a variation of a classifier parameter t that can be either a threshold that influences the classification (i.e. threshold classifiers) or, as it will be discussed later, a resampling rate or, in the case of DT the criterion for leaves labelling. The ROC curve, depicted in Fig. 1, puts into evidence the benefits in terms of TP_R and the costs in terms of FP_R that can be achieved by modifying t. One of the main characteristics of the ROC curve is that it can be used to find the optimal value of the t parameters and to compare different classifiers over a range of parameters value (t).

The Area Under the Curve (AUC) is a measure derived from the ROC and that is often used to summarize it. AUC measures the actual area under the ROC curve: the larger the area the higher the classifier performance.

4 Approaches for the Classification of Unbalanced Datasets

Given the strategic importance of the problem in most industrial and real world frameworks, many methods have been developed for the classification of unbalanced dataset, focusing on the detection of the interesting and rare patterns. These methods are traditionally grouped into two main families with respect to the way they operate to counterbalance the effects of class unbalance within the training dataset in order to favour the correct detection of rare patterns. The *internal* methods consist on techniques expressly designed for facing this specific problem whilst the *external* ones operate at data level by modifying the training dataset and reducing the unbalance ratio. The main ideas and techniques belonging to both these groups are depicted in the following sections.

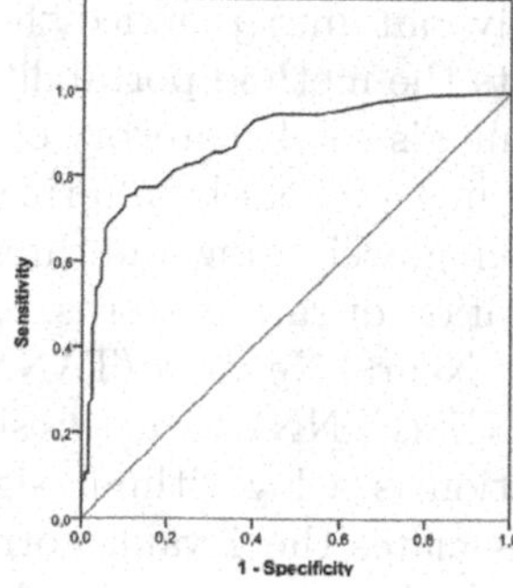

Fig. 1. Example of ROC curve

4.1 Internal Methods

Internal methods include not only the algorithms expressly developed for unbalanced data classification but also standard algorithms that are modified for this purpose.

Algorithms in this group are often designed for facing a specific problem and/or data distribution and - due to their specificity - the results they achieve are in most cases extremely satisfactory. Nevertheless this characteristic is also the cause of the main drawback of this class of approaches that lies in their scarce suitability to be successfully applied to other problems than that for which they are originally developed. Most known techniques in this group challenged the problem of portability and led to almost general purpose algorithms that require a minimal tuning from the user in order to achieve successful performance.

Cost–Sensitive Learning. The Cost Sensitive Learning (CSL) methods are based on the idea of directly tackling the tendency of standard methods of being biased toward frequent patterns that is due to the fact that the misclassification of a rare pattern is as penalized (i.e. has the same cost) as the misclassifiaction of a frequent one. CSL techniques operate by giving a different weight to these two types of errors by emphasizing the missed detection of rare patterns during the training of the classifier. This operation promotes their detection and is suitable for those problems where both class distribution and misclassification costs are unbalanced. One of the main advantages of these techniques is the possibility of combining it with standard techniques. An altered cost matrix or function can be used for instance within the ANN training algorithm or, as in [18], within DT for guiding the choice of the best attribute to be associate at tree nodes.

The main problem related to CSL is the determination of a suitable cost–matrix. In some early approaches costs are set inversely proportional with respect to class rates within the training dataset, theoretically re–balancing the effect of class unbalance during classifier tuning. In [29] it was shown that this approach leads to severe over–fitting problems. In some cases the cost–matrix is arbitrarily set by the users that directly put into practice the peculiar requirements of handled problems but it limits the method portability.

Cost–sensitive meta–learning is a sub–category of CSL and works at a higher level with respect to the learning process by altering the output of the trained classifier on the basis of sampling, weighting and thresholding operations aiming at decreasing the misclassification of rare patterns.

The Thresholded Artificial Neural Network (TANN) [25] combines a standard two–layers Feed–Forward ANN (FFNN) to a threshold operator. The FFNN output layer activation function is a logarithmic sigmoid and is connected to a threshold operator that associates the 1 value corresponding to positive/rare sample to the FFNN outputs higher than the threshold t and 0 otherwise. The threshold t determines the sensitivity of the TANN to rate patterns: the lower t the more TANN is encouraged to classify an arbitrary pattern as belonging to the minority class. The tuning of t implements the asymmetric cost of misclassification and simultaneously takes into account the overall classifier accuracy (ACC), the rate of minority class samples detected (TP_R) as well as the rate of false positives (i.e. false alarms) (FP_R). Once the FFNN within the TANN architecture is trained the optimal t value is selected among a set of candidate thresholds spanning in [0;1] according to the following merit function that formalizes the requirements of unbalanced datasets classification (the higher the better):

$$E(t) = \frac{\gamma TP_R(t) - FA_R(t)}{ACC(t) + \mu FA_R(t)} \quad (1)$$

and where γ and μ are two empirical parameters set by users according to the specific task targets.

One Class Learning. One class learning approaches (OCL) (also known as *recognition based*) include a set of methods for the training of some types of standard classifiers by exploiting only the samples belonging to the minority class. This idea aims at facing the issues encountered by discriminative learners (ANN, SVM, Fuzzy Systems) that base their training on the exploitation of multiple classes and discriminate among them: in the case of unbalanced datasets these classifiers tend to recognize with satisfactory precision majority classes as they are designed to achieve an optimal global performance to which minority samples marginally contribute. OCL can be used for the training of different types of classifiers, although it cannot be employed for some others such as DT and Naive Bayes. In [13] for instance this approach is used for the ANN– autoencoder training. The autoencoder reconstructs minority samples in the output layer and is subsequently used to recognize novel instances. Classification is possible, after training, since minority instances are expected to be reconstructed accurately while negative instances are not. Further, in [21] a OCL version of SVMs, the v-SVM, has been developed in a similar way for unbalanced classification. The v-SVM is trained by processing only minority samples. In this case the v-parameter, which acts as a similarity threshold, is suitably tuned in order to favour the detection of rare patterns. In [12] it was empirically put into evidence that for some problems, especially those characterized by extremely unbalanced ratios and where the generation of a higher rate of FP is acceptable, OCL outperforms discriminative methods.

The LASCUS Method. This method [23] merges different techniques and concepts related to statistical learning for unbalanced dataset classification. LASCUS (LAbelled Clusters for Unbalanced Sets) is designed to be general purpose and is particularly suitable for those problems where the identification of minority patterns is fundamental, accepting the risk of the generation of an acceptable amount of false positives. In many industrial applications such as, for instance, machinery faults prediction the generation of some false alarms is not a problem while the missed detection of a fault can be harmful and costly. The basic idea of LASCUS is to partition efficiently the input space and to suitably label the formed clusters in order to promote the minority patterns output. Clusters to which a *relatively high ratio* of rare patterns is associated are assigned through labelling to the minority class in order to favour its classification. The key issues of LASCUS concern the data partition and the determination of the rare patters concentration that determines the assignment of a cluster to the classes. The LASCUS training process can be subdivided into two main phases as depicted in Fig. 2.

The first step employs a Self Organizing Maps (SOM) for the creation of the clusters to which the rare and frequent patterns are assigned according to the euclidean distance metric: each sample is associated to the nearest cluster. Subsequently the *relative density* of minority patterns is calculated for each cluster. The determination of a critical minority samples density defines the criterion for which clusters are associated to the minority class: clusters characterized by a minority samples density higher than the critical one are associated to rare patterns, otherwise they are associated to majority class. For this purpose the rates $ACC(d)$, $TP_R(d)$ and $FP_R(d)$ are calculated for varying values of

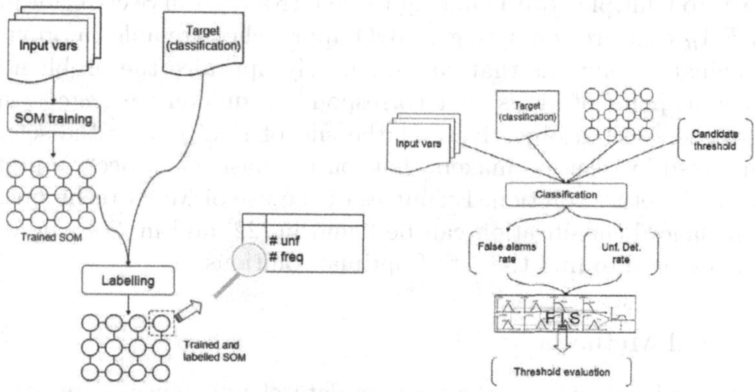

Fig. 2. The two steps of the LASCUS training procedure.

candidate densities d: these features are then fed to a fuzzy inference system which implements a human driven criterion for the LASCUS performance evaluation. The best rated density is then selected and all the clusters assignments are calculated by consequence. LASCUS has been successfully used within several industrial applications.

Ensemble Based Approaches. Ensemble methods combine the output of multiple learners to determine the overall output. Single learners are normally simple and their individual performance is globally only moderate so that they are named *weak*. Nevertheless the combination of the weak learners outputs can get the ensemble to extremely good performance. The most known ensemble paradigms are boosting and bagging. The first one builds progressively the ensemble, adding at each step a new weak learner that is trained by using only a subset of the training data. This subset is formed so as to favour the samples on which the ensemble performs worst in order to improve the system accuracy on these instances. Due to this characteristics boosting has been used with unbalanced datasets. AdaCost [8] for instance selects the samples to be included in the weak learners training datasets according to performance and user–defined misclassification costs. Rare–Boost [15] in an analogous way assigns a different cost that is calculated according to each cell of the confusion matrix. Bagging ensembles train simultaneously different weak classifiers that exploiting different training dataset that slightly overlap. In the case of unbalanced data this method creates a set of re–balanced datasets for the weak learners training so as to create a set of classifiers with limited performance but not biased toward majority class [19]. Ensemble methods have been often used in synergy with resampling techniques in the so–called hybrid approaches some of which are analysed in Sect. 4.3.

Multi–objective Optimization. Multi–Objective Optimization techniques (MOO) are suitable for coping with unbalances datasets classifications. Since in this context it is generally not possible to achieve a solution that is optimal with respect to multiple criteria among those introduced in Sect. 3 (for instance, TP_R and FA_R that are conflicting), MOO approaches are able to provide a set of non–dominated solutions that simultaneously optimize the problem according to different point of views that correspond to different *optimal* conditions in terms of the selected objectives. At the end of this process the set of solutions is analysed by decision makers that, on the basis of subjective preference, select one or the other solution. Examples of the use of MOO techniques in the field of unbalanced classification can be found in [22] and in [20] where genetic algorithms are used to find the set of optimal solutions.

4.2 External Methods

External methods counter act the training dataset unbalance by re–sampling it and creating a more balanced version of the training dataset that is then fed to a standard classifier for learning. In most cases datasets are not totally rebalanced but rather the relative frequency of the minority class is increased up to a pre–determined ratio. Unfortunately, since neither a fix unbalance ratio that suits for any classifier and problems nor a rationale for its determination exist, such ratio is often established experimentally. One of the main advantages of external approaches with respect to internal ones is their portability, in facts they can be used for training any type of classifier without requiring any modification of the standard algorithms or data structures. Datasets can be re–balanced either by removing majority class samples or by improving the number of samples belonging to the minority class until the desired unbalance ratio is reached. These two approaches that operate in opposite direction are called *under–sampling* and *over–sampling* respectively. Various studies investigated on the impact of these opposite approaches on classifiers performance without reaching clear and stable conclusions that can be addressed in order to outline an optimal re–sampling strategy [1,6]. The simplest approach to re–sampling consists in the random selection of the instances to be removed or replicated in the cases of under–sampling and over–sampling respectively. This trivial approach, yet fruitful in some cases, present some drawbacks. Random under–sampling can potentially remove instances with high informative content and being detrimental for the overall performance of the classifier. Random over–sampling can lead to over-fitting by creating compact *clusters* of positive samples that reduce instead of expanding the area of the input space associated to minority class.

Advanced Re–sampling Techniques. The limitations and risks connected to random re–sampling pushed the scientific community to develop more and more sophisticated techniques for the selection of the instances subjected to removal or replication. The idea behind these methods is to *globally* select a set of instances whose removal or replication (or their combination) is most beneficial for the classifier performance. For this reason such methods aim at conditioning the decision boundaries determined by classifiers.

In [16] focused under–sampling is achieved by selecting noisy and border–line samples in order to reduces the dimension of domain areas associated to frequent patterns, creating a more specific concept of the corresponding class. In [17] data are under–sampled by removing majority samples from compact homogeneous clusters of negative samples in order to reduce redundancy. In [11] focused under–sampling replicates the positive samples that lie on the boundary regions between minority and majority classes in order to spread the regions that the classifier associates to the positive samples and to limit eventual classification conflicts that standard classifiers would solve in favour of majority class.

SMOTE. SMOTE (Synthetic Minority Oversampling TEchnique) is an innovative oversampling approach originally presented in [4]. SMOTE aims at avoiding the overfitting problems that may arise when over–sampling is achieved by replicating existing minority class samples that do not actually add informative content to the training dataset but just rebalance it.

The main element of novelty of SMOTE is the synthetic creation of new positive samples that are not yet present in the training dataset. The synthetic samples are placed where they *probably* could be (and maybe are not in the dataset due to lack of data), for instance along the lines connecting existing minority samples. Depending on the number of synthetic samples required to rebalance the dataset, different numbers of couples of minority samples are selected for the generation of the new ones. Synthetic creation, contrary to replication, broadens the regions of the domain that are associated with minority class. Unfortunately the risk SMOTE assumes is the generation of misleading information since there is no control on the positioning of synthetic positive samples with respect to existing negative ones.

SUNDO. SUNDO is an advanced re–sampling method that synergistically combines oversampling and under–sampling for a re–balance of the training dataset [3]. The resampling procedure is based on several strategies devoted to the selection of most informative patterns to be maintained in the training dataset and the synthetic creation of minority samples non–conflicting and redundant with respect to the others. The method requires the specification of the target unbalance ratio so as to determine the number of synthetic samples to be created through oversampling (n_{over}) and to be eliminated through under–sampling (n_{under}).

The first step consists on the generation of a number $k \cdot n_{over}(k > 1)$ of synthetic minority samples higher than the number of positive instances actually required for the achievement of the set unbalance ratio. Synthetic instances are generated by using the original spatial distribution of the minority samples in the training dataset in order to locate them where they actually may be. Among these samples, n_{over} are selected on the basis of their distance from frequent observations: the closest to frequent observations are eliminated in order to limit inter-class interferences during the training. The under–sampling step is devoted to the removal of redundant positive samples in order to rebalance the dataset without loosing informative content. The n_{under} minority samples to be removed are selected on the basis of the relative distances among couples of frequent observations: once these distances are calculated, the n_{under} closest couples are selected and one out of the two samples is randomly selected and eliminated from each of these couples. This procedure reduces the redundancy among frequent samples.

Smart Undersampling. The focused under–sampling together with the automatic determination of the *optimal* unbalance ratio is proposed in [27]. The basic idea of Smart–Undersampling is to remove from the training dataset the optimal ratio of majority samples whose inclusion in the dataset is most detrimental according to a set of criteria that are calculated for each single negative sample. Utilized indicators take into consideration the relative distances among positive and negative instances and redundancy among majority samples and, once calculated, are used to create distinct rankings of the negative samples. At the top of these rankings the samples whose removal is most beneficial are located. A Genetic Algorithms (GAs) based optimization procedure is utilized for determining the optimal removal percentages associated to the each ranking.

These percentages are used to select the frequent patterns to be pruned from the training dataset. More in detail, each one represents the rate of samples, picked from the associated ranking, to be removed from the dataset according to their rating (i.e. starting from the top of the ranking) and managing eventual intersections. The fitness function employed by the GAs engine, given an arbitrary candidate solution expressed in terms of removal percentages according to each ranking, creates the corresponding reduced training dataset, trains an arbitrary classifier and evaluates its performance in terms of Eq. 1 in order to find the optimal removal rates according to an unbalanced classification framework. The main advantage of this approach is that it grants the elimination of majority samples whose removal is not detrimental for the classifier performance and, as a side effect, the automatic determination of the unbalance ratio that directly derives from the optimal removal rates provided by GAs optimization.

4.3 Hybrid Methods

Among all internal and external approaches there is no one that overcomes the others in any application but the choice is rather dependent on the handled problems, data distributions and evaluation criteria. As mentioned in the previous sections, internal and external methods have both advantages and weak points that led the scientific community to search for a *fusion* of these two families into *hybrid* approaches that maintain the pros and discard the cons of both of them. Some of the most representative examples of hybrid approaches are introduced in this section.

Combination of SMOTE and Ensemble Approaches. The SMOTE over–sampling method has been widely used in literature in combination with internal approaches due to its innovative characteristics. In particular SMOTE has been successfully utilized together with ensemble techniques.

SMOTE-Bagging [28] employs SMOTE for the over sampling of the training datasets utilized by the weak learners included within a bagging–type ensemble. More in details, for each base learner a part of negative samples is included in the training dataset so as to include, throughout all the weak learners, each negative instance at least once. Further original positive instance are included together with an additional quantity of minority samples (according to a pre–defined unbalance ratio) that are constituted in (small) part by originally replicated minority samples and SMOTE generated samples. The main advantage of this approach is that all the weak learners are trained by using a well balanced training dataset and that diversity among weak–learners - which is fundamental for ensemble performance - is granted by the different ratios of samples that are either replicated or synthetically created for each learner.

Smote-Boost [5] employs SMOTE within a boosting ensemble with the aim of improving the ensemble classification accuracy as far as the minority class concerns while the boosting ensemble architecture aims at keeping the overall classifier performance satisfactory. Smote-Boost provides at each step of the ensemble construction not only those minority samples (and eventual majority samples) that are not correctly classified by the ensemble, but also the new synthetic samples created by SMOTE in order to broaden at each step the representation of rare instances. Smote-Boost has been proven to outperform standard boosting techniques in terms of performance on unbalanced classification tasks.

Dynamic Resampling. The Dynamic Resampling is a hybrid approach that combines a FFNN based classifier characterized by a modified training procedure and a re–sampling operation that takes place dynamically during the training of the network itself [26]. The FFNN used as a classifier is a single–output network whose structure can be arbitrarily determined. The output neuron is activated by a logistic function that fires in the range [0;1]. For the classification purpose a threshold set to 0.5 is used to assign the FFNN output to positive or negative classes.

The basic idea of this approach is to mitigate the effect of class unbalance by using a different resampled versions of the training dataset through distinct phases of the network training process. The training dataset is resampled throughout the learning process so as that in each phase all minority samples are exploited but only part of the majority ones according to a resampling rate to be defined by the user. The training process is divided into *blocks of epochs* $[B_1, B_2, .., B_N]$ and for each of them a different training dataset is set up. At each block of epochs the required amount of majority samples is probabilistically selected according to two criteria: probability is higher for those instances that have been less frequently employed in previous blocks; probability of selecting a sample s is proportional to the classification performance achieved by the classifier during the training within the blocks of epochs that actually involved s. This approach presents two main interesting characteristics: on one hand it allows a balanced training of the FFNN that exploits, at each epoch of the training, a smaller but balanced (as defined by the user) dataset, on the other hand this training procedure is informative content loss–less since the samples selection

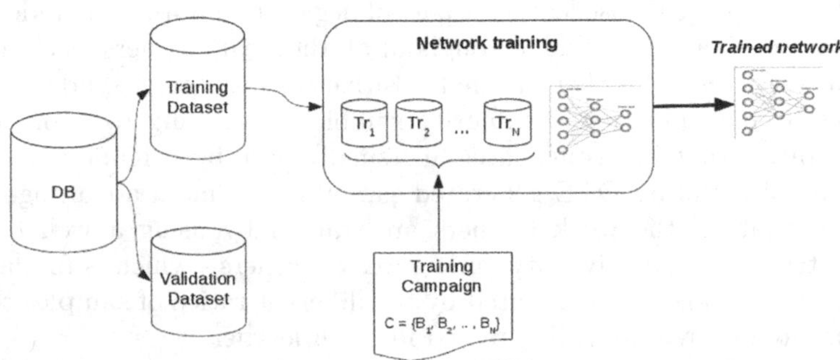

Fig. 3. Flow chart depicting the Dynamic Resampling training procedure. The number of epochs within each training block B_i exploits a different re–balanced training set Tr_i.

algorithm avoids it and moreover tends to privilege the samples whose use is mostly advantageous (Fig. 3).

5 Discussion

In this paper the main aspects related to the classification of unbalanced datasets and the associated problem of detection of rare patterns have been dealt, focusing on an analysis of the multiple and interacting factors that prevent standard classifiers to perform satisfactorily on these tasks that play a fundamental role in many practical and industrial environments. The issues on unbalanced datasets classification are associated both to the intrinsic training dataset characteristics (unbalance ratio, lack of data, instances distribution) and to the inclination of standard classifier to better recognize majority instances due to the search of the optimal overall performance. The analysis of techniques and ideas behind main methods developed for this task put into evidence that all approaches try to counterbalance in different ways the detrimental effect of class unbalance that biases the classifiers toward the majority class by assigning only limited regions of the domain to the minority class. Actually no method or family of approaches is able to achieve universally satisfactory results on any problem since solutions are deeply related to the peculiarities and requirements of single problems: it rather seems that it is up to the users to select and eventually combine the methods that best suited to the peculiar exigences.

The research activities in this field reached a satisfactory maturity level and led to the development of algorithms able to achieve good performance and an appreciable flexibility. Both the internal and external approaches put into evidence positive characteristics that complement each other. The future lines of the research in this framework could address the integration - already started - of these two families of methods.

References

1. Batista, G., Prati, R.C., Monard, M.C.: A study of the behavior of several methods for balancing machine learning training data. SIGKDD Explor. Newsl. **6**(1), 20–29 (2004)
2. Borselli, A., Colla, V., Vannucci, M., Veroli, M.: A fuzzy inference system applied to defect detection in flat steel production (2010)
3. Cateni, S., Colla, V., Vannucci, M.: A method for resampling imbalanced datasets in binary classification tasks for real-world problems. Neurocomputing **135**, 32–41 (2014)
4. Chawla, N.V., Bowyer, K.W., Hall, L.O., Kegelmeyer, W.P.: Smote: synthetic minority over-sampling technique. J. Artif. Intell. Res. **16**(1), 321–357 (2002)
5. Chawla, N.V., Lazarevic, A., Hall, L.O., Bowyer, K.W.: SMOTEBoost: improving prediction of the minority class in boosting. In: Lavrač, N., Gamberger, D., Todorovski, L., Blockeel, H. (eds.) PKDD 2003. LNCS (LNAI), vol. 2838, pp. 107–119. Springer, Heidelberg (2003)

6. Chawla, N.: C4.5 and imbalanced data sets: investigating the effect of sampling method, probabilistic estimate, and decision tree structure. In: Proceedings of ICML03 Workshop on Class Imbalances (2003)
7. Estabrooks, A., Jo, T., Japkowicz, N.: A multiple resampling method for learning from imbalanced data sets. Comput. Intell. **20**(1), 18–36 (2004)
8. Fan, W., Stolfo, S.J., Zhang, J., Chan, P.K.: Adacost: misclassification cost-sensitive boosting. In: Proceedings of the Sixteenth International Conference on Machine Learning, ICML 1999, pp. 97–105. Morgan Kaufmann Publishers Inc., San Francisco (1999)
9. García-Pedrajas, N., Ortiz-Boyer, D., García-Pedrajas, M.D., Fyfe, C.: Class imbalance methods for translation initiation site recognition. In: García-Pedrajas, N., Herrera, F., Fyfe, C., Benítez, J.M., Ali, M. (eds.) IEA/AIE 2010, Part I. LNCS, vol. 6096, pp. 327–336. Springer, Heidelberg (2010)
10. He, H., Garcia, E.A.: Learning from imbalanced data. IEEE Trans. Knowl. Data Eng. **21**(9), 1263–1284 (2009)
11. Japkowicz, N.: The class imbalance problem: significance and strategies. In: Proceedings of the 2000 International Conference on Artificial Intelligence, ICAI, pp. 111–117 (2000)
12. Japkowicz, N.: Concept-learning in the presence of *Between−Class* and *Within−Class* imbalances. In: Stroulia, E., Matwin, S. (eds.) Canadian AI 2001. LNCS (LNAI), vol. 2056, pp. 67–77. Springer, Heidelberg (2001)
13. Japkowicz, N., Myers, C., Gluck, M.: A novelty detection approach to classification. In: Proceedings of 14th International Joint Conference on Artificial Intelligence, pp. 518–523 (1995)
14. Japkowicz, N., Stephen, S.: The class imbalance problem: a systematic study. Intell. Data Anal. **6**(5), 429–449 (2002)
15. Joshi, M., Kumar, V., Agarwal, R.: Evaluating boosting algorithms to classify rare classes: comparison and improvements, pp. 257–264 (2001)
16. Kubat, M., Matwin, S.: Addressing the curse of imbalanced training sets: one-sided selection. In: Proceedings of the Fourteenth International Conference on Machine Learning, pp. 179–186. Morgan Kaufmann (1997)
17. Laurikkala, J.: Improving identification of difficult small classes by balancing class distribution. In: Quaglini, S., Barahona, P., Andreassen, S. (eds.) AIME 2001. LNCS (LNAI), vol. 2101, pp. 63–66. Springer, Heidelberg (2001)
18. Ling, C.X., Yang, Q., Wang, J., Zhang, S.: Decision trees with minimal costs. In: Proceedings of the Twenty-first International Conference on Machine Learning, ICML 2004, p. 69. ACM, New York (2004)
19. Liu, Y., Chawla, N., Harper, M., Shriberg, E., Stolcke, A.: A study in machine learning from imbalanced data for sentence boundary detection in speech. Comput. Speech Lang. **20**(4), 468–494 (2006)
20. Maheta, H.H., Dabhi, V.K.: Classification of imbalanced data sets using multi objective genetic programming. In: 2015 International Conference on Computer Communication and Informatics (ICCCI), pp. 1–6, January 2015
21. Schölkopf, B., Smola, A.J., Williamson, R.C., Bartlett, P.L.: New support vector algorithms. Neural Comput. **12**(5), 1207–1245 (2000)
22. Soda, P.: A multi-objective optimisation approach for class imbalance learning. Pattern Recogn. **44**(8), 1801–1810 (2011)
23. Vannucci, M., Colla, V.: Novel classification method for sensitive problems and uneven datasets based on neural networks and fuzzy logic. Appl. Soft Comput. J. **11**(2), 2383–2390 (2011)

24. Vannucci, M., Colla, V., Nastasi, G., Matarese, N.: Detection of rare events within industrial datasets by means of data resampling and specific algorithms. Int. J. Simul. Syst. Sci. Technol. **11**(3), 1–11 (2010)
25. Vannucci, M., Colla, V., Sgarbi, M., Toscanelli, O.: Thresholded neural networks for sensitive industrial classification tasks. In: Cabestany, J., Sandoval, F., Prieto, A., Corchado, J.M. (eds.) IWANN 2009. LNCS, vol. 5517, pp. 1320–1327. Springer, Heidelberg (2009)
26. Vannucci, M., Colla, V., Vannocci, M., Reyneri, L.: Dynamic resampling method for classification of sensitive problems and uneven datasets. In: Greco, S., Bouchon-Meunier, B., Coletti, G., Fedrizzi, M., Matarazzo, B., Yager, R.R. (eds.) IPMU 2012. CCIS, vol. 298, pp. 78–87. Springer, Heidelberg (2012)
27. Vannucci, M., Colla, V.: Smart under-Sampling for the detection of rare patterns in unbalanced datasets. Springer International Publishing, Cham (2016)
28. Wang, S., Yao, X.: Diversity analysis on imbalanced data sets by using ensemble models. In: IEEE Symposium on Computational Intelligence and Data Mining, CIDM 2009, pp. 324–331, March 2009
29. Weiss, G.M.: Mining with rarity: a unifying framework. SIGKDD Explor. Newsl. **6**(1), 7–19 (2004)

Variable Selection for Efficient Design of Machine Learning-Based Models: Efficient Approaches for Industrial Applications

Silvia Cateni and Valentina Colla(✉)

TeCIP Institute, Scuola Superiore Sant'Anna, via G. Moruzzi, 1, 56124 Pisa, Italy
{s.cateni,colla}@sssup.it

Abstract. In many real word applications of neural networks and other machine learning approaches, large experimental datasets are available, containing a huge number of variables, whose effect on the considered system or phenomenon is not completely known or not deeply understood. Variable selection procedures identify a small subset from original feature space in order to point out the input variables, which mainly affect the considered target. The identification of such variables leads to very important advantages, such as lower complexity of the model and of the learning algorithm, savings of computational time and improved performance. Moreover, variable selection procedures can help to acquire a deeper knowledge of the considered problem, system or phenomenon by identifying the factors which mostly affect it. This concept is strictly linked to the crucial aspect of the stability of the variable selection, defined as the sensitivity of a machine learning model with respect to variations in the dataset that is exploited in its training phase. In the present review, different categories of variable section procedures are presented and discussed, in order to highlight strengths and weaknesses of each method in relation to the different tasks and to the variables of the considered dataset.

1 Introduction

In machine learning and statistics, Variable Selection (VS) (or feature selection), is the process of selecting a subset of significant input variables considering the words variable or feature as synonymous. Actually *variable* is the term to address the "raw" input variable, while *feature* is a variable which can result from a preprocessing of the input variables. Concerning VS the two terms are usually used without distinction, as there is no impact on the selection algorithms [32]. VS is particularly important when dealing industrial datasets, especially when the amount of possible input variables is significant with respect to the number of available instances. In effect, in the industrial field, usually a large number of variables related to a single process is collected through several sensors distributed along the whole production chain [16,20,21]. The issue of VS has been deeply analysed in literature for purposes such as prediction [10,44,56], classification [11–13] or clustering [60]. VS also improves phenomena interpretation,

C. Jayne and L. Iliadis (Eds.): EANN 2016, CCIS 629, pp. 352–366, 2016.
DOI: 10.1007/978-3-319-44188-7_27

as it allows to ignore not significant effects as well as irrelevant, noisy or unreliable variables, by thus reducing noise, enhancing the machine learning system performance and speeding up the design time. Moreover, VS can be an interesting tool to reduce the risk of over-fitting of machine learning models or to lower their computational burden [7].

In literature three main approaches to VS can be found: *filters*, which are independent on the learning machine performed; *wrappers*, that consider the learning algorithm as a black box; *embedded* approaches, which execute VS as component of the learning phase. The most trivial variables selection approach consists in the analysis of all combinations of variables, also called *exhaustive search* or *brute force approach* but its computational time complexity is exponential, thus it is not viable when the number of input variables is high.

A further issue to be carefully considered when dealing with VS procedures is the so-called *stability*: a VS algorithm is said to be stable when the exploitation of different training data sets related to the same phenomenon leads to the selection of the same variable subsets. This concept is crucial in real applications where VS is designed to interpret the behaviour of the considered process.

The proposed tutorial provides a survey of recent and relevant literature results in the field of VS, especially related to industrial applications. The paper is organised as follows: a survey of the basic concepts inherent the VS task is proposed in Sect. 2, while the importance of VS for the design of Neural Networks (NNs) is highlighted in Sect. 3. In Sect. 4 the classification task for VS is treated, while Sect. 5 introduces the problem of the stability. Finally, Sect. 6 provides some concluding remarks.

2 Variable Selection

The complexity of selecting the most relevant input variables of a machine learning system can be mainly due to the large size of the initial inputs set, the correlations among variables which lead to redundancy and, finally, the presence of variables which do not affect the considered phenomenon or process [43]. In order to choose the optimal subset of input variables the following considerations should be taken into account:

- **Relevance.** The selected variables must convey all or most of the significant information inherent to the considered phenomenon.
- **Computational efficiency.** The number of selected input variables should be not too high in order to limit the computational burden. This fact is particularly relevant in the filed of NNs: the inclusion of redundant and irrelevant input variables adds noise and slows down the training of the network.
- **Knowledge improvement.** The optimal selection of input variables allows a deeper understanding of the behaviour of the considered/modelled process.

To sum up, the optimal set of input variables should include only the variables needed to explain the behaviour of the considered process or phenomenon with no or minimum redundancy, in order to build an accurate, efficient, inexpensive and more easy interpretable model. In literature VS techniques are categorized into three main classes: filter, wrapper and embedded methods.

2.1 Filter Methods

Filter approaches can be considered as a pre-processing phase; they exploit statistical tests to carry out the selection by computing a pertinence score for each variable: the variables which have a high pertinence score are finally selected. Filter are independent on the learning algorithm adopted in the developed machine learning system and their computational burden is reasonable, but they are subjected to over-fitting. Filter variables ranking is mostly suitable when the knowledge about the discriminative value of individual variables is searched. Filter variables ranking evaluates performance of each variable individually, being an univariate method. When interaction between variables is of interest, univariate methods are unsuccessful.

A popular example of filters is the *correlation-based approach* which computes the correlation coefficient between each variable and the considered target [62, 63]; inputs are then ranked and a subset is selected including the variables with the highest correlation coefficient. This approach is fast and the removal of variables with a low correlation coefficient decreases the redundancy of the initial variable set. However the linear correlation approach is not very adequate when dealing with real-world datasets, where the variables are usually not linearly correlated with the target. Other widely applied filter approaches are the *Fisher criterion* [31] and the *T-test* [47], which are used to evaluate the importance of each variable in the case of binary classification, taking into account the mean value and the standard deviation of the samples belonging to the two classes. *Single Variable Classifier* (SVC) is an alternative feature ranking approach where variables are ranked according to their individual predictive power, i.e. variables are ranked considering the performance of a classifier designed with a single variable [32].

A common filter approach is the one based on the concept of *Mutual Information* (MI). MI evaluates the amount of information included in a variable or a group of variables, in order to forecast the dependent one. The main advantage of MI is that it is model-independent and nonlinear at the same time. It is model-independent, as no assumption is made on the adopted model; it is also nonlinear as it evaluates the nonlinear relationships between variable and target. Also MI is a non-linear measure of dependency between variables and it is a general correlation measure that, contrarily to the correlation coefficient, can be generalized to all probability distributions [53]. MI can be computed by exploiting entropy measurements [23] as follows:

$$MI(\mathbf{x}, y) = H(y) + H(x_1) + H(x_2) + ... + H(x_n) - H(y, x_1, x_2, ..., x_n) \quad (1)$$

where $\mathbf{x}$ is the input vector and y is the output of the considered system. The MI of y and $\mathbf{x}$ represents the amount of information that $\mathbf{x}$ contains related to y and H is the *Shannon Entropy, Cover91*. The Shannon entropy of one generic variable z is defined as:

$$H(z) = -\int_z f(z) \log [f(z)] dz \quad (2)$$

The Shannon entropy in the case of n generic variables $\mathbf{z}$ is defined as:

$$H(z) = -\int_{z_1} \dots \int_{z_n} f(z_1, \dots, z_n) \log [f(z_1, \dots, z_n)] dz_1 \dots dz_n \quad (3)$$

where *f(z)* is the probability density function of z and *f(z)* is the joint Probability Density Function (PDF) of $\mathbf{z}$. The main complexity of this approach lies in the fact that the densities are all unknown and should be estimated from the data. To this aim, distribution fitting techniques can be applied both belonging to traditional statistics [6,26,48] and exploiting NNs [22,46] or evolutionary computation [18,19]. The entropy concept has been generalised in [42], where MI is also applied.

Finally, another filter feature selection method is the one based on the computation of the *relative entropy*, also known as *Kullback-Leibler divergence* [40]. This index was originally used to measure of how different two probability distributions are and has been used to rank the input variables maximizing the Kullback-Leibler distance between the target classes in the binary classification task [17]. The Kullback-Leibler divergence between two probability distribution p and q can be defined for distributions of a discrete variable as follows:

$$KL(p, q) = \sum_i p_i \log \frac{p_i}{q_i} \quad (4)$$

where p represents the probability distribution of the samples belonging to the unitary class while q is the probability distribution of the samples belonging to the null class. For distributions of continuous variables the summation is substituted by the integral. Once the KL-distance is calculated, the variables are ranked according to the KL-distance and the ones associated to the highest KL-distance values are selected.

2.2 Wrappers

The wrapper approach that was introduced in 1997 by Kohavi and John [39] exploits the performance of the learning machine in order to select a subset of variables evaluating their predictive power. Wrappers consider the learning algorithm as a black box capable of learning from examples and making predictions once trained. This fact makes wrappers remarkably universal, as they can be applied using different kind of learning algorithms. An obvious wrapper method is the *exhaustive search* also called *brute force method* which analyses all combinations of available variables. When the number of input variables are considerable the exhaustive approach becomes impractical. In fact, if there are n potential input variables, there are 2^n possible subsets to test and, consequently 2^n training procedures to perform.

A common wrapper strategy is the so called *Greedy Search* strategy which progressively creates the variables subset by adding or removing single variables from an initial set. Greedy search in fact can work into two directions: *Sequential Forward Selection* (SFS) and *Sequential Backward Selection* (SBS). SFS starts with an empty set of variables and the other variables are iteratively included until a fixed stopping criterion is reached. Usually the adopted performance index is the accuracy of the learning machine performed. SBS acts in the opposite sense: the procedure begins including all features and progressively removes the least relevant ones. A variable is considered relevant and is kept, if the accuracy of the learning machine tends to decreases when it is removed. Greedy search strategies need at maximum $n(n+1)/2$ training procedures and, as SFS starts with a small variables set, it is less expensive than SBS, if it stops early. In comparison, in filters there is no search and only n training procedures are necessary.

Wrappers are thus heavier from the computational point of view than filters, as the induction algorithm is trained for each tested subset, but it is more efficient in terms of accuracy.

2.3 Embedded Methods

Embedded methods execute the VS as part of the learning stage and are usually specific of a particular learning machine. Common examples of embedded approaches include classification trees, random forests [3] and methods based on regularization techniques, which make this combination very efficient in terms of both computational cost and suitable selection of variables [4]. The main advantage of embedded methods lies in their connection to the learning algorithm. Embedded approaches also utilize all the variables to create a model and then analyses it to deduce the importance of the variables. A simple example is represented by a decision tree which gives an embedded score of variable importance that can be achieved by the number and the quality of splits that are generated from a predictor variable.

In conclusion, filter methods are appropriate to deal with very high-dimensional datasets because they are computationally simple and fast and independent on the algorithm used. Wrapper methods utilize the learning algorithm as a black box evaluating its performance for the selection of variables. On the other hand wrappers have a high risk of over-fitting and they are very weighty from a computational point of view, in particular if the datasets are huge or if the constructed classifier has a high computational cost. Finally, embedded methods have a lower computational cost than the wrappers but they are too definite for a given classifier.

In the last years several hybrid VS methods have been proposed in order to jointly exploit their advantages by overcoming their drawbacks. An example of hybrid VS approach is proposed in [13] where the set of available variables is firstly reduced through a combination of filter selection methods and then exhaustive search (belonging to the wrappers category) is performed in order to obtain a sub-optimal set of variables in a reasonable time. A similar hybrid algorithm for VS is described in [9], where a combination of four filter methods with a popular sequential selection method is applied in order to obtain a more informative subset in a reasonable time. This approach can be applied to several kind of datasets without any a priori hypothesis on the data and also it is suitable to large or imbalanced datasets, commonly present in industrial context.

3 Variable Selection in NNs Applications

NNs are widely used in industrial application when highly complex and non-linear phenomena are considered. The set of input variables must contain only the most relevant variables with a low degree of correlation. Previous background on the considered application can provide some information for VS, but such information can be insufficient and, in some cases, even misleading. On the other hand, a good selection is recommended, as every unnecessary input lower the system robustness, adds noise and increases the computational effort.

While for standard parametric mathematical models the complexity of the VS task can be mitigated by the a priori hypothesis of the functional form of the model based on several physical interpretation of the system to be processed, for NNs or other similar statistical modelling approaches, the very generic and highly non linear structure of the model if makes the application of standard VS methods quite difficult. This complexity becomes greater when the variables selection are performed in the learning phase. The idea that an NN is effectively able of detecting redundant and noise variables during training, and that the trained network will adopt only the relevant variables can be utopic, especially is some restrictions are present on the time required for the learning. Moreover, especially naive NN users tend to design the network using all available input variables believing that their redundancy will lead to a more robust model. Consequently, NN models are still too often developed without considering the effect that the choice of input variables has on model complexity, efficiency of the learning procedure and performance of the trained NN.

Expert NN modellers recognise the importance of input VS techniques in the network design as a preliminary phase. Thus the choice of the VS method represents an important step: techniques that are appropriate for linear regression are not suitable for the highly non-linear neural model. Wrappers and embedded approaches are commonly applied when the number of training instances is quite small and the number of input variables is large. The search strategy is a compromise between the number of considered unique solutions and the sustainable computational burden. The SBS approach trains NN models of limited dimensions and can be highly efficient. On the other hand, SFS performs an efficient search provided that appropriate redundancy checking is included into the statistical analysis of variables. This guarantees that the approach selects the most informative input subset with the lowest number of variables. Brute force search is impracticable in most real cases, while evolutionary search approaches can provide an appropriate compromise that allows to cover a large combinations of input variables [10]. Other approaches allow to perform simultaneous weights, parameters and features optimization in an NN, such as proposed, for instance, in [28] and [49].

Filter algorithms represent a fast, model-free approach to VS and are principally appropriate to applications where independence from a specific NN architecture is mandatory or where computational efficiency is required. The capability to detect an optimal subset of input variables before training an NN avoids the computational burden related to the training and model selection, which can effectively reduce the overall effort of NN development. Moreover, simple ranking procedures select more input variables than needed, without taking into account the redundancy of variables, and also are not appropriate to multivariate NN regression. MI provides a generic measure of the relevance of a candidate input variable and is highly appropriate for NN modelling because it measures both linear and non-linear interactions. MI is also less sensitive to data transformations than correlation, which makes it more suitable, even for linear data analysis applications. However, evaluation of MI is more computationally intensive than estimation of correlation, due to the need to estimate density functions of the data [43].

An exemplar industrial application concerning the classification of particular surface defects of flat steel products, called *inclusions*, on the basis of parameters related to the steel chemistry and to the processes proceeding the cold rolling stage was proposed in [7]. VS was performed to discover the most relevant input variables among the 18 available ones, by applying a combined approach [13]. Five popular binary classifiers were tested and their accuracy on the real dataset has been calculated. The main objective of this work was to evaluate the importance of the preliminary VS step by exploiting the validation dataset to compute the performance of the classifiers. Finally a 10-cross validation approach was adopted to evaluate the performance of the classifiers in term of Balanced Classification Rate (BCR) [51]. The variables selection approach selects 4 variables (Blowed Oxygen, the Nitrogen content in the liquid steel, the first speed of the casting and the ladle temperature) which actually represent the parameters which mostly affect the defects classification as inclusion. Table 1 depicts the classification results which were obtained by considering all the available variables and only the subset selected through the VS procedure: noticeably the accuracy of all five classifiers is significantly improved after the application of the VS procedure.

Such results highlight the advantages which can be achieved coupling a VS procedure and AI-based methods, NNs included.

Table 1. Results of defects classification for flat steel products

Classifier	BCR 18 vars.	BCR 4 vars
Bayesian	0.59	0.78
Decision tree	0.57	0.72
Support vector machine	0.58	0.78
Multi layer perceptron	0.53	0.77
Self organizing maps	0.57	0.76

4 Variable Selection for Classification Task

Several works had been proposed in order to demonstrate the effectiveness of the VS task in classification problems, which are very relevant for real-world applications. In particular, binary classification has a relevant importance from the practical point of view in many real world applications; for instance, anomalies detection and forecasting are addressed as binary classification problems. Moreover intrusion detection, which occupies an important function in the protection of communication networks, is often approached as a binary classification problem faced by pattern recognition systems, whose accuracy is highly dependent on the variables which are fed in inputs. Binary classifiers are commonly applied in the medical field to detect some diseases diagnosys problems [29,30,37] as well as in industrial contexts for faults and anomalies diagnosis or forecasting and defective products detection in quality monitoring in the industrial applications are often approached through binary classifiers.

In the industrial field, the developments of the sensing technologies allow to disseminate industrial plants with a high number of sensors collecting and storing a huge quantity of information that can be exploited to detect or even forecast anomalies and faults. However, mainly for large and very complex processes, for instance where a series of chemical, physical and thermo-mechanical reactions simultaneously happen, such as in process industry, it is not easy to discover the reasons which affect faults and quality problems. In all these applications irrelevant and redundant variables tend to reduce classification performance and this explains the large quantity of VS approaches developed for that task.

Eid et al. [25] propose an approach for the selection of optimal variable subset based on the analysis of the Pearson correlation coefficients, which improves the performance of classifiers applied to decide if a considered system activity is intrusive or legitimate. Recent works based on the application of VS for intrusion detection performing binary classifiers can be found in [38,50].

A recent filter VS method which exploit the binary classification is proposed in [14], where a combination of four popular filters using a fuzzy logic based approach is proposed. The main element of originality of this method lies in the joint exploitation of four indexes deriving from popular VS procedures to be calculated on the variables to select in order to study their relevance with respect to the classification task. The fuzzy inference system is used to combine the evaluation provided by the four scores in a natural way in order to detect a subset of variables which is smaller or at least equal to the one selected by each single method without decreasing the classification performance. The method is applied to the design of binary classifiers and the effectiveness of the approach has been confirmed by the development of several classic binary classifiers by exploiting different datasets coming from both public repositories and real industrial applications. This method is a filter method and it could be applied with large datasets, as it is very fast and suitable for all binary classifiers. For each input variable 4 indexes derived from 4 common filter VS techniques are evaluated. These indexes are computed considering the distribution of the two classes; thus the method is suitable for binary classification.

Other literature methods are suitable also for multi-class classification and are applied to industrial datasets. For instance, in [11] VS is realized by means of the use of GAs during the selection process based on the assessment of the performance of the possible variable subsets adopted to train a decision-tree classifier. Furthermore the method is able to optimize some parameters of the employed classifier. The proposed approaches, which are characterized by different initialization and fitness functions of the GAs has been tested on a real industrial problem obtaining very satisfactory results.

In [12,15] the task of the variables selection to be fed in input to a labelled Self Organizing Map (SOM) has been faced using GAs. The GA-based optimization procedure is used to identify not only the subset of input variables which maximizes the performance of the classifier but also to determine the optimal parameters set of the SOM. These approaches are not appropriate for large dataset, as their computational burden is quite high; they belong to the wrapper approaches and their efficiency is related to their capability to optimize the adopted classifier.

Even though a lot of VS methods have been proposed, an optimal method, i.e. an approach which is suitable on all VS problems, has not been identified. Some experiments [33,64] demonstrated that there are considerable differences of performance, in terms of classification accuracy, among different VS approaches over a specific data set. This consideration leads to a very important question: which VS methods should be chosen for a specified dataset? The most obvious answer is to perform all candidate VS methods to the dataset and to prefer one with the best performance using, for example, the cross-validation approach. Nevertheless, this solution could be very expensive in terms of time, especially for high-dimensional data [5]. In order to address this problem in a more efficient way, automated VS methods (e.g. based on application of computational intelligence and meta-learning) are recommended [1,2,34,52,59]. In [59] efficient meta learning based VS algorithms are proposed. Their approach is based on the assumption that the performance of an algorithm on a given dataset depends on the characteristics of the dataset itself. The meta-features, which are commonly adopted in meta-learning [58] are introduced to differentiate the datasets. Moreover, a multi-criteria metric, which considers not only the performance of a classifier obtained by applying a VS method but also the time complexity and the length of the selected is introduced. The k-Nearest Neighbour based method is used to select appropriate VS algorithm for a unseen dataset. The main objective of the proposed meta-learning algorithm lies in the automatic selection of appropriate VS methods; the effectiveness of the approach is confirmed on the results obtained on 115 real world datasets, performing 22 different VS approaches and 5 representative classification algorithms.

5 The Stability Problem

The stability notion was introduced in 1995 by Turney [57]. Stability is defined as the sensitivity of a VS procedure with respect to variations in the training dataset. Several works demonstrate that the exploitation of different training sets can lead to select different variable subsets also performing the same VS algorithm. The stability measures are defined on the basis on the typology of target to be considered and can be classified as follows:

- a weighting-scoring $\boldsymbol{w} = (w_1, w_2, ..., w_n)$, $w \in W \subseteq R^n$;
- a ranking vector: $\boldsymbol{r} = (r_1, r_2, ..., r_n)$, $1 \leq r_k \leq n$, where r_k is the rank of variable k;
- an n-dimensional binary vector where each component is associated to a variable and its null or unitary value represents, respectively, absence or presence of a variable in the selected subset: $\boldsymbol{b} = (b_1, b_2, ..., b_n)$, $b_k \in [0, 1]$

In the first definition, in order to compute the similarity between two weighting vectors $\boldsymbol{w}^1 = (w_1^1, w_2^1, ..., w_n^1)$ and $\boldsymbol{w}_2 = (w_1^2, w_2^2, ..., w_n^2)$ the *Pearson's correlation coefficient* [55] can be computed as:

$$S_w(\boldsymbol{w}^1, \boldsymbol{w}^2) = \frac{\sum_{k=1}^{n} (w_k^1 - \mu_1) \cdot (w_k^2 - \mu_2)}{\sqrt{\sum_{k=1}^{n} (w_k^1 - \mu_1)^2 \cdot \sum_k (w_k^2 - \mu_2)^2}} \tag{5}$$

where $\mu_1 = 1/n \sum_{k=1}^{n} w_k^1$ and $\mu_2 = 1/n \sum_{k=1}^{n} w_k^2$. S_w belongs to the range $[-1, 1]$: the null value is related to the absence of correlation while unitary values indicate that $\boldsymbol{w}_1$ and $\boldsymbol{w}_2$ are perfectly (positively or negatively) correlated.

Similarly, in order to quantify the similarity between two rankings $\boldsymbol{r}^1 = (r_1^1, r_2^1, ..., r_n^1)$ and $\boldsymbol{r}^2 = (r_1^2, r_2^2, ..., r_n^2)$, the *Spearman's rank correlation coefficient* [54] which is also also known as "the Pearson correlation coefficient between the ranked variables", is calculated as in 6:

$$S_R(\boldsymbol{r}^1, \boldsymbol{r}^2) = 1 - 6 \cdot \frac{\sum_{k=1}^{n} (r_k^1 - r_k^2)}{m(m^2 - 1)} \tag{6}$$

S_R lies in the range $[-1, 1]$: the unitary value indicates that the two rankings are identical, the null value means absence of correlation and the value -1 means that the two rankings have inverse orders.

Finally the similarity between two binary vectors $\boldsymbol{b}^1$ and $\boldsymbol{b}^2$ is evaluated through the *Tanimoto distance* [24], which is defined as in (7):

$$S_B(\boldsymbol{b}^1, \boldsymbol{b}^2) = \frac{|\boldsymbol{b}^1 \cdot \boldsymbol{b}^2|}{|\boldsymbol{b}^1| + |\boldsymbol{b}^2| - |\boldsymbol{b}^1 \cdot \boldsymbol{b}^2|} \tag{7}$$

where $|\cdot|$ indicates the norm of the binary vector and $\boldsymbol{b}^1 \cdot \boldsymbol{b}^2$ represents the scalar product of $\boldsymbol{b}^1$ and $\boldsymbol{b}^2$. S_B lies in the range $[0, 1]$ where the null value indicates that there is no overlap between the two sets while an unitary value means that the two sets are identical [35].

The motivation for studying the stability of VS methods lies in the need to offer to the experts in the considered application field a quantified evidence that the selected variables are rather robust with respect to variations in the training data. This request is mainly crucial in real applications where the knowledge of the phenomenon under consideration is an important outcome of the VS and a stable identification of the variables that mostly affect the target is essential.

The stability of a VS method is commonly defined as the robustness of the variables selected relative to the differences in the training sets created from the same generating distribution [36]. In [45] a new stability index based on the Shannon entropy to assess the overall occurrence of individual variables in selected subsets of possibly varying cardinality has been proposed. In [35] different similarity measures to compute the stability of variable preferences via executing several tests with several VS algorithms has been studied. Moreover, in [41,61] stable VS approaches that cluster variables and select representatives of each cluster for the final subset has been dealt. In [8] a novel method that improves the stability of the wrapper VS algorithms while preserving and possibly improving the classification performance has been proposed. The main idea of the approach lies in the consideration of the mutual interaction between couples of variables which is commonly ignored by traditional VS algorithms. An important advantage of this approach is that it is generic and can be performed on any kind of binary classifier, on any kind of wrapper strategy and finally on any kind of real dataset. Authors show the method effectiveness and universality by applying it to three different binary classifiers, with datasets coming from real word applications and by comparing its performance with different wrapper procedures. The obtained results show that the proposed approach is effective independently on the type of VS method, on the type of classifier and on the database. The stability is also measured using the Tanimoto distance, which is more appropriate for binary classification.

In [27], instead of varying the sample population, authors change the classifier used in the SVCs feature ranking and study the consequence of this alteration on the stability of the final selected variables. The paper analyses how unstable the SVCs feature ranking results could be using different classifiers and several datasets is shown, as well as the bias of classifiers in SVCs ranking . Moreover the similarity and correlation of the results obtained using different classifiers is calculated using the Spearman's rank correlation coefficient.

6 Conclusions

The paper present recent and relevant approaches to VS, a task which is particularly important in many industrial applications where often the considered process is unknown or difficult to model and the dimensionality of data collected by sensors can be very high. A survey of the main VS approaches is presented in order to underline advantages and disadvantages of different type of approaches, especially for applications of neural networks. The application of VS in the field of classification, which is very relevant from the industrial point of view, is anaylsed in detail. Moreover the notion of stability is provided to introduce the problem that the variation of different training data sets can provide to the selection of different variable subsets. Such concept is essential in real applications where VS is applied to study the behaviour of the considered process.

References

1. Ali, S., Smith, K.: On learning algorithm selection for classification. Appl. Soft Comput. **6**(2), 119–138 (2006)
2. Brazdil, P., Carrier, C., Soares, C., Vilalta, R.: Metalearning: Applications to Data Mining. Springer, Heidelberg (2008)
3. Breiman, L.: Random forests. Mach. Learn. **45**, 5–32 (2001)
4. Breiman, L., Friedman, J.H., Olshen, R.A., Stone, C.J.: Classification and Regression Trees. Wadsworth and Brooks, Belmont (1984)
5. Brodley, C.E.: Addressing the selective superiority problem: automatic algorithm/model class selection. In: Proceedings of 10th International Conference on Machine Learning, pp. 17–24 (1993)
6. Cam, L.L.: Maximum likelihood an introduction. ISI Rev. **58**(2), 153–171 (1990)
7. Cateni, S., Colla, V.: The importance of variable selection for neuralnetworks-based classification in an industrial context. In: Advances in Neural Networks (Computational Intelligence for ICT). Smart Innovation, Systems and Technologies (2016)
8. Cateni, S., Colla, V.: Improving the stability of wrapper variable selection applied to binary classification. Int. J. Comput. Inf. Syst. Ind. Manage. Appl. (2016) (in press)
9. Cateni, S., Colla, V.: A hybrid variable selection approach for nn-based classification in industrial context. In: Smart Innovation, Systems and Technologies (in press)
10. Cateni, S., Colla, V., Vannucci, M.: General purpose input vvariable extraction: a genetic algorithm based procedure give a gap. In: Proceedings of the 9th International Conference on Intelligence Systems design and Applications ISDA09 (2009)
11. Cateni, S., Colla, V., Vannucci, M.: Variable selection through genetic algorithms for classification purpose. In: IASTED International Conference on Artificial Intelligence and Applications (2010)
12. Cateni, S., Colla, V., Vannucci, M.: A genetic algorithm-based approach for selecting input variables and setting relevant network parameters of a som-based classifier. Int. J. Simul. Syst. Sci. Technol. **12**(2), 30–37 (2011)
13. Cateni, S., Colla, V., Vannucci, M.: A hybrid feature selection method for classification purposes. In: 8th European Modeling Symposium on Mathematical Modeling and Computer simulation EMS2014, Pisa, Italy, vol. 1, pp. 1–8 (2014)
14. Cateni, S., Colla, V., Vannucci, M.: A fuzzy system for combining filter features selection methods. Int. J. Fuzzy Syst. (In press)
15. Cateni, S., Colla, V., Vannucci, M.: A genetic algorithms-based approach for selecting the most relevant input variables in classification tasks. In: Proceedings of the 4th European Modelling Symposium on Mathematical Modelling and Computer simulation EMS2010, Pisa, pp. 63–67, 17–19 November 2010
16. Cateni, S., Colla, V., Vannucci, M., Vannocci, M.: A procedure for building reduced reliable training datasets from real-world data. In: 13th IASTED International Conference on Artificial Intelligence and Applications AIA 2014, Innsbruck, Austria, 17–19 Febbraio 2014
17. Coetzee, F.: Correcting the kullback-leibler distance for feature selection. Pattern Recogn. Lett. **26**(11), 1675–1683 (2005)
18. Colla, V., Nastasi, G., Cateni, S., Vannucci, M., Vannocci, M.: Genetic algorithms applied to discrete distribution fitting. In: Proceedings - UKSim-AMSS 7th European Modelling Symposium on Computer Modelling and Simulation, EMS 2013, pp. 30–35 (2013)

19. Colla, V., Nastasi, G., Materese, N.: Gadf - genetic algorithms for distribution fitting. In: Proceedings of the 2010 10th International Conference on Intelligent Systems Design and Applications, ISDA 2010, pp. 6–11 (2010)
20. Colla, V., Valentini, R., Bioli, G.: Mechanical properties prediction for aluminium-killed and interstitial-free steels, Revue de Métalurgie Special Issue JSI, pp. 100–101, Dicembre 2004
21. Colla, V., Vannucci, M., Fera, S., Valentini, R.: Ca-treatment of al-killed steels: inclusion modification and application of artificial neural networks for the prediction of clogging. In: Proceedings of 5th European Oxygen Steelmaking Conference EOSC 2006, vol. 1, pp. 387–394, 26–28 June 2006
22. Colla, V., Vannucci, M., Reyneri, L.: Artificial neural networks applied for estimating a probability density function. Intell. Data Anal. **19**(1), 29–41 (2014)
23. Cover, T., Thomas, A.: Elements of Information Theory. Telecommunications and Signal Processing. Wiley Series, Hoboken (1991)
24. Duda, R., Hart, P., Stork, D.: Pattern Classification. Wiley, New York (2001)
25. Eid, H., Hassanien, A., Kim, T.H., Banerjee, S.: Linear correlation-based feature selection for network intrusion detection model. Commun. Comput. Inf. Sci. **381**, 240–248 (2013)
26. Epanechnikov, V.: Non-parametric estimation of a multivariate probability density. Theory Probab. Appl. **14**, 239–250 (1969)
27. Fakhraei, S., Zadeh, H.S., Fotouhi, F.: Bias and stability of single variable classifiers for feature ranking and selection. Expert Syst. Appl. **14**(15), 6945–6958 (2014)
28. Fiasché, M., Taisch, M.: On the use of quantum-inspired optimization techniques for training spiking neural networks: a new method proposed. Smart Innov. Syst. Technol. **37**, 359–368 (2015)
29. Froelich, W., Wrobel, K., Porwik, P.: Diagnosing parkinson's disease using the classification of speech signals. J. Med. Inform. Technol. **23**, 187–194 (2014)
30. Ghumbre, S., Patil, C., Ghatol, A.: Heart disease diagnosis using support vector machine. In: International Conference on Computer Science and Information Technology, ICCSIT 2011, Pattaya (2011)
31. Golub, T., Slonim, D., Tamayo, P., Huard, C., Gaasenbeek, M., Mesirov, J., Coller, H., Loh, M., Downing, J., Caligiuri, M., Lander, C.B.E.: Molecular classification of cancer: class discovery and class prediction by gene expression monitoring. Science **286**, 531–537 (1999)
32. Guyon, I., Elisseeff, A.: An introduction to variable and feature selection. Mach. Learn. **3**, 1157–1182 (2003)
33. Hall, M.A.: Correlation-based feature selection for machine learning. Ph.D. thesis, The University of Waikato (1999)
34. Kalousis, A., Gama, J., Hilario, M.: On data and algorithms: understanding inductive performance. Mach. Learn. **54**(3), 275–312 (2004)
35. Kalousis, A., Prados, J., Hilario, M.: Stability of feature selection algorithms. In: Proceedings of 5th IEEE International Conference on Data Mining, ICDM 2005, pp. 218–225 (2005)
36. Kalousis, A., Prados, J., Hilario, M.: Stability of feature selection algorithms: a study on high-dimensional spaces. Knowl. Inf. Syst. **12**, 95–116 (2007)
37. Khashei, M., Eftekhari, S., Parvizian, J.: Diagnosing diabetes type ii using a soft intelligent binary classification model. Rev. Bioinform. Biometrics (RBB) **1**, 9–23 (2012)
38. Koc, L., Carswell, A.D.: Network intrusion detection using a hnb binary classifier. In: 17th UKSIM-AMSS International Conference on Modelling and Simulation (2015)

39. Kohavi, R., John, G.: Wrappers for feature selection. Artif. Intell. **97**, 273–324 (1997)
40. Kullback, S., Leibler, R.: On information and sufficiency. Ann. Math. Stat. **22**, 79–86 (1951)
41. Loscalzo, S., Yu, L., Ding, C.: Consensus group stable feature selection. In: Proceedings of ACM SIGKDD International Conference on Knowledge Discovery and Data Mining, vol. 1, pp. 567–575 (2009)
42. Mammone, N., Fiasché, M., Inuso, G., La Foresta, F., Morabito, C.F.C., Versaci, M.: Information theoretic learning for inverse problem resolution in bioelectromagnetism. In: Apolloni, B., Howlett, R.J., Jain, L. (eds.) KES 2007, Part III. LNCS (LNAI), vol. 4694, pp. 414–421. Springer, Heidelberg (2007)
43. May, R., Dandy, G., Maier, H.: Review of input variable selection methods for artificial neural networks. In: Suzuki, K. (ed.) Artificial Neural Networks Methodological Advances and Biomedical Applications. INTECH, Rijeka (2011)
44. Mitchell, T., Toby, J., Beauchamp, J.: Bayesian variable selection in linear regression. J. Am. Stat. Assoc. **83**, 1023–1032 (1988)
45. Novovicova, J., Somol, P., Pudil, P.: A new measure of feature selection algorithms stability. In: IEEE International Conference on Data Mining Workshops, vol. 1, pp. 382–387 (2009)
46. Reyneri, L., Colla, V., Vannucci, M.: Estimate of a probability density function through neural networks. In: Cabestany, J., Rojas, I., Joya, G. (eds.) IWANN 2011, Part I. LNCS, vol. 6691, pp. 57–64. Springer, Heidelberg (2011)
47. Rice, J.A.: Mathematical Statistics and Data Analysis, 3rd edn. Duxbury Advanced, Duxbury (2006)
48. Rosenblatt, M.: Remarks on some nonparametric estimates of a density function. Ann. Math. Stat. **27**(3), 443–471 (1956)
49. Schliebs, S., Defoin-Platel, M., Worner, S., Kasabov, N.: Integrated feature and parameter optimization for an evolving spiking neural network: exploring heterogeneous probabilistic models. Neural Netw. **22**(5–6), 623–632 (2009)
50. Shetty, M., Shekokar, N.M.: Data mining techniques for real time intrusion detection systems. Int. J. Sci. Eng. Res. **3**(4), 1–7 (2012)
51. Sokolova, M., Lapalme, G.: A systsystem analysis of performance measures for classification tasks. Inf. Process. Manage. **45**, 427–437 (2009)
52. Song, Q.B., Wang, G., Wang, C.: Automatic recommendation of classification algorithms based on data set characteristics. Pattern Recogn. **45**(7), 2672–2689 (2012)
53. Souza, F., Araujo, R., Soares, S., Mendes, J.: Variable selection based on mutual information for soft sensors application. In: Proceedings of the 9th Portuguese Conference on Automatic Control, Controlo 2010, At Coimbra, Portugal (2010)
54. Spearman, C.: The proof and measurement of association between two things. Am. J. Psychol. **15**, 72–101 (1904)
55. Stigler, S.: Francis galtons account of the invention of correlation. Stat. Sci. **4**(2), 73–79 (1989)
56. Sun, Y., Robinson, M., Adams, R., Boekhorst, R., Rust, A.G., Davey, N.: Using feature selection filtering methods for binding site predictions. In: Proceedings of 5th IEEE International Conference on Cognitive Informatics, ICCI 2006 (2006)
57. Turney, P.: Techncal note: bias and the quantification of stability. Mach. Learn. **20**, 23–33 (1995)
58. Vitalta, R., Drissi, Y.: A perspective view and survey of meta-learning. Artif. Intell. Rev. **18**(2), 77–95 (2002)
59. Wang, G., Song, Q., Sun, H., Zhou, X.: A feature subset selection algorithm automatic reccomendation method. J. Artif. Intell. Res. **47**, 1–34 (2013)

60. Wang, S., Zhu, J.: Variable selection for model-based high dimensional clustering and its application on microarray data. Biometrics **64**, 440–448 (2008)
61. Yu, L., Ding, C., Loscalzo, S.: Stable feature selection via dense feature groups. In: Proceedings of the ACM SIGKDD International Conference on Knowledge Discovery and Data Mining, pp. 803–811 (2008)
62. Yu, L., Liu, H.: Feature selection for high-dimensional data: a fast correlation basedfilter solution. In: Proceedings of the 20th International Conference on Machine Learning ICML, vol. 1, pp. 856–863 (2003)
63. Zhang, K., Li, Y., Scarf, P., Ball, A.: Feature selection for high-dimensional machinery fault diagnosis data using multiple models and radial basis function networks. Neurocomputing **74**, 2941–2952 (2011)
64. Zhao, Z., Liu, H.: Searching for interacting features. In: Proceedings of the 20th International Joint Conference on Artifical Intelligence Morgan Kaufmann Publishers Inc., pp. 1156–1161 (2007)

Author Index

Printed in the United States
By Bookmasters